Advanced Two-Dimensional Material-Based Heterostructures in Sustainable Energy Storage Devices

Advanced Two-Dimensional Material-Based Heterostructures in Sustainable Energy Storage Devices provides a detailed overview of advances and challenges in the development of 2D materials for use in energy storage devices. It offers deep insight into the synthesis, characterization, and application of different 2D materials and their heterostructures in a variety of energy storage devices, focusing on new phenomena and enhanced electrochemistry.

This book:

- Introduces 2D materials, synthesis methods, and characterization techniques.
- Discusses application in a wide range of batteries and supercapacitors.
- Offers perspectives on future investigations necessary to overcome existing challenges.

This comprehensive reference is written to guide researchers and engineers working to advance the technology of energy-efficient energy storage devices.

Advanced Two-Dimensional Material-Based Heterostructures in Sustainable Energy Storage Devices

Edited by
Srikanth Ponnada
Susmita Naskar

CRC Press
Taylor & Francis Group
Boca Raton London New York

CRC Press is an imprint of the
Taylor & Francis Group, an **informa** business

Designed cover image: Shutterstock

First edition published 2025
by CRC Press
2385 NW Executive Center Drive, Suite 320, Boca Raton FL 33431

and by CRC Press
4 Park Square, Milton Park, Abingdon, Oxon, OX14 4RN

CRC Press is an imprint of Taylor & Francis Group, LLC

ISBN: 978-1-032-51974-6 (hbk)
ISBN: 978-1-032-51975-3 (pbk)
ISBN: 978-1-003-40472-9 (ebk)

DOI: 10.1201/9781003404729

Typeset in Times
by MPS Limited, Dehradun

Contents

Preface

The dwindling of fossil fuels and climate change are two of the most critical problems that confront us every day. One of the most significant attempts to reduce the effects of climate change and reduce carbon footprint is to switch to renewable energy sources. Despite an unprecedented rise in their use over the last few decades, renewable energy sources continue to confront challenges, including high costs and inconsistent availability. Efficient energy storage and management are more important than ever in the constantly changing world of modern energy solutions. Innovations in energy storage are of paramount importance due to the proliferation of renewable energy sources and the increasing demand for electric cars and portable electronics. Strategies for energy storage can be of assistance to us in addressing the intermittent nature of renewable energy sources. There has been a proliferation of research into and production of various batteries since Gaston Planté invented the first rechargeable battery, and John Goodenough's development of lightweight high-energy-density rechargeable lithium batteries sparked an uptick in interest in their potential for use in portable devices and for electric mobility.

The inexorable march towards a sustainable energy future demands innovative solutions that transcend the conventional boundaries of materials science and engineering. In this quest, two-dimensional (2D) materials with their exceptional performance and unique characteristics have risen to the forefront of game-changing technology. This book, *Advanced Two-Dimensional Material-Based Heterostructures in Sustainable Energy Storage Devices*, delves into the forefront of research and development where the convergence of nanotechnology, electrochemistry, and sustainable energy meets. The urgent need for affordable, environmentally friendly, and scalable solutions is driving a fundamental transformation in the energy storage industry. Harnessing the unique qualities of 2D materials, such as their atomic-scale thickness and exceptional electrical, mechanical, and thermal capabilities, opens previously unimagined possibilities. This book unravels the intricate tapestry of 2D material heterostructures, where the synergy of distinct materials at the nanoscale heralds a new era in energy storage device design. Our journey begins with a comprehensive exploration of the fundamental principles underpinning 2D materials and their heterostructures. We traverse the rich terrain of graphene, transition metal dichalcogenides, and other exotic 2D materials, understanding their electronic structures, synthesis methods, and scalable production. Armed with this foundational knowledge, we embark on an odyssey into the design, fabrication, and optimization of heterostructures, unlocking the full potential of these materials for energy storage applications. As we navigate through the chapters, we will see diverse applications for 2D material heterostructures in sustainable energy storage systems. The flexibility of these heterostructures is seen in applications ranging from improved batteries to supercapacitors and beyond. The story is intertwined with cutting-edge scientific results, technical achievements, and rising commercialization potential.

The authors of this book, experts in the field, shared their views and experiences with readers, helping them through the complexities of 2D material heterostructures. This book is not merely an assortment of research articles; it is a road map for academics, engineers, and enthusiasts keen on contributing to the growing landscape of sustainable energy storage. Finally, our book, *Advanced Two-Dimensional Material-Based Heterostructures in Sustainable Energy Storage Devices*, demonstrates the promise of 2D materials in determining the future of energy storage. We hope that this book will serve as a source of inspiration and information, encouraging collaboration and creativity in the pursuit of a cleaner and more sustainable energy paradigm.

About the Editors

Dr Srikanth Ponnada (MRSC) is currently a postdoctoral fellow at (Prof. Andy Herring's Group) Chemical and Biological Engineering Department, Colorado School of Mines, USA. Previously, he worked as a postdoctoral research associate at the Indian Institute of Technology Jodhpur, Rajasthan, India. He received his Dual Degree Masters in Applied Chemistry from the Department of Engineering, Chemistry from the Andhra University College of Engineering (A), Andhra University, India, and his PhD is in the area of "Functional materials and their Electrochemical Applications in Batteries and Sensors". As a part of his PhD, he worked at the Indian Institute of Science, Bengaluru, India. His current research area includes functional materials synthesis, polymer electrolyte membranes, fuel cell assembly and engineering, energy storage (Li-S, Li-ion and Li-free batteries, all-solid-state battery electrolytes), conversion devices (AEM and PEM fuel cells), electrochemical sensors, electrocatalysis and electrochemical engineering for net zero emissions and carbon neutrality. He previously worked at CSIR-Central Electrochemical Research Institute, India, as a Project Assistant Level-III in Lead-free perovskites-based photovoltaics and electrocatalysis and at the Indian Institute of Technology (ISM) Dhanbad, India, as a project research fellow in gold nanoparticle-assisted heterogeneous catalysis in alcohol and hydrocarbon oxidation reactions.

Prof Susmita Naskar is an assistant professor in the Faculty of Engineering and Physical Sciences at the University of Southampton. She worked as a postdoctoral research fellow at the Whiting School of Engineering of Johns Hopkins University in the collaboration of Army Research Lab, USA. She moved to the USA after completing her doctoral degree from the University of Aberdeen. Susmita's research interests and expertise broadly lie in the field of multi-scale structural mechanics and multi-physics analysis, focusing on engineered materials and advanced composites involving the intersection of additive manufacturing, material characterization through computational design and experiments in engineering. In addition, she is working on advanced manufacturing techniques that are relevant to the fabrication of engineered materials like composites and metamaterials.

Contributors

Cyril A Andrews
Indian Institute of Technology
Jodhpur, Rajasthan, India

Lakshman Kumar Anisetty
NTPC Energy Technology Research
   Alliance
Greater Noida, India

Trapa Banik
Tennessee Technological University
Cookeville, Tennessee, United States

Indranil Bhattacharya
Tennessee Technological University
Cookeville, Tennessee, United States

Rapaka S Chandra Bose
Centre for Materials for Electronic
   Technology
Thrissur, Kerala, India

Shankara Devi
Pondicherry University
Pondicherry, India

Ayan Dey
Indian Institute of Technology
Jodhpur, Rajasthan, India

Ismail Fidan
Tennessee Technological University
Cookeville, Tennessee, United States

Demudu Babu Gorle
Indian Institute of Science
Bengaluru, Karnataka, India

Ankit Gupta
California State University
Los Angeles, California, United States

Rimpa Jaiswal
Bhabha Atomic Research Centre
Mumbai, Maharashtra, India

Maryam Sadat Kiai
University College of Dublin
Dublin, Ireland

Praveen Kumar
Colorado School of Mines
Golden, Colorado, United States

Raj Kumar
University of Delhi
Delhi, India

Susmita Naskar
University of Southampton
Southampton, United Kingdom

Anjali Palariya
Jai Narain Vyas University
Jodhpur, Rajasthan, India

Ravinder Pawar
National Institute of Technology
Warangal, Telangana, India

Tirupathi Rao Penki
Bar-Ilan University
Ramat Gan, Israel

Srikanth Ponnada
Colorado School of Mines
Golden, Colorado, United States

Varun T S
Centre for Materials for Electronic
   Technology
Thrissur, Kerala, India

Kirankumar Venkatesan Savunthari
Marshall University
Huntington, West Virginia,
    United States

Srijan Sengupta
Indian Institute of Technology
Jodhpur, Rajasthan, India

Mohit K Sharma
Changwon National University
Changwon, Republic of Korea

Rakesh K Sharma
Indian Institute of Technology
Jodhpur, Rajasthan, India

Ghanshyam Varshney
Indian Institute of Technology
Jodhpur, Rajasthan, India

Krishna Harika Villa
Pondicherry University
Pondicherry, India

Neetu Yadav
Indian Institute of Technology
Roorkee, Uttarakhand, India

Sarita Yadav
National Institute of Technology
Warangal, Telangana, India

1 Introduction and Characterization of Two-Dimensional Materials

Anjali Palariya and Srikanth Ponnada

1.1 INTRODUCTION

Nanoscience is the branch of material science that deals with the study of science, technology, and engineering of materials at the nanoscale. At this scale, the number of constituting atoms/molecules and dimensions plays a crucial role in reactivity and physicochemical properties. Nanomaterials are categorized as zero-dimensional (these include nanoparticles, fullerenes, and quantum dots where all the dimensions are in the range of a few nanometres), one-dimensional (1D) (these include nanotubes, nanorods, and nanowires where these are in nanoscale in two dimensions), and two-dimensional (2D) (these include graphene, borophene, hexagonal boron nitride (BN), 2D carbon nitride-layered double hydroxide, and transition metal chalcogenides with layered structure where only one dimension is restricted [1]). 2D materials typically refer to layered solids with strong in-plane bonds and weaker interplanar interactions (viz. van der Waals forces). A thinner overall structure with a limited number of layers can be yielded from these layered materials with different exfoliation and stabilization techniques. A more rigid definition of 2D materials is restricted to only monoatomic layers however; few layered 2D materials are more versatile in terms of ease of synthesis and hence these are more application-focused [2,3].

2D nanomaterials display unique physicochemical properties resulting from the quantum-size effect, higher aspect ratio, and improved surface chemistry. During the 16th century, graphite was used for drawing and marking sheep and after its first synthesis in the late 1800s, the industrial uses of graphite were commenced. Metal-free 2D materials are primarily studied as carbon-based, boron-based, and phosphorous-based materials. Various bottom-up and top-down methods are used for the synthesis of these materials. Micromechanical exfoliation, surface-assisted in-situ growth, and exfoliation into colloidal solution are some novel methods for thickness control of metal-free 2D materials [1–6].

Diverse 2D metallic nanomaterials have appealed immense applications in catalysis, solar cells, sensing, bioimaging, and so on owing to their fascinating surface properties and anisotropic structure [7]. Metals usually have a highly

DOI: 10.1201/9781003404729-1

1

symmetric lattice structure and therefore 2D growth is thermodynamically unfavourable [8,9]. In these synthesis processes, the total free energy of the nanostructure is decreased, or the addition of atoms is delayed facilitating kinetic control for 2D anisotropic growth [10]. Different synthetic strategies often consist of bottom-up growth with chemical reduction or thermal decomposition of the metal precursor where the symmetry of isotropic nuclei is broken to induce anisotropic growth. Capping agents are generally used to achieve adequate lattice energy by decreasing the surface energy for the synthesis of these materials [7,11]. Both bottom-up and top-down methods are used for the synthesis of these materials. Some of the frequently used wet-chemical methods are organic ligand-assisted growth, templated-confined growth, photochemical synthesis, polyol method, particle assembly, seeded growth, hydro/solvothermal synthesis, biological synthesis, and small molecules/ions-mediated synthesis. Examples of a few important top-down methods are mechanical compression, nanolithography, and liquid-phase exfoliation method [7].

The properties of 2D materials obtained by different methodologies primarily depend on lattice structure, layer control, surface-to-volume ratio, and inter/intralayer interactions. The physical, chemical, electronic, optical, and morphological characteristics of 2D nanomaterials dictate their potential applications. These features can be monitored with characterization techniques of varying sensitivity, sample preparation, and sample preparation technique [12].

The intriguing physicochemical properties comprising stability, structural, electronic, magnetic, and plasmonic properties have appealed enormous interdisciplinary attention in fields of catalysis as a high surface-to-volume ratio enables abundant exposure of catalytically active sites. The development of heterogeneous catalysts is crucial for sustainable water splitting, N_2 fixation, CO_2 capture/conversion, oxygen reduction reaction, biomass conversion, industrially important catalytic transformations, and so on. The field of energy conversion and storage has also developed tremendously with the fabrication of 2D nanomaterials for fuel cells, solar cells, batteries, and electrochemical capacitors. 2D materials have also shown potential in fascinating arenas of gas storage/separation, photochemical/electrochemical sensing, bioimaging, optoelectronic devices, and nanomedicines [1,7,13–15].

In this chapter, a concise structural overview of different metallic and metal-free 2D nanomaterials along with a few essential characterization techniques is presented.

1.2 TYPES OF 2D NANOMATERIALS AND TYPICAL CHARACTERIZATION TECHNIQUES

1.2.1 METAL-FREE 2D NANOMATERIALS

1.2.1.1 Graphene and Related C-Based Materials

Carbon is one of the most abundant elements and it bears the foundation of life. The pristine graphene comprises a symmetric honeycomb lattice of hexagonal C_6 rings with sp^2 hybridization. The sp^2 hybridized C monolayer of graphene displays superior heat conductivity, optical transparency, and quantum Hall effects owing to its Dirac-like band structure nearing the Fermi level. It is a prototypical 2D material due to its chemical inertness, charge carrier efficiency, and high mechanical strength [16]. Different monoatomic layers of carbon atoms consisting of C_4, C_5, C_7, C_8, C_9, and C_{10} rings can be accumulated to form allotropes of graphene [17]. Another sp^2 hybridized C structure is graphenylene which was first reported in 2013 [18]. Graphynes include carbonaceous 2D materials with sp hybridization. These materials show semiconductor properties with band gaps of 0.5–0.6 eV [19]. Elemental doping by chemical reaction or adsorption method can modulate the electronic configuration of graphene for various applications.

Graphene oxide (GO) consists of hexagonal carbon rings with various oxygen functional groups, viz., C-OH, C=O, C-OOH, and C-O-C groups. These oxygen functionalities assist GO in overcoming the low dispersion ability of graphene, making it favourable for catalysis and formation of heterojunctions. The presence of the hydroxyl, carboxyl, and epoxy groups imparts adequate hydrophilicity and thus facilitates exfoliation to monolayer GO [1]. Reduced GO is prepared by chemical reduction or thermal annealing which partially restores sp^2 hybridization and high conductivity of graphene [20].

Carbon nitrides have seven phases with band gaps in the range of 2.7–5.5 eV among which graphitic carbon nitride (g-C_3N_4) with triazine and heptazine units is thermodynamically most favoured. This sp^2 hybridized 2D porous material is synthesized with thermal polymerization of N-rich precursors like urea, thiourea, melamine, cyanamide, and dicyandiamide. The physicochemical properties can be easily tuned with techniques like nanostructure design, elemental/molecular doping, and heterojunction construction [21,22].

2D covalent organic frameworks (COFs) are polymers consisting of two dimensionally linked organic monomers with covalent bonds. These are further stacked on one another through non-covalent interactions [23]. These porous and highly crystalline nanosheets are synthesized with methods like surfactant-assisted synthesis, interfacial synthesis, and ultrasonic/mechanical/electrochemical exfoliation. COF nanosheets are also synthesized using modulators to regulate coordination equilibrium and thus implement 2D morphology. These materials have shown tremendous potential in catalysis electrochemical processes and optoelectronic devices [23,24].

The thorough characterization of carbonaceous materials consists of basic structural and morphological analysis like X-ray diffraction (XRD), infrared spectroscopy (IR), scanning electron microscopy (SEM), and transmission electron microscopy (TEM). Crucial characteristics such as thickness, phase, number of layers, and their stacking configuration are analysed with facile and non-destructive Raman spectroscopy. Figure 1.1a, b, and c displays the layered structure of graphene, GO, and g-C_3N_4. Figure 1.1d illustrates the Raman spectra of various graphene derivatives. The effects of structural changes in the respective Raman spectra are visualized through the changes in intensity, sharpness, and position of the characteristic D and G bands [25]. Figure 1.1e and f demonstrate the effects of a number of layers and their stacking configuration in g-C_3N_4, where exfoliation dramatically reduces the peak intensities in bulk g-C_3N_4. Again, the heteroatom doping drastically alters the shape and peak intensities of the Raman spectra [26]. Optoelectronic properties of g-C_3N_4 and 2D COF are analysed with UV-diffuse reflectance spectroscopy (UV-DRS), photoluminescence (PL), and time-resolved

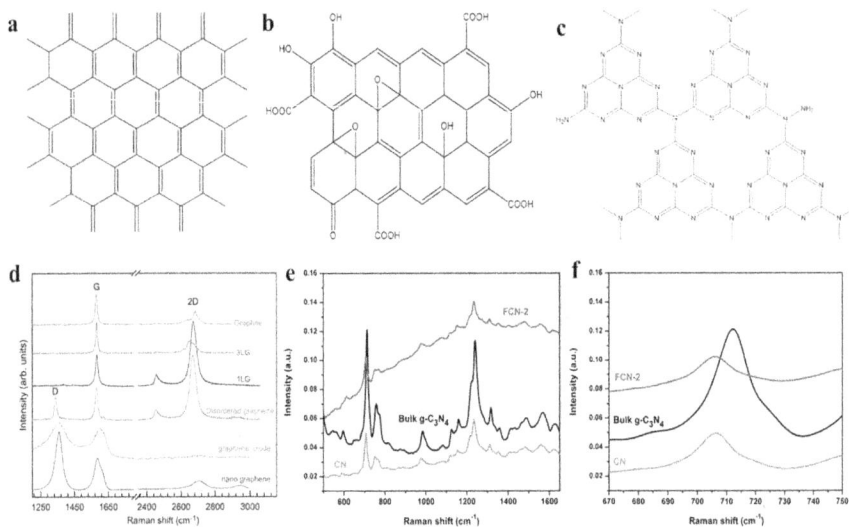

FIGURE 1.1 Structures of (a) graphene, (b) GO, (c) g-C_3N_4, and (d) Raman spectra of graphite, monolayer graphene, three-layered graphene, GO, and nanographene. Reproduced with permission copyright © 2018 RSC publishing [25]. (e, f) Raman spectra of bulk g-C_3N_4, exfoliated g-C_3N_4, and Fe-doped g-C_3N_4. Image reproduced with permission copyright © 2014 RSC publishing [26].

photoluminescence TRPL techniques. As the surface-to-volume ratio and nature of porosity play a significant role in the activity of the metal-free 2D catalysts, N_2 adsorption-desorption analysis for specific surface area and porosity is essential for the detailed characterization of materials.

1.2.1.2 Boron-Based Materials

The trivalent electronic configuration prevents Boron from satisfying the octet rule resulting in an electron-deficient and delocalized (among three or more atoms) bonding configuration [27]. 2D boron sheets were first theoretically proposed in the early 2000s and their synthesis on metallic Au, Ag, and Cu surfaces was first predicted in 2013 [28]. BN contain covalently bonded B and N atoms in crystalline arrays as shown in Figure 1.2a. These layered inorganic materials are synthesized using several methods like chemical vapour deposition, electrodeposition, molecular beam epitaxy, and atmospheric pressure catalytic flux methods. BN exist in different crystal structures, viz., cubic BN, wurtzite BN, hexagonal BN (h-BN), and rhombohedral BN. These materials usually exhibit high thermal and chemical stability, and cubic BN is the thermodynamically most favourable phase. However, hexagonal BN has been conventionally more explored as a ceramic material for lubrication and heat shielding, and as a filler material in structural composites. Recently, a monolayer of hexagonal BN has been explored as an insulating support due to its wide band gap (~6.5 eV) in nano-electronic devices [29,30].

The overall quality including crystallinity, morphology, and the number of layers of hexagonal BN is analysed using accretive sophisticated characterization

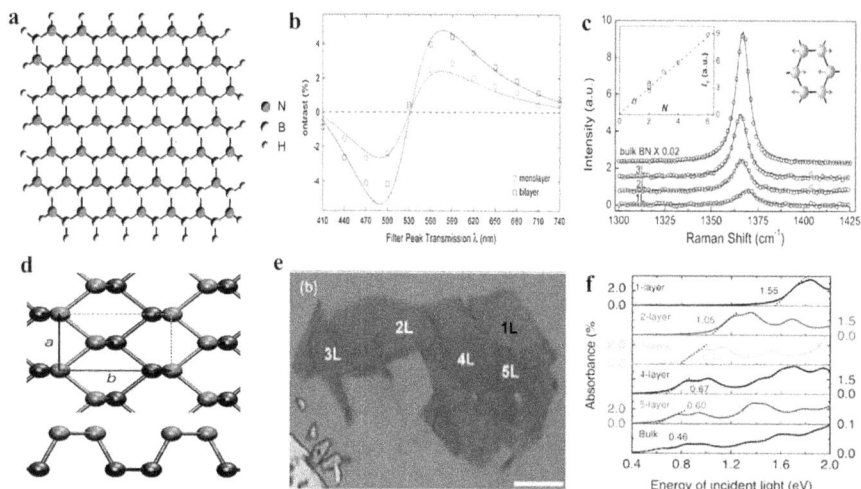

FIGURE 1.2 (a) Structures of h-BN. Image reproduced with permission copyright © 2017 RSC publishing [31]. (b) Difference in optical contrast of mono and bilayer h-BN with respect to wavelength of light. Image reproduced with permission copyright © 2017 Elsevier [32]. (c) Raman spectra of h-BN. Image reproduced with permission copyright © 2017 RSC publishing [31]. (d) Structure of bulk phosphorene reproduced with permission copyright © 2016 RSC publishing [35]. (e) AFM image of phosphorene. (f) Optical absorption spectra of phosphorene. Image reproduced with permission copyright © 2017 Elsevier [36].

techniques. Mono and few layered BN are not easily distinguishable with optical microscopy; however, subtle changes in optical contrast in monochromatic light can differentiate the change in thickness of 2D materials as shown in Figure 1.2b. Raman spectroscopy is a direct tool for the detection of h-BN due to the prominent E_{2g} Raman vibration mode of h-BN at 1366 cm^{-1}. The intensity of the Raman E_{2g} band is also associated with the number of layers and decreasing the number of layers proportionally decreases the intensity as shown in Figure 1.2c. A high-angled annular dark field scanning transmission electron microscope (HAADF-STEM) is used to analyse individual atoms, defects, and grain size of h-BN. Again, the topography is precisely analysed by tapping mode in an atomic force microscope (AFM) [31,32].

1.2.1.3 Phosphorous-Based Materials

Elemental phosphorous exhibits multiple bulk phases, viz., white, red, purple, and black phosphorous. 2D phosphorene as shown in Figure 1.2d is usually synthesized using exfoliation of black phosphorous. Different bottom-up methods like chemical vapour deposition and wet-chemical synthesis are also used to synthesize atomically thin phosphorene. Phosphorene exhibits an anisotropic orthorhombic lattice structure ensuing incredible mechanical, optical, electronic, and transport properties. This semiconductor has a moderate and tunable band gap of 0.3–2.0 eV which is crucial for optoelectronic devices, sensors, and solar cells. Defect-induced, doped, and 2D heterojunctions of phosphorene have attained growing applications in catalysis, nanoelectronic devices, sensors, energy storage, and thermoelectric applications [33,34].

Similar to borophene, Raman spectroscopy and HAADF-STEM are essential for the characterization of phosphorene. The topography and number of layers can be visualized with AFM as shown in Figure 1.2e [35]. The high anisotropic nature of phosphorene affects its light absorption and this nature is used to experimentally identify the orientation of layered phosphorene as depicted in Figure 1.2f [36].

1.2.2 2D Metallic Nanomaterials

1.2.2.1 2D Metals and Alloys

2D metals are a relatively new addition to the family of 2D materials and a number of metallic nanosheets have been successfully synthesized and studied till date. 2D metals and alloys are primarily prepared as atomistically thin films as well as freestanding ultrathin films with varying thicknesses. The strong non-directional metallic bonding creates thermodynamic instability in their 2D forms, particularly in freestanding geometry [37]. However, Nevaletia et al. in 2019 explored the liquid drop model with DFT to analyse the energetic stability of 45 elemental metal candidates. This study demonstrated the role of pores in covalent 2D templates for the stabilization of 2D metal patches in respective sizes [38]. Top-down methods used for the fabrication of 2D metals are mechanical compression and polymer surface buckling-enabled exfoliation; however, these methods yield large lateral dimensions. Wet chemical methods like small molecule-guided growth and ligand-assisted growth usually result in well-controlled geometry, surface-to-volume ratio, and crystalline structure [7]. The high electrical mobility, high surface-to-volume ratio, and localized surface plasmon resonance of these novel materials have resulted in fascinating physical, chemical, and optical properties. 2D metals and alloys have found applications in electrocatalysis, energy storage, biomedical imaging, and magnetic memory devices [7,37].

The morphology of 2D metals and alloys is analysed by SEM and TEM techniques; however, precise experimental visualization of elemental distribution is achieved by HAADF-STEM and electron energy loss spectroscopy (EELS). Figure 1.3 a, b displays HRTEM and SAED patterns of Pd nanosheets synthesized using small molecules and ions-mediated method [39]. The HAADF-STEM analysis in Figure 1.3c depicts Sn segregation in Al-Cu-Sn alloy which infers low solid solubility of Sn in Al matrix [40]. Again, the remarkable magnetic anisotropy of nano-Co as depicted in Figure 1.3d, e indicates 2D growth [41].

1.2.2.2 2D Metal Oxides

Based on their structural characteristics, 2D metal oxides are categorized as layered, lamellar, and non-layered metal oxides. MoO_3, V_2O_5, and layered double hydroxides (LDH) have a layered structure. Here, each atomic layer indicates octahedral metal and oxygen atoms [42]. LDH are composed of stacked brucite-type layers of divalent and trivalent metals with intercalated anions [43]. 2D perovskites have a lamellar structure where weak electrostatic forces bond the metals to the oxide layer. In 2018, Chen et al. reported a similar structure for Bi_2O_2Se [44]. CeO_2,

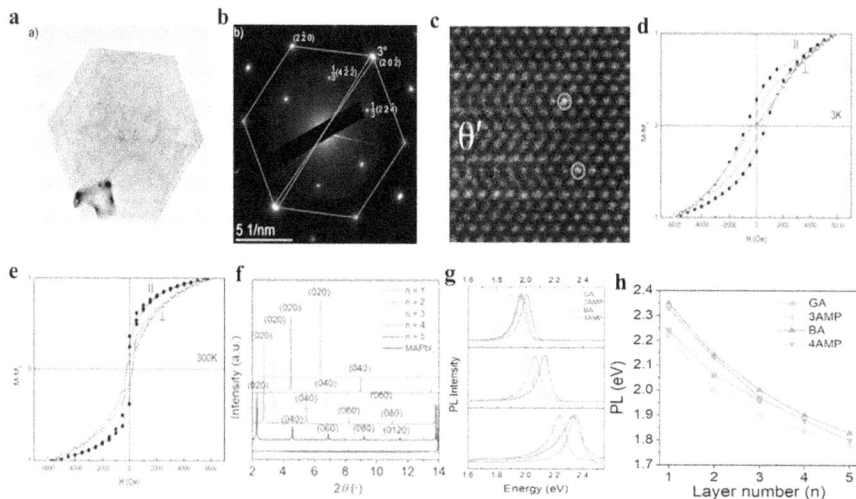

FIGURE 1.3 (a) HRTEM image and (b) SAED pattern of multilayer Pd nanosheets. Image reproduced with permission copyright © 2014 ACS publishing [39]. (c) HAADF-STEM image of Sn rich column in Al-Cu-Sn alloy. Reproduced with permission copyright © 2017 IOP Science [40]. (d, e) Hysteresis loops with magnetic field applied parallel and perpendicular to 2D Co nanoparticles. Image reproduced with permission copyright © 2007 ACS publishing [41]. (f) High-resolution PXRD of a perovskite (n = 1 − 5), (g, h). Evaluation of PL emission of different perovskites (n = 1 − 3). Image reproduced with permission copyright © 2018 ACS publishing [45].

SnO_2, and WO_3 have a non-layered structure. Mechanical exfoliation, ultrasonic-assisted liquid phase exfoliation, intercalation-assisted exfoliation, and electrochemical exfoliation techniques are frequently used to obtain 2D metal oxides from bulk. Chemical vapour deposition, solvothermal method, and atomic layer deposition techniques are used for high-quality synthesis of 2D metal oxides [42]. These materials have shown bright potential in the fields of catalysis, sensing, energy storage, and smart devices.

The characterization of 2D metal oxides follows the same route as 2D metals where XRD, SEM, TEM, HAADF-STEM, EELS, and XPS analysis are essential for understanding at the atomic level. Characterization of photo-active 2D metal oxides can be achieved by UVDRS, PL, and TRPL analysis. Figure 1.3f depicts the X-ray diffraction pattern of a perovskite in relation to the size of the constituent 2D slabs. Again, the comparison of PL emission of diverse hybrid halide perovskites and the relation of the size of the constituent 2D slabs to PL emission energy is evident in Figure 1.3g, h [45].

1.2.2.3 2D Metal Carbide/Nitride and MXenes

Since the discovery of Ti_3C_2 in 2011, transition metal carbides, nitrides, and carbonitrides (family of MXenes) have gained enormous attention due to their versatile and tunable chemistry. MXenes are synthesized with the etching of selected layers from the precursor (MAX phase), where aqueous fluoride containing acidic solution is typically used for selective etching. 2D metal carbides and nitrides are also synthesized using bottom-up methods like chemical vapour deposition and pulsed layer deposition [46,47]. MXene films are transparent, conductive, and have shown exquisite mechanical strength [47,48]. Based on the M, X, and surface termination MXene shows metallic to semiconductor properties. MXene flakes are not quite stable in the presence of oxygen, moisture, and light. The family of MXenes has displayed applications in energy generation/storage, reinforcement for composites, bio/gas sensors, water purification/desalination, and catalysis [47].

The diversity of MXenes structures and compositions demands the use of sophisticated analysis techniques to understand the precursor (MAX phase), track and verify successful synthesis, and evaluate structure, composition, and properties. The XRD pattern in Figure 1.4a of $Mo_XV_{X-4}AlC_3$ shows the purity of the MAX phase for MXenes synthesis where multiple low-intensity peaks for impurities are observed. Figure 1.4b displays close-ups of (002) and (100) peak shifts as a function of a slight change in the composition of the MAX phase. This assists the researchers in evaluating the stoichiometry of the precursors with lattice parameters. Figure 1.4b shows high-resolution STEM (HRSTEM) of different MAX phases where distinct M layers are observed with little to no merging which is favourable for MXene synthesis. Various stages of MXene synthesis were monitored with XRD patterns in Figure 1.4 f. The SEM images in Figure 1.4g–j analyse the effect of the concentration of HF acid in the etching process. Apart from the above-described techniques, X-ray absorption spectroscopy (XAS), AFM, XPS, and several other characterization techniques are frequently explored [49,50].

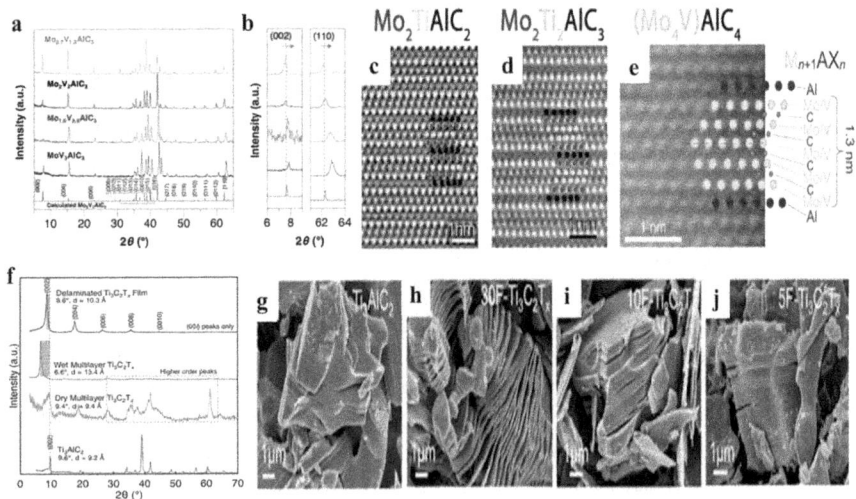

FIGURE 1.4 (a, b) XRD pattern of $Mo_XV_{X-4}AlC_3$ pressed powders. Images reproduced with permission copyright © 2020 RSC publishing [50]. (c, d, e) HRSTEM of different MAX phases. (f) XRD pattern of Ti_3AlC_2 MAX phase during different stages of synthesis. SEM images of (g) Ti_3AlC_2 MAX phase, (h, i. j) MXenes etched with different percentages of HF. Images reproduced with permission copyright © 2021 Elsevier [49].

1.2.2.4 Transition Metal Dichalcogenides

These 2D layered materials with unique optical and electronic properties include transition metal sulphides, selenides, and tellurides. Mechanical and chemical exfoliation processes are employed to synthesize 2D flakes of transition metal dichalcogenides (TMD). However, bottom-up growth methods like chemical vapour deposition, molecular beam epitaxy, and wet chemical methods yield a precise layered structure. These materials exhibit hexagonal or trigonal antiprism arrangements. Depending on their chemical composition, these materials show diverse electronic properties ranging from semiconductors to superconductors. As their bandgap lies in the visible region of the electromagnetic spectrum, UV-Vis spectroscopy, fluorescence, and Raman spectroscopy are used to study the optical properties [51,52]. MoS_2 is one of the most investigated TMD for various applications including water electrolysis, where it displays comparable efficiency to noble metal catalysts [53]. First-row transition metal selenides are increasingly favoured for the design of electrocatalysts and electrode materials due to the metallic nature of Se imparting enhanced electronic conductivity [54]. First-row transition metal sulphides and selenides and their heterostructures with other 2D materials is a promising strategy for the design of sensors, energy storage devices, optoelectronic devices, and catalysis.

The above-described structural and morphological analyses are used to characterize TMDs; the initial growth and defect density of the TMD layers over the substrate are visualized with scanning tunnelling microscopy as displayed in Figure 1.5a.

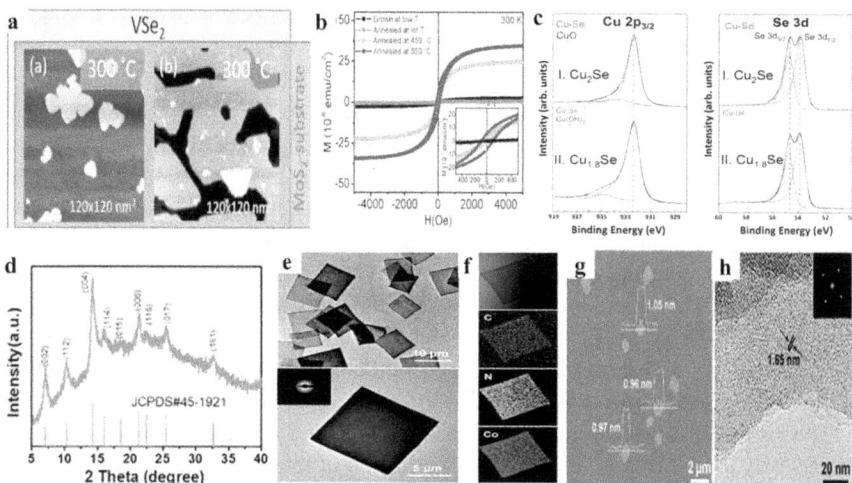

FIGURE 1.5 (a) Comparison of initial growth of VSe_2 over MoS_2 substrate at 300 °C growth temperature. (b) Magnetometry measurements of VSe_2. Image reproduced with permission copyright © 2021 Elsevier [55]. (c) HR-XPS of Cu 2p and Se 3d regions of Cu_2Se and $Cu_{1.8}Se$. Image reproduced with permission copyright © 2020 MDPI [56]. (d) PXRD of Co-MOF, (e) HRTEM images of Co-MOF nanosheets and corresponding SAED pattern, and (f) dark field TEM image and elemental mapping of Co-MOF. Image reproduced with permission copyright © 2018 RSC publishing [62]. (g) AFM image, (h) HRTEM and FTT pattern of ZnPd-MOF nanosheet. Image reproduced with permission copyright © 2017 ACS publishing [63].

Low-energy electron diffraction analysis is also used to monitor the growth of TMDs over different substrates. The magnetometry measurements displayed in Figure 1.5b depict that the VSe_2 grown in low temperature shows no magnetic moment, whereas strong ferromagnetic hysteresis is observed parallel to post-growth annealing [55]. The activity of the TMDs is highly dependent on the precise stoichiometry and surface chemical environment of the elements present. XPS analysis is used to monitor stoichiometry, surface chemical environment, as well as heteroatom doping in TMDs. Figure 1.5c demonstrates the subtle differences in intensity and binding energies of Cu_2Se and $Cu_{1.8}Se$ materials [56].

1.2.2.5   2D Metal-Organic Framework

This next-generation material is distinctive in terms of high surface-to-volume ratio and ultrathin layered structure. Top-down methods, e.g., ball-milling, sonication, chemical exfoliation, and Li-intercalation exfoliation, are used to prepare 2D MOF. However, the yields of these methods are 15–20% high yield and even exfoliation is a challenge. Some combination approaches involving sonication/Li-intercalation exfoliation and chemical reduction have been explored which significantly increased the yield. Straight-forward synthesis using bottom-up methods typically restricts vertical stacking of the layers which encourages 2D growth [57]. Interfacial synthesis extensively used the method where the MOF grow only in the confined interface, ensuing 2D nanosheets [58]. To slow down the diffusion and growth rate, a three-layer crystal growth strategy is employed. Two miscible solvents with different densities are used where the low-density solvent with the metal ions goes to top and the high-density solvent with the organic linker occupies the bottom layer with a middle buffer layer containing equal amounts of both solvents. In the static condition, the migration of the metal ions and organic linker to the middle layer is decelerated, encouraging the synthesis of 2D MOF nanosheets [59]. Surfactants and some small organic molecules are used to promote 2D growth during wet chemical synthesis [57]. Sonication during the synthesis is a novel and environmentally benign method for the direct synthesis of MOF nanosheets [60]. These 2D materials have shown potential as highly selective and ultra-permeable membranes for gas separation and sensing owing to their nanometre thickness and distinct pore structure [61]. The abundance of exposed metal sites and high surface-to-volume ratio have inspired applications in catalysis, energy conversion, and storage. Biodegradability and controllable surface functionalities have displayed potential in biomedicine and drug delivery [57].

The results of structural, morphological, textural, and physicochemical characterizations play a vital role in the targeted application of these nanosheets. The XRD of MOF nanosheets resembles their 3D counterpart; however, a slight decrease in crystalline nature is usually observed. Figure 1.5d depicts the powder XRD pattern of Co-MOF synthesized using the bottom-up method. Figure 1.5e shows overlapping Co-MOF nanosheets with a thickness ~40 nm and a lateral dimension of ~8 μm, whereas Figure 1.5f shows the uniform presence of the elements on the nanosheet [62]. The thickness of the ZnPd-MOF nanosheet was observed to be ~1.0 nm with AFM as shown in Figure 1.5g. The FFT pattern in Figure 1.5h showed a four-fold symmetry, which infers an intact crystalline structure after exfoliation [63].

1.3 CONCLUSION

The research on the fabrication of 2D materials has skyrocketed in the past few decades benefitting from rising demands for future technologies and advancement in sophisticated analysis tools. In this chapter, we have summarized research advancements made in synthesis strategies, characterization tools, and material-specific interpretation of analysis results for a thorough understanding of properties and assessment of potential applications. Metal-free 2D functional nanomaterials are highly desirable due to their inexpensive and environment-friendly nature; therefore, these materials have been extensively studied for the past several decades. Owing to the tunable morphological/electronic nature, fascinating surface properties, and anisotropic structure, these materials and their composites have attracted diverse areas of applications in catalysis, energy generation/storage, sensing, and bio-imaging. These materials also perform as host materials or templates for functional material design. The development of sophisticated analytical tools has allowed better insights into their fabrication and chemistry to unlock the full potential of these materials.

The inherent isotropic nature of metals imposes serious challenges towards fabrication and stability of 2D metal nanomaterials with surface energy in comparison to their 0D, 1D, and 3D counterparts. This chapter discusses different methodologies to prepare stable 2D metal nanomaterials as well as their alloys with well-defined geometry, composition, and lattice structure. The structure and stability of different metals in 2D form were analysed with ab initio calculations. The characterization of these nanomaterials at the atomistic level is also stated here. A simple overview of metal-containing compounds, viz., oxides/hydroxides, carbides, nitrides, and dichalcogenides, are elaborated here in terms of fabrication, characterization, and potential applications. Novel 2D nanomaterials like MXene and 2D MOFs are also detailed here in terms of synthetic strategies along with the challenges, stability, and potential applications. The sophisticated characterization tools used for these materials are cited in the respective sections. Based on the aforementioned discussion, we believe that the research on 2D nanomaterials has opened leeway for mesmerizing physicochemical properties as well as improvement in characterization tools.

ACKNOWLEDGEMENTS

Dr Srikanth Ponnada would like to thank his postdoctoral supervisor, Prof Andy Herring and Colorado School of Mines, USA, for the postdoctoral funding and resources. Also, the authors would like to thank the Department of Science and Technology, India, and University Grants Commission, India.

CONFLICTS OF INTEREST

The authors declare no competing interest.

REFERENCES

1. Mas-Balleste, R., Gomez-Navarro, C., Gomez-Herrero, J. and Zamora, F., 2011. 2D materials: to graphene and beyond. *Nanoscale, 3*(1), pp.20–30.
2. Geim, A.K., 2009. Graphene: status and prospects. *Science, 324*(5934), pp.1530–1534.
3. Rosso, C., Filippini, G., Criado, A., Melchionna, M., Fornasiero, P. and Prato, M., 2021. Metal-free photocatalysis: two-dimensional nanomaterial connection toward advanced organic synthesis. *ACS Nano, 15*(3), pp.3621–3630.
4. Novoselov, K.S., Jiang, D., Schedin, F., Booth, T.J., Khotkevich, V.V., Morozov, S.V. and Geim, A.K., 2005. Two-dimensional atomic crystals. *Proceedings of the National Academy of Sciences, 102*(30), pp.10451–10453.
5. Chen, Z.G., Zou, J., Liu, G., Li, F., Wang, Y., Wang, L., Yuan, X.L., Sekiguchi, T., Cheng, H.M. and Lu, G.Q., 2008. Novel boron nitride hollow nanoribbons. *ACS Nano, 2*(10), pp.2183–2191.
6. Sasaki, T., Watanabe, M., Hashizume, H., Yamada, H. and Nakazawa, H., 1996. Macromolecule-like aspects for a colloidal suspension of an exfoliated titanate. Pairwise association of nanosheets and dynamic reassembling process initiated from it. *Journal of the American Chemical Society, 118*(35), pp.8329–8335.
7. Chen, Y., Fan, Z., Zhang, Z., Niu, W., Li, C., Yang, N., Chen, B. and Zhang, H., 2018. Two-dimensional metal nanomaterials: synthesis, properties, and applications. *Chemical Reviews, 118*(13), pp.6409–6455.
8. Wang, F., Wang, Z., Shifa, T.A., Wen, Y., Wang, F., Zhan, X., Wang, Q., Xu, K., Huang, Y., Yin, L. and Jiang, C., 2017. Two-dimensional non-layered materials: synthesis, properties and applications. *Advanced Functional Materials, 27*(19), p.1603254.
9. Liz-Marzán, L.M. and Grzelczak, M., 2017. Growing anisotropic crystals at the nanoscale. *Science, 356*(6343), pp.1120–1121.
10. Hu, H., Zhou, J., Kong, Q. and Li, C., 2015. Two-dimensional Au nanocrystals: shape/size controlling synthesis, morphologies, and applications. *Particle & Particle Systems Characterization, 32*(8), pp.796–808.
11. Lohse, S.E. and Murphy, C.J., 2013. The quest for shape control: a history of gold nanorod synthesis. *Chemistry of Materials, 25*(8), pp.1250–1261.
12. Shelke, N.T. and Late, D.J., 2021. Synthesis and characterization of 2D materials. In *Fundamentals and Supercapacitor Applications of 2D Materials* (Elsevier), pp. 77–104.
13. Kumar, R., Joanni, E., Singh, R.K., Singh, D.P. and Moshkalev, S.A., 2018. Recent advances in the synthesis and modification of carbon-based 2D materials for application in energy conversion and storage. *Progress in Energy and Combustion Science, 67*, pp.115–157.
14. Ouyang, J., Rao, S., Liu, R., Wang, L., Chen, W., Tao, W. and Kong, N., 2022. 2D materials-based nanomedicine: from discovery to applications. *Advanced Drug Delivery Reviews*, p.114268.
15. Tan, T., Jiang, X., Wang, C., Yao, B. and Zhang, H., 2020. 2D material optoelectronics for information functional device applications: status and challenges. *Advanced Science, 7*(11), p.2000058.
16. Mannix, A.J., Kiraly, B., Hersam, M.C. and Guisinger, N.P., 2017. Synthesis and chemistry of elemental 2D materials. *Nature Reviews Chemistry, 1*(2), p.0014.
17. Jana, S., Bandyopadhyay, A., Datta, S., Bhattacharya, D. and Jana, D., 2021. Emerging properties of carbon based 2D material beyond graphene. *Journal of Physics: Condensed Matter, 34*(5), p.053001.
18. Song, Q., Wang, B., Deng, K., Feng, X., Wagner, M., Gale, J.D., Müllen, K. and Zhi, L., 2013. Graphenylene, a unique two-dimensional carbon network with nondelocalized cyclohexatriene units. *Journal of Materials Chemistry C, 1*(1), pp.38–41.

19. Baughman, R.H., Eckhardt, H. and Kertesz, M., 1987. Structure-property predictions for new planar forms of carbon: layered phases containing sp 2 and sp atoms. *The Journal of Chemical Physics*, *87*(11), pp.6687–6699.
20. Dreyer, D.R., Todd, A.D. and Bielawski, C.W., 2014. Harnessing the chemistry of graphene oxide. *Chemical Society Reviews*, *43*(15), pp.5288–5301.
21. Ong, W.J., Tan, L.L., Ng, Y.H., Yong, S.T. and Chai, S.P., 2016. Graphitic carbon nitride (g-C3N4)-based photocatalysts for artificial photosynthesis and environmental remediation: are we a step closer to achieving sustainability?. *Chemical Reviews*, *116*(12), pp.7159–7329.
22. Zhu, J., Xiao, P., Li, H. and Carabineiro, S.A., 2014. Graphitic carbon nitride: synthesis, properties, and applications in catalysis. *ACS Applied Materials & Interfaces*, *6*(19), pp.16449–16465.
23. Alahakoon, S.B., Diwakara, S.D., Thompson, C.M. and Smaldone, R.A., 2020. Supramolecular design in 2D covalent organic frameworks. *Chemical Society Reviews*, *49*(5), pp.1344–1356.
24. Wang, J., Li, N., Xu, Y. and Pang, H., 2020. Two-dimensional MOF and COF nanosheets: synthesis and applications in electrochemistry. *Chemistry – A European Journal*, *26*(29), pp.6402–6422.
25. Wu, J.B., Lin, M.L., Cong, X., Liu, H.N. and Tan, P.H., 2018. Raman spectroscopy of graphene-based materials and its applications in related devices. *Chemical Society Reviews*, *47*(5), pp.1822–1873.
26. Tonda, S., Kumar, S., Kandula, S. and Shanker, V., 2014. Fe-doped and-mediated graphitic carbon nitride nanosheets for enhanced photocatalytic performance under natural sunlight. *Journal of Materials Chemistry A*, *2*(19), pp.6772–6780.
27. Sergeeva, A.P., Popov, I.A., Piazza, Z.A., Li, W.L., Romanescu, C., Wang, L.S. and Boldyrev, A.I., 2014. Understanding boron through size-selected clusters: structure, chemical bonding, and fluxionality. *Accounts of Chemical Research*, *47*(4), pp.1349–1358.
28. Liu, Y., Penev, E.S. and Yakobson, B.I., 2013. Probing the synthesis of two-dimensional boron by first-principles computations. *Angewandte Chemie International Edition*, *52*, pp.3156–3159.
29. Naclerio, A.E. and Kidambi, P.R., 2023. A review of scalable hexagonal boron nitride (h-BN) synthesis for present and future applications. *Advanced Materials*, *35*(6), p.2207374.
30. Molaei, M.J., Younas, M. and Rezakazemi, M., 2021. A comprehensive review on recent advances in two-dimensional (2D) hexagonal boron nitride. *ACS Applied Electronic Materials*, *3*(12), pp.5165–5187.
31. Wang, J., Ma, F. and Sun, M., 2017. Graphene, hexagonal boron nitride, and their heterostructures: properties and applications. *RSC Advances*, *7*(27), pp.16801–16822.
32. Khan, M.H., Liu, H.K., Sun, X., Yamauchi, Y., Bando, Y., Golberg, D. and Huang, Z., 2017. Few-atomic-layered hexagonal boron nitride: CVD growth, characterization, and applications. *Materials Today*, *20*(10), pp.611–628.
33. Chaudhary, V., Neugebauer, P., Mounkachi, O., Lahbabi, S. and EL FATIMY, A., 2022. Phosphorene – an emerging two-dimensional material: recent advances in synthesis, functionalization, and applications. *2D Materials*, *9*(2022), p.032001.
34. Akhtar, M., Anderson, G., Zhao, R., Alruqi, A., Mroczkowska, J.E., Sumanasekera, G. and Jasinski, J.B., 2017. Recent advances in synthesis, properties, and applications of phosphorene. *npj 2D Materials and Applications*, *1*(1), p.5. npj 2D Mater Appl 1, 5 (2017).
35. Vierimaa, V., Krasheninnikov, A.V. and Komsa, H.P., 2016. Phosphorene under electron beam: from monolayer to one-dimensional chains. *Nanoscale*, *8*(15), pp.7949–7957.

36. Khandelwal, A., Mani, K., Karigerasi, M.H. and Lahiri, I., 2017. Phosphorene – the two-dimensional black phosphorous: properties, synthesis and applications. *Materials Science and Engineering: B, 221*, pp.17–34.

37. Wang, T., Park, M., Yu, Q., Zhang, J. and Yang, Y., 2020. Stability and synthesis of 2D metals and alloys: a review. *Materials Today Advances, 8*, p.100092.

38. Nevalaita, J. and Koskinen, P., 2019. Stability limits of elemental 2D metals in graphene pores. *Nanoscale, 11*, pp.22019–22024.

39. Yin, X., Liu, X., Pan, Y.T., Walsh, K.A. and Yang, H., 2014. Hanoi tower-like multilayered ultrathin palladium nanosheets. *Nano Letters, 14*(12), pp.7188–7194.

40. Nie, J.F., 2017, July. Applications of atomic-resolution HAADF-STEM and EDS-STEM characterization of light alloys. In IOP Conference Series: Materials Science and Engineering (Vol. 219, No. 1, p. 012005). IOP Publishing.

41. Xu, R., Xie, T., Zhao, Y. and Li, Y., 2007. Single-crystal metal nanoplatelets: cobalt, nickel, copper, and silver. *Crystal Growth & Design, 7*(9), pp.1904–1911.

42. Xie, H., Li, Z., Cheng, L., Haidry, A.A., Tao, J., Xu, Y., Xu, K. and Ou, J.Z., 2022. Recent advances in the fabrication of 2D metal oxides. *iScience, 25*(1), p.103598.

43. Mishra, G., Dash, B. and Pandey, S., 2018. Layered double hydroxides: a brief review from fundamentals to application as evolving biomaterials. *Applied Clay Science, 153*, pp.172–186.

44. Chen, C., Wang, M., Wu, J., Fu, H., Yang, H., Tian, Z., Tu, T., Peng, H., Sun, Y., Xu, X. and Jiang, J., 2018. Electronic structures and unusually robust bandgap in an ultrahigh-mobility layered oxide semiconductor, Bi2O2Se. *Science Advances, 4*(9), p.eaat8355.

45. Mao, L., Stoumpos, C.C. and Kanatzidis, M.G., 2018. Two-dimensional hybrid halide perovskites: principles and promises. *Journal of the American Chemical Society, 141*(3), pp.1171–1190.

46. Verger, L., Xu, C., Natu, V., Cheng, H.M., Ren, W. and Barsoum, M.W., 2019. Overview of the synthesis of MXenes and other ultrathin 2D transition metal carbides and nitrides. *Current Opinion in Solid State and Materials Science, 23*(3), pp.149–163.

47. Anasori, B., Lukatskaya, M.R. and Gogotsi, Y., 2017. 2D metal carbides and nitrides (MXenes) for energy storage. *Nature Reviews Materials, 2*(2), pp.1–17. *Nat Rev Mater* **2**, 16098 (2017).

48. Ling, Z., Ren, C.E., Zhao, M.Q., Yang, J., Giammarco, J.M., Qiu, J., Barsoum, M.W. and Gogotsi, Y., 2014. Flexible and conductive MXene films and nanocomposites with high capacitance. *Proceedings of the National Academy of Sciences, 111*(47), pp.16676–16681.

49. Shekhirev, M., Shuck, C.E., Sarycheva, A. and Gogotsi, Y., 2021. Characterization of MXenes at every step, from their precursors to single flakes and assembled films. *Progress in Materials Science, 120*, p.100757.

50. Pinto, D., Anasori, B., Avireddy, H., Shuck, C.E., Hantanasirisakul, K., Deysher, G., Morante, J.R., Porzio, W., Alshareef, H.N. and Gogotsi, Y., 2020. Synthesis and electrochemical properties of 2D molybdenum vanadium carbides–solid solution MXenes. *Journal of Materials Chemistry A, 8*(18), pp.8957–8968.

51. Manzeli, S., Ovchinnikov, D., Pasquier, D., Yazyev, O.V. and Kis, A., 2017. 2D transition metal dichalcogenides. *Nature Reviews Materials, 2*(8), pp.1–15. Nature Review Materials 2, 17033 (2017).

52. Joseph, S., Mohan, J., Lakshmy, S., Thomas, S., Chakraborty, B., Thomas, S. and Kalarikkal, N., 2023. A review of the synthesis, properties, and applications of 2D transition metal dichalcogenides and their heterostructures. *Materials Chemistry and Physics, 297*, p.127332.

53. Cao, Y., 2021. Roadmap and direction toward high-performance MoS2 hydrogen evolution catalysts. *ACS Nano*, *15*(7), pp.11014–11039.
54. Mondal, A. and Vomiero, A., 2022. 2D transition metal dichalcogenides-based electrocatalysts for hydrogen evolution reaction. *Advanced Functional Materials*, *32*(52), p.2208994.
55. Lasek, K., Li, J., Kolekar, S., Coelho, P.M., Zhang, M., Wang, Z. and Batzill, M., 2021. Synthesis and characterization of 2D transition metal dichalcogenides: recent progress from a vacuum surface science perspective. *Surface Science Reports*, *76*(2), p.100523.
56. Hamawandi, B., Ballikaya, S., Råsander, M., Halim, J., Vinciguerra, L., Rosén, J., Johnsson, M. and S. Toprak, M., 2020. Composition tuning of nanostructured binary copper selenides through rapid chemical synthesis and their thermoelectric property evaluation. *Nanomaterials*, *10*(5), p.854.
57. Zhao, M., Huang, Y., Peng, Y., Huang, Z., Ma, Q. and Zhang, H., 2018. Two-dimensional metal–organic framework nanosheets: synthesis and applications. *Chemical Society Reviews*, *47*(16), pp.6267–6295.
58. Maeda, H., Sakamoto, R. and Nishihara, H., 2017. Interfacial synthesis of electrofunctional coordination nanowires and nanosheets of bis (terpyridine) complexes. *Coordination Chemistry Reviews*, *346*, pp.139–149.
59. Rodenas, T., Luz, I., Prieto, G., Seoane, B., Miro, H., Corma, A., Kapteijn, F., Llabrés i Xamena, F.X. and Gascon, J., 2015. Metal–organic framework nanosheets in polymer composite materials for gas separation. *Nature Materials*, *14*(1), pp.48–55.
60. Zhao, S., Wang, Y., Dong, J., He, C.T., Yin, H., An, P., Zhao, K., Zhang, X., Gao, C., Zhang, L. and Lv, J., 2016. Ultrathin metal–organic framework nanosheets for electrocatalytic oxygen evolution. *Nature Energy*, *1*(12), pp.16184.
61. Peng, Y., Li, Y., Ban, Y., Jin, H., Jiao, W., Liu, X. and Yang, W., 2014. Metal-organic framework nanosheets as building blocks for molecular sieving membranes. *Science*, *346*(6215), pp.1356–1359.
62. Cheng, J., Chen, S., Chen, D., Dong, L., Wang, J., Zhang, T., Jiao, T., Liu, B., Wang, H., Kai, J.J. and Zhang, D., 2018. Editable asymmetric all-solid-state supercapacitors based on high-strength, flexible, and programmable 2D-metal–organic framework/reduced graphene oxide self-assembled papers. *Journal of Materials Chemistry A*, *6*(41), pp.20254–20266.
63. Ding, Y., Chen, Y.P., Zhang, X., Chen, L., Dong, Z., Jiang, H.L., Xu, H. and Zhou, H.C., 2017. Controlled intercalation and chemical exfoliation of layered metal–organic frameworks using a chemically labile intercalating agent. *Journal of the American Chemical Society*, *139*(27), pp.9136–9139.

2 Introduction to Batteries and Supercapacitors

Ghanshyam Varshney, Ayan Dey, and Srijan Sengupta

2.1 INTRODUCTION TO BATTERIES

Batteries use regulated chemical processes, where the intended reaction happens electrochemically and all other reactions, such as corrosion ones, are ideally absent or very tightly kinetically controlled. The precise selection of the chemical components, including their morphology and structure, is necessary to achieve the appropriate selectivity. The articles in this section discuss electrochemical energy storage in batteries, a topic that is becoming increasingly important for supplying power to high-tech gadgets and enabling a more sustainable lifestyle for permitting renewable energy sources like solar and wind, as well as a less energy-intensive transportation sector. A smart grid, which among other things, can safeguard the infrastructure in the event of future natural disasters. Despite this, there are less expensive ways to store electrical energy. There are different electrochemical devices and battery is one of them, as it converts chemical energy to electrical energy. They are an essential part of our modern life and are used in laptops, electric vehicles, as well as renewable energy systems. In this chapter, we will introduce batteries, including their history, basic principles, and types [1].

2.1.1 HISTORY OF BATTERIES

The history of batteries dates to the late 18th century when Italian physicist Alessandro Volta invented the first true battery known as the Voltaic pile. This early battery consisted of alternating layers of zinc and copper discs separated by cloth soaked in salt water or acid. When the two metals were connected by a wire, current flowed, producing a small but steady stream of electricity [2].

Over the next few decades, scientists and inventors around the world experimented with different materials and designs for batteries. In 1836, British chemist John Frederick Daniell invented the Daniell cell, which used copper and zinc electrodes immersed in separate solutions of copper sulphate and zinc sulphate. The Daniell cell provided a more stable and reliable source of electricity than the Voltaic pile, and it became the standard battery for telegraph and telephone systems for many years [3].

In the 19th and 20th centuries, several new battery technologies were developed, including the lead-acid battery and the alkaline battery. These batteries were smaller,

DOI: 10.1201/9781003404729-2

lighter, and more efficient than earlier designs, and helped to power the growth of new technologies like automobiles and portable electronics.

Today, batteries are used in different applications, from powering satellites and medical devices to storing renewable energy for homes and businesses.

2.1.2 BASIC PRINCIPLES OF BATTERIES

Batteries operate based on the principle of electrochemistry, which involves the transfer of electrons between different materials. In a battery, two electrodes are immersed in an electrolyte solution, which contains ions that can carry an electrical charge [1], [4].

When the two electrodes are connected by a wire, electrons flow from one electrode (the anode) to the other (the cathode), producing an electrical current. At the same time, ions in the electrolyte flow from the cathode to the anode, maintaining an electrical balance.

The chemical reactions that take place at the electrodes and in the electrolyte determine the voltage and capacity of the battery. In most batteries, the chemical reactions are reversible, meaning that the battery can be recharged by applying a current in the opposite direction [5].

2.1.3 TYPES OF BATTERIES

There are many different types of batteries, each with its own strengths and weaknesses. Some of the most common types of batteries are discussed below.

2.1.3.1 Primary Batteries

These batteries are those types of cells that can be used only once because they generate an irreversible reaction. They are mostly used in portable devices as they generate current instantly. Generally, in (Figure 2.1) primary batteries have a very less self-discharge rate and can be used for long-term storage purpose. Primary batteries have a better capacity and prime voltage compared to secondary batteries, a high energy density, and a sloping discharge curve according to the ratio weight/volume [6] (Figure 2.1).

Primary cells are disposable because their electrochemical reaction cannot be reversed. They contain only a fixed amount of the reacting compounds and can be discharged only once so when the full reaction of the material is completed, the new generation of energy is not possible at that time. We have to throw that particular battery so that means it is not rechargeable [7]. The reacting compounds are consumed by discharging and the cell cannot be used again. Mostly these cells are used in smoke detectors, flashlights, and remote controls. Dry cell is an example of primary-type batteries made up of an outer zinc container that work like an anode and the carbon rod or graphite acting as the cathode surrounded by powdered MnO_2 and carbon. Binding the cathode and anode, a mixture paste of NH_4Cl and $ZnCl_2$ is present as an electrolyte almost the same as zinc carbon batteries; it's also called the 'Leclanche cell'.

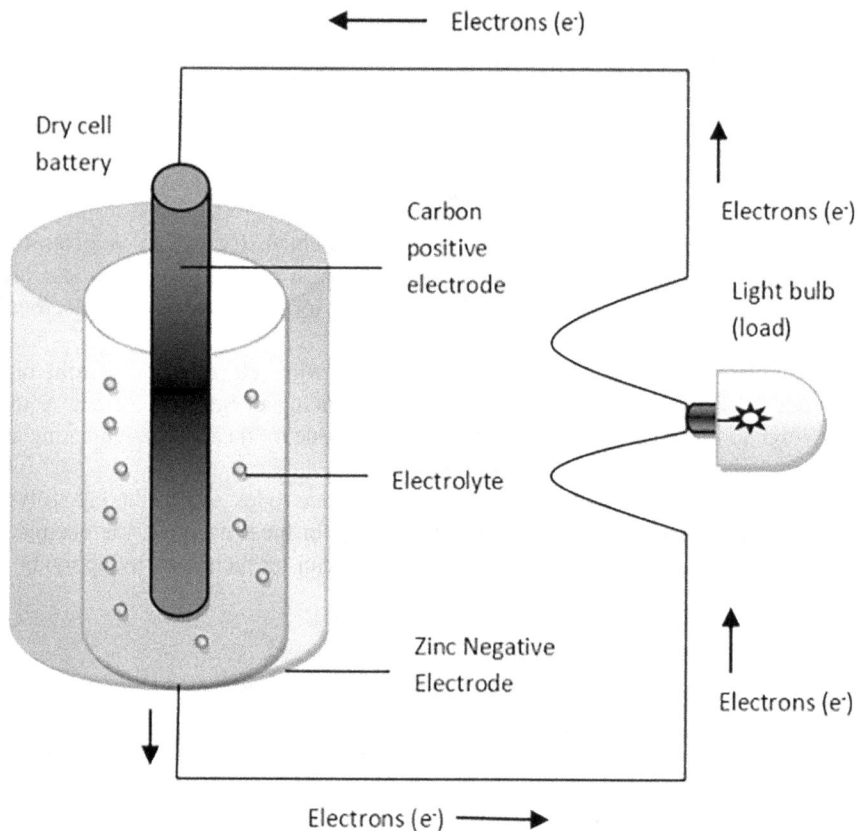

FIGURE 2.1   Scheme of a primary dry cell.

2.1.3.2 Secondary Batteries

Secondary-type cells are galvanic cells that aim to provide cheaper large-scale battery storage and help accelerate the shift to a renewable future. These cells must be charged before they can be used; they can also be recharged many times and this process is a spontaneous feasible reaction that produces electrical energy that is reversed; in other words, the charging process occurs. Reversible reactions occur in such types of batteries, which means the reactant is converted into the product and the product is converted to the reactant. Both kinds of reactions happen in secondary-type energy storage batteries [8].

During the charging condition, as the surplus of electrons is there on anode, that is why it is termed as negative electrode and electron deficit observerd at the cathode side, that is why it is designated as positive terminal.

During Discharging, as the electrons flow negative terminal (anode) to cathode (positive electrode through an external circuit and produce electricity. Furthermore this spontaneous reaction helps to create the potential difference between the two electrode cathode and anode, that is termed as electromotive force (emf).

The most commonly used secondary-type energy storage devices are mentioned below.

2.1.3.2.1   Lead-Acid Batteries

Lead-acid batteries are the oldest type of rechargeable batteries and are still widely used today in automobiles, boats, and other applications. They consist of lead plates immersed in sulphuric acid, with the lead acting as the anode and the lead dioxide acting as the cathode. Lead-acid batteries are relatively cheap and reliable but are heavy and have a relatively short lifespan. Lead acid batteries are the best examples of secondary-type energy storage devices which can be recharged a number of times and can be used for longer output. So first we are going to specify the composition of lead-acid storage cells and the amount of voltage generated by this type of battery [9]. Express the chemical reaction in half anodic and cathodic cells and define when it works like a galvanic cell and how it works like an electrolytic cell because mostly energy storage devices work like both types of cells. In the forward direction, it works like a galvanic cell and in reverse it works like an electrolytic cell [10] (Figure 2.2).

In lead-acid storage devices, pure lead metal was taken as anode and cathode is made up of lead oxide (PbO_2). Here, sulphuric acid (38% H_2SO_4 which shows the presence of water) is taken for electrolyte purpose. So, this is the statistics about the composition of lead-acid energy storage cells [11]. The types of operation we will discuss in lead acid batteries are mentioned below.

In the anodic half-cell, always at anode, pure lead involved in the loss of electrons is used to release two electrons and this process is called oxidation

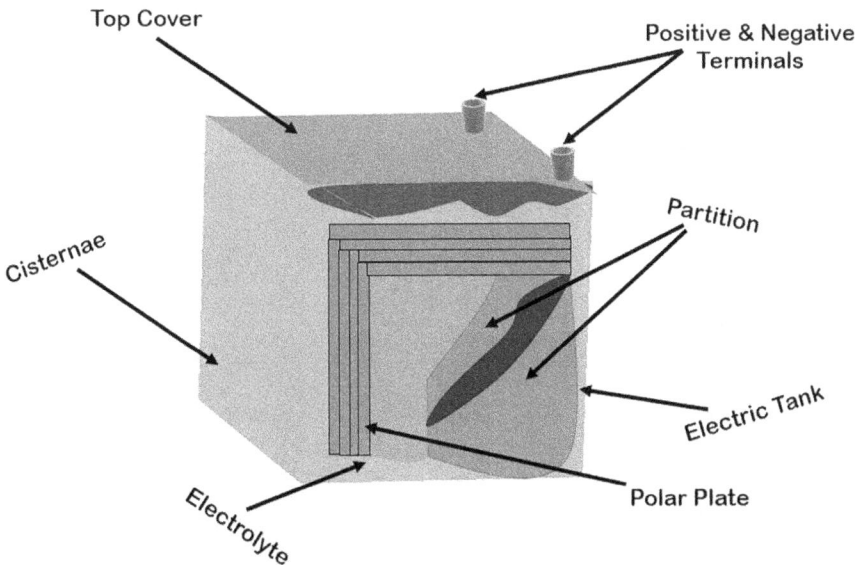

FIGURE 2.2   Schematic diagram of a lead-acid battery.

$$Pb \rightarrow Pb^{+2} + 2(e^-)$$

As it released Pb^{+2} and there is an existence of radical SO_4^- ,Which cannot exist alone and forms $PbSO_4$ combining with the Pb.

$$Pb^{+2} + SO_4^{-2} \rightarrow PbSO_4$$

As H^+ ion presents in the solution from H_2SO_4 and it will start accept electrons from the PbO_2 cathode and SO_4^- ions, thereafter it wil donate electrons to Pb plate, as a result imbalance of electron phenomenon occurs. This drives the current through external circuit to balance it, which is termed as discharging phenomenon of this Lead-acid battery. The possible reaction are shown below.

$$Pb^{+4}O_2 + 2(e^-) + 4H^+ \rightarrow Pb^{+2} + 2H_2O$$

So, the net final reaction product is

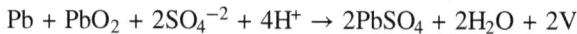

$$Pb + PbO_2 + 2SO_4^{-2} + 4H^+ \rightarrow 2PbSO_4 + 2H_2O + 2V$$

In this reaction, consider 2 (volts) of current is generated during the conversion of shorting material into product.

2.1.3.2.2 Nickel-Cadmium Batteries

Nickel-cadmium batteries are commonly used rechargeable batteries but have largely been replaced by newer technologies. They consist of nickel oxide and cadmium electrodes immersed in a potassium hydroxide electrolyte [12]. Nickel-cadmium batteries are lightweight and have a long lifespan, but they are relatively expensive and have a high environmental impact due to the toxic cadmium used in their manufacture (Figure 2.3).

The active element of rechargeable Ni-Cd batteries used nickel hydroxide as cathode and cadmium as anode; as an electrolyte, usually potassium hydroxide can

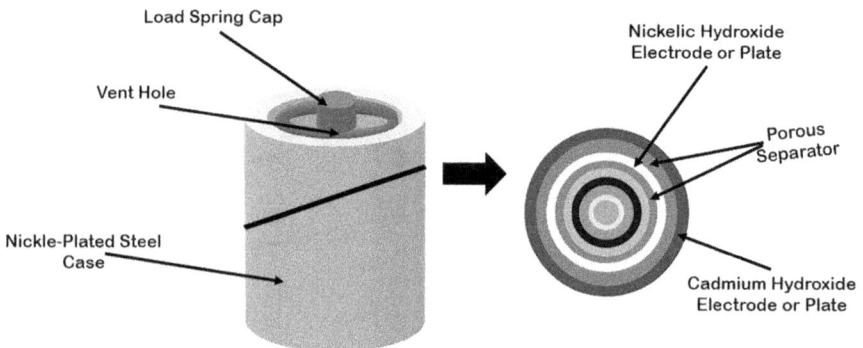

FIGURE 2.3   Scheme of Ni-Cd batteries.

be utilized due to its standard conducting property, lower internal resistance, very high supply for extremely high concern, and rapid recharge capabilities.

At the negative plate reaction, $Cd + 2OH \rightarrow Cd\ (OH)_2 + 2(e^-)$

At the positive plate reaction, $NiOOH + H_2O + 2(e^-) \rightarrow Ni\ (OH)_2 + OH$

Overall cell reactions, $2NiOOH + Cd + 2H2O \rightarrow 2Ni\ (OH)_2 + Cd\ (OH)_2$

2.1.3.2.3 Silver Zinc Energy Storage Device

In this type of energy storage device, zinc oxide is chosen for and for cathode we use the silver oxide and some content of graphite is mixed to improve its conductivity. In this rechargeable battery water cycle takes place during charging. When the cycle starts water dissociates at the positive terminal and this dissociated hydroxide ion travels to the anode side and generate water after recombination. This gives a cell voltage of 1.7V. So, zinc is going to loose mass and the silver cathode is gain mass [13]. High cost of silver is the main disadvantage of this type of energy storage batteries.

2.1.3.2.4 Lithium-Ion Batteries

Nowadays, lithium-ion batteries are the most widely used in different applications, powering everything from smartphones to electric vehicles [14]. These batteries use lithium cobalt oxide as cathode and graphite anode immersed in an electrolyte solution. Lithium-ion batteries (Figure 2.4) are lightweight, have a high energy density, and are rechargeable, making them ideal for portable electronics and electric vehicles [15]. However, these batteries can be relatively expensive and if damaged or overheated can lead to a fire hazard [16] (Figure 2.4).

2.1.3.2.5 Alkaline Batteries

Alkaline batteries are a type of disposable batteries that are commonly used in devices like flashlights and toys. They use zinc as anode and MnO_2 as cathode is

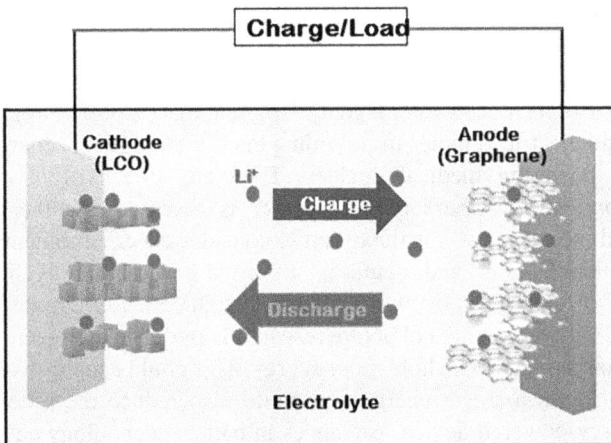

FIGURE 2.4 Schematic diagram of a LCO/graphene cell.

immersed in an alkaline electrolyte. Alkaline batteries are relatively cheap and have a long shelf life, but they are not rechargeable and not as efficient as other battery types [17].

2.1.3.2.6   Flow Batteries

Flow batteries are a type of rechargeable batteries that are used in energy storage systems for renewable energy sources. They consist of two electrolyte solutions that are stored in separate tanks and circulated through the battery cells by pumps. When the battery is charged, the electrolyte solutions flow through the cells, producing an electrical current. When the battery is discharged, the electrolyte solutions are pumped back into the tanks [8]. Flow batteries are relatively efficient and can be scaled up to provide large amounts of energy storage, but they are relatively expensive and require a large amount of space.

2.1.3.2.7   Solid-State Batteries

Solid-state batteries are a type of batteries that use a solid electrolyte instead of a liquid or gel electrolyte. This makes them safer, more efficient, and more durable than other battery types, but they are currently more expensive to manufacture and are not yet widely available [18].

2.1.4   CONCLUSION

Batteries are an essential part of modern life, providing the power for everything from portable electronics to electric vehicles and renewable energy systems. Batteries operate based on the principles of electrochemistry, and there are many different types of batteries, each with its own strengths and weaknesses. As technology advances, batteries will continue to play an important role in shaping the future of energy and transportation. However, it is important to consider the environmental impact of batteries, and to continue to improve battery recycling and the sustainability of battery production. Researchers and manufacturers are continually exploring new materials, designs, and manufacturing processes to improve battery performance and reduce costs. Solid-state batteries, alternative chemistries, and 3D printing are just a few examples of the areas of active research in battery technology.

Batteries also have a wide range of applications, from powering portable electronics and electric vehicles to providing energy storage for renewable energy systems and powering medical devices. They are also used in military and aerospace applications where reliable power is essential. As the demand for batteries continues to grow, it is important to consider the environmental impact of battery production, use, and disposal. Improving battery recycling and the sustainability of battery production can help to reduce the environmental impact of batteries [19]. Another area of active research is in the development of batteries that can be charged faster and hold more energy. This could enable electric vehicles to travel further on a single charge and could also reduce the time required to charge a battery-powered device. Advances in battery technology could also lead to the development of more efficient and reliable energy storage systems for homes and businesses.

Finally, as the demand for batteries continues to grow, there may be new challenges to overcome, such as the need for more efficient and sustainable methods of battery production, as well as the need for more effective battery recycling programs. Addressing these challenges will require collaboration among researchers, manufacturers, policymakers, and the public.

Furthermore, the development of batteries with higher energy densities, lower costs, and improved safety will be essential for the continued growth of battery-powered devices and renewable energy systems. This will require continued investment in battery research and development, as well as collaboration between academia, industry, and government. However, it is important to consider the environmental and social impacts of battery production, use, and disposal. The mining of raw materials such as lithium, cobalt, and nickel can have significant environmental and social impacts, and the disposal of batteries can also pose environmental and health risks. To mitigate these risks, efforts are underway to develop more sustainable methods of mining, to improve battery recycling and reuse, and to ensure that the benefits of battery technology are shared equitably. It is worth mentioning that batteries have the potential to revolutionize the transportation sector. Electric vehicles powered by batteries offer a cleaner and more sustainable alternative to vehicles powered by fossil fuels and petroleum [20]. In addition, electric vehicles powered by batteries have lower operating costs than vehicles powered by internal combustion engines, making them a more cost-effective option in the long term.

As battery technology continues to improve, we can expect to see more electric vehicles on the roads, as well as increased adoption of other forms of electric transportation such as buses, trains, and boats. These developments have the potential to significantly reduce emissions from the transportation sector, which is one of the biggest sources of greenhouse gas emissions globally [21]. The development of battery technology has been driven by a variety of factors, including the demand for portable electronics, the need for energy storage for renewable energy systems, and the desire to reduce greenhouse gas emissions from the transportation sector. As battery technology continues to improve, we can expect to see even more applications for batteries, as well as increased adoption of battery-powered devices.

However, the use of batteries also poses significant environmental and social challenges such as the mining of raw materials, the disposal of batteries, and the potential for unequal access to the benefits of battery technology. To address these challenges, it is important to invest in sustainable and responsible battery production and use, as well as to promote battery recycling and reuse [22].

In addition, researchers use battery modelling and simulation to better understand battery behaviour and optimize battery design and performance. Battery modelling and simulation can be used to predict battery behaviour under different operating conditions, to optimize battery design and operation, and to improve battery management systems [23]. Designing and optimizing batteries for specific applications is a complex and multidisciplinary process that requires a deep understanding of the physics and chemistry of batteries, as well as expertise in materials science, engineering, and modelling and simulation. By developing new materials, optimizing battery structures and compositions, using modelling and simulation, and testing and validating battery performance, researchers and engineers can continue to improve battery performance and unlock new applications and markets for battery technology [24].

2.1.5 FUTURE DEVELOPMENTS

As the demand for portable electronics, electric vehicles, and renewable energy systems continues to grow, the need for more efficient, reliable, and affordable batteries will also increase. To meet this demand, researchers and manufacturers are exploring new materials, designs, and manufacturing processes for batteries [25].

One area of active research is the development of solid-state batteries. Solid-state batteries use a solid electrolyte instead of a liquid or gel electrolyte, which can provide several advantages over traditional batteries such as higher safety, as well as higher energy density, and cycle life. However, solid-state batteries are currently more expensive to manufacture and have not yet been scaled up for mass production.

Another area of research is the use of new materials for battery electrodes. For example, silicon has been shown to have a high capacity for lithium-ion batteries, which could significantly increase their energy density. However, silicon electrodes are prone to expansion and contraction during charging and discharging, which can cause damage to the battery.

Researchers are also exploring the use of alternative chemistries for batteries such as sodium-ion and magnesium-ion batteries. These battery types could offer lower costs and improved safety compared to lithium-ion batteries, but they are still in the early stages of development and have not yet been widely adopted.

Manufacturers are also working to improve battery manufacturing processes to reduce costs and improve efficiency. For example, some manufacturers are using 3D printing to create battery components, which can reduce the amount of material waste and improve the accuracy and consistency of the manufacturing process [26].

In conclusion, the future of batteries is likely to involve continued improvements in efficiency, safety, and affordability. As new materials and manufacturing processes are developed, the performance and cost of batteries will continue to improve, making them an even more important part of modern life.

2.1.6 ENVIRONMENTAL CONSIDERATIONS

As the demand for batteries increases, it is important to consider the environmental impact of battery production, use, and disposal. Batteries can contain toxic chemicals and heavy metals, such as lead, cadmium, and mercury, which can pose a risk to human health and the environment if not properly handled.

One approach to reducing the environmental impact of batteries is to improve battery recycling. Batteries can be recycled to recover valuable materials, such as lithium, cobalt, and nickel, which can be used to manufacture new batteries. However, battery recycling is still relatively low, with many batteries ending up in landfills or being improperly disposed of.

Another approach to reducing the environmental impact of batteries is to improve the sustainability of battery production. This can involve using renewable energy sources, such as solar and wind power, to power battery factories, as well as reducing the use of toxic chemicals and heavy metals in battery manufacturing [27,28].

Finally, it is important to consider the environmental impact of battery use. For example, electric vehicles powered by batteries can reduce greenhouse emission compared to petroleum vehicles, but they can also have a higher carbon footprint if the electricity used to charge the batteries comes from fossil fuels. Renewable energy sources, such as solar and wind power, can help to reduce the carbon footprint of battery-powered devices.

2.1.7 APPLICATIONS OF BATTERIES

Batteries are used in different types of applications, from powering portable electronics and vehicles to providing energy storage for renewable energy systems. Some of the most common applications of batteries are discussed below.

2.1.7.1 Portable Electronics

Batteries are an essential part of portable electronics such as smartphones, laptops, and tablets. Lithium-ion batteries are commonly used in these devices due to their high energy density, lightweight, and rechargeability. Other battery types, such as alkaline and zinc-carbon batteries, are also used in some portable electronics.

2.1.7.2 Electric Vehicles

Batteries are also used to power electric vehicles, which are becoming increasingly popular as a more sustainable alternative to gasoline-powered vehicles. Lithium-ion batteries are commonly used in electric vehicles owing to higher energy density and rechargeability. However, other batteries, such as solid-state batteries and Na-ion batteries, are also being developed for use in electric vehicles.

2.1.7.3 Renewable Energy Systems

Batteries are an essential part of renewable energy systems such as solar and wind power. These systems rely on batteries to store energy when it is generated, so that it can be used when it is needed. Lithium-ion and flow batteries are commonly used in renewable energy systems owing to their higher energy density and ability to be scaled up to provide large amounts of energy storage.

2.1.7.4 Medical Devices

Batteries are also used in medical devices, such as pacemakers and hearing aids, which require a reliable source of power. Zinc-air batteries and silver oxide batteries are commonly used in medical devices due to their long shelf life and stable voltage output.

2.1.7.5 Military and Aerospace

Batteries are also used in military and aerospace applications, where reliable power is essential. Lithium-ion and nickel-metal hydride batteries are commonly used in these applications due to their high energy density and ability to operate in extreme temperatures.

2.2 INTRODUCTION TO CAPACITORS

Fossil fuel and different natural resources are the primary sources of energy. But the main problem with this is regulating and maintaining the supply chain of these resources. The sources can be converted into different forms and stored in grid storage for peak saving. Also, the main issue nowadays is that the demand for energy has increased day by day due to the overwhelming population around the world. As we depend on the combustion of fossil fuel which severely affects our ecology and economy of the environment, we need to think beyond that source that can fulfil the energy demand. That is why there is an increasing demand for economical, environment-friendly electrochemical energy storage devices. There are several types of devices used such as (i) batteries ii) supercapacitors, and (iii) fuel cells. Electrochemical energy storage devices are the best portfolio for futuristic clean energy systems [29]. Capacitors and supercapacitors are the most important and interesting parts among them. A great example of this is that they can act as a portable source of electricity. Another important aspect is that its potential demand will witness an increase in the next 20 years. A proper practical example of such electrochemical systems was first introduced by Professor Alessandro Volta at the University of Pavia, Italy, in 1800. In 1840, there was a big discovery in the field of electrochemical energy storage; for the first time, a high-current electroplating battery was developed, which paved the way for a new dimension. This new development led to the development of consumer market batteries and also the manufacture of electric bells for domestic and industrial applications. In the end of 1870, flashlight was invented, and after 20 years a big invention was made by Sir Edison. The idea of large-scale introduction of such types of sources has opened a new path for electricity generation and storage [30]. This is the principal motivation for the development of such technology (e.g., secondary storage systems). The global market of batteries has crossed billion. Besides those batteries, capacitors also attracted attention due to their application in portable electronics. Such electrolytic capacitors are very common in circuits. One very common example of that is in an E-Vehicle, where it has been used in combination with a battery. One of the major applications of capacitors is that they can provide peak power while travelling through steep areas, whereas batteries provide continuous supply as it operates in a low power mode [31]. This combined technology helped to improve the performance of E-vehicles. Apart from the basic introduction, in the next section, we are going to elaborate on the capacitors and how the idea of supercapacitors came from the capacitors.

2.2.1 CLASSIFICATION AND WORKING PRINCIPLE

2.2.1.1 Capacitors

A capacitor is basically a two-terminal electrical device that stores energy in the form of an electrical charge, also known as capacitance. The capacitance can be expressed by the following equation:

$$C = \frac{Q}{V}$$

where Q = Electric charge in Coulombs

C = Amount of charge stored (Capacitance) in Farad.

V = Voltage over the parallel plates.

Capacitor is a passive electronic element that consists of two or more conducting materials that is separated.

This invention of capacitor technology goes to Leyden Jar, who made this technology with the help of glass vessels and metal foils, where metal foils work as working electrodes and glass vessels act as dielectric. The dielectric is the main storage of charge that is accumulated by the capacitor and is accumulated in two ways: the positive charge is accumulated on one side of the electrode used and the negative charge is on the other side. Discharging is an important phenomenon to circulate the accumulated charge for powering any device or electronic equipment. When a external wire is connected to the electrode, this discharging phenomenon happens [29]. If we look back into the history of capacitor technology, we can see the first ever electrolytic capacitor was invented in the 1920s. There are different types of capacitors such as (i) ceramic, (ii) polymer, and (iii) electrolytic capacitors. There is a short portrayal of the different types of capacitors and their properties.

2.2.1.2   Ceramic Capacitors

Capacitors are used to store charge in the dielectric field [32]. It is one of those types of capacitors that use ceramic material which acts as a dielectric medium where the charge accumulates [33]. It is a non-polarized device, i.e., it has no specific polarity [34], so we can connect it in any direction [31]. Due to this, it is more advantageous than the common electrolyte capacitors [32], [35]. Some common compounds that are used for the ceramic capacitors (Figure 2.5) include titanium dioxide and Barium$^+$ (Figure 2.5).

2.2.1.3   Polymer Capacitors

Polymer capacitors (Figure 2.6) are mainly using conductive polymers as their electrolyte. These electrolytes are basically solid in form; this solid form of

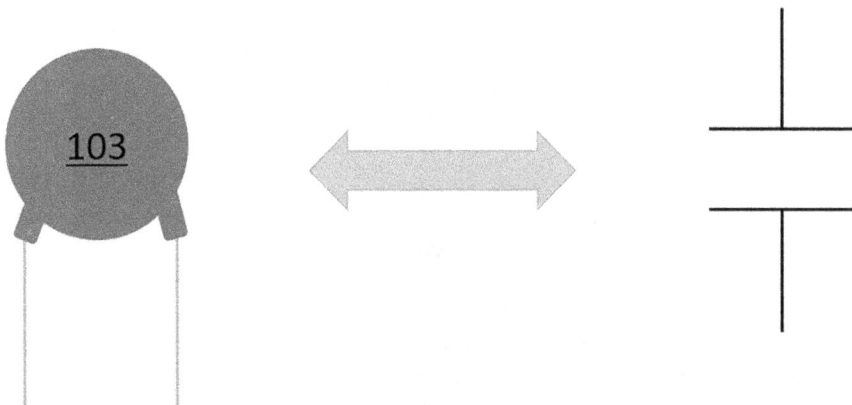

FIGURE 2.5   Ceramic capacitor.

FIGURE 2.6 Polymer capacitors.

electrolyte gives an extra advantage over the liquid one [36]. The liquid electrolyte may have a chance of drying. The dryness of the electrolyte hampers the lifetime of capacitors, for example, electrolytic capacitors showed a short lifetime due to the liquid electrolyte [37]. Polymer capacitors are superior in the life cycle, stability as well as low equivalent series resistance (ESR), which is why polymer capacitors are a good [38] replacement for electrolytic or conventional ones [39]. It can also withstand higher temperatures and its operating voltage is around 100 V DC (Figure 2.6).

Some important characteristics are listed below.

i. Lifetime – Its lifetime exceeded even after using at very high temperatures.
ii. Voltage rate and capacitance: It shows a maximum operating rate of 100 V and a capacitance rate between 10 µF and 1 mF.
iii. ESR: Its ESR does not change with the operating temperature; it seems to be constant.

The most common application of this type of capacitor is in the field of converters like DC-DC converters and buck-boost converters.

2.2.1.4 Electrolytic Capacitors

Electrolytic capacitor is a type of capacitor that uses electrolytes to get a higher amount of capacitance. So, what is an electrolyte? Electrolyte is in the form of liquid or gel, which contains high concentrations of ions that help to transport the charge carriers [40].

One of the important characteristics is, that this is a polarized capacitor, which means the voltage from the positive terminal should be higher from the negative terminal. One of the important characteristics of the electrolytic capacitor is the capacitance drift so that it can tolerate large tolerance values. Some important applications are as follows:

i. They are extensively used to reduce voltage fluctuation in various devices.
ii. They are also used for noise-filtering purposes in power supplies.

2.2.2 Energy Storage Mechanism in Capacitors

There is a difference in charge storage mechanism between batteries and capacitors. Battery basically works on the mechanism of electrochemical charge transfer reaction, whereas capacitor stores energy by separation of charge. Simple capacitor can store the charge in a thin layer of dielectric material, on which the charge has been stored. The stored energy can be calculated by $\frac{1}{2} * CV^2$. The stored energy depends upon the dielectric material that is used [41].

In battery, the charge stored mechanism totally depends upon the active material of the electrode. Here, the energy is released in the form of electricity, when an external load or wire is connected through it. The stored charge in battery is voltage-dependent, which in turn depends on the active materials and also close to the open circuit potential for those materials [42].

Electrochemical capacitors (Figure 2.7) are also known as ultracapacitor; its construction is also the same like a battery, where the electrodes are immersed into an electrolyte with a separator used to separate the two electrodes. The main criteria of choosing the electrode are that it should have a higher surface area and porous structure, should be in the range of 500–2000 m²/g, and higher than the battery. Charge is stored in the micropores of the electrode and the interface of the

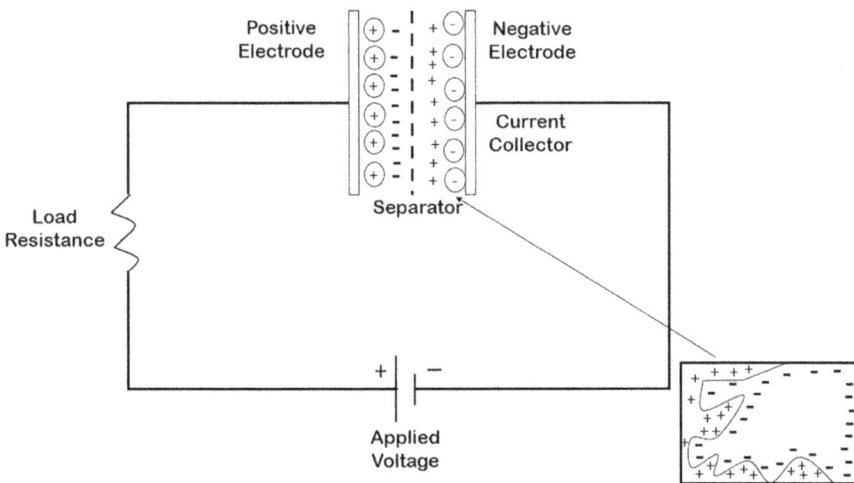

FIGURE 2.7 Schematic diagram of electrochemical capacitors.

electrode–electrolyte. Based on the different charge storage mechanisms and activities, it is quite important to discuss those processes like double layer and pseudo-capacitances [43]. The following sections explain briefly about those processes (Figure 2.7).

2.2.2.1 Double-Layer Capacitors

The charge storage mechanism of this double-layer formation works based on the interface between the electrolyte and the electrode material. A schematic diagram is also shown earlier. The energy can be calculated by $\frac{1}{2}*CV^2$ and also the accumulation of charge can be calculated by $C*V$. The capacitance mainly depends upon some parameters such as specific surface area of the active material [44] and as well as the pore size distribution. Specific capacitance of an electrode is calculated by

$$C/gm = (F/cm^2)_{act} * (cm^2/gm)_{act}$$

As the capacitance is surface area dependent, so its active pores help to accumulate the charge [45], which in turn helps to develop the double-layer capacitance [46]. Simply, it can be derived by

$$(F/cm^2)_{act} = (K/\text{thickness of double layer})_{eff}$$

The thickness of the double layer is quite small, so the surface area is higher, which results in an increase in specific capacitance from 150 to 300 F/g. That is why in most of the cases carbon materials are being used to develop the ultracapacitor, due to its higher pore size as well as surface area with different electrolytes like organic and aqueous [47]. But for aqueous, it shows less capacitance compared to organic, as the size of ions for organic electrolytes is higher than aqueous electrolytes, due to which the pore is not fully accessible to the electrolyte. Also, material with higher pores helps to enhance the capacitance [48]; lower pore size results in a large fall of capacitance. Besides pore size and surface area, the electrolyte plays an important role in cell voltage. Aqueous-based systems show a cell voltage of 1 V, whereas organic electrolyte shows 3–3.4 V [49]. Table 2.1 shows such a comparison of the following types of electrodes with their performance.

TABLE 2.1
Comparison of Electrode Performance

Choice of Electrode Material	Density (g/cm³)	Electrolyte	Capacitance (F/g)
Carbon cloth	0.35	Potassium hydroxide; organic	200
Activated carbon	0.7	Potassium hydroxide; organic	160
Anhydrous RuO_2	2.7	H_2SO_4	150
Hydrous RuO_2	2.0	H_2SO_4	650
Doped polymer	0.7	Organic	315

2.2.2.2 Charge Storage as Pseudo-Capacitance

In the case of an ideal double-layer mechanism, no such faradaic reaction happens between the interface of electrode and electrolyte, as the dQ/dv is constant and not dependent upon the voltage. But in the case of pseudo-capacitance charge storage mechanism, there is a faradaic reaction, i.e., charge transfer takes place between the solid electrode and the electrolyte interface, for that the charge transfer became voltage dependent [50]. Three important steps are involved during the development of ultracapacitor for the mechanism of pseudo-capacitance; they are as follows: (1) surface adsorption of ions from the electrolyte, (2) doping or undoping of the electrode material for the enhancement of conductivity, (3) redox reaction involved at the interfacial region of electrode and electrolyte [51]. The first and third mechanisms basically dependent on surface mechanisms, which is why it depends upon the specific surface area. The second process involves doping, un-doping, which is basically to increase the concentration of holes or electrons via adding some elements like boron or arsenic, which is why it is independent of the surface area. But it should be noted that the designed electrode must have higher electronic conductivity to accumulate or trap the electron [52]. The total mechanism can be well understood by using cyclic voltammetry, as well as EIS (electrochemical impedance spectroscopy) methods [53]. For good understanding of its characteristics, it is more prominent to use the average capacitance value as follows:

$$C_{av} = Q_{tot} / V_{tot}$$

2.2.2.3 Hybrid Capacitors

Another example of capacitors is hybrid capacitors (Figure 2.8), where two electrodes are of different types: one electrode is a double layer of carbon-based material and another electrode is made of pseudo-capacitance material. Those devices are termed as hybrid capacitors [40]. Pseudo-capacitance materials are being used as positive electrode. Metal oxide and metal oxide frameworks (MOFs) are also used as hybrid capacitors [54]. Due to this hybrid structure, its energy density is higher than the conventional one [11], [47]. Its charge storage mechanism is in one electrode with capacitive features and another with faradaic characteristics [55] (Figure 2.8).

2.2.3 ELECTRODE MATERIALS: CAPACITORS

There are different types of materials used as electrode material, some of which are enlisted below with their signifying properties.

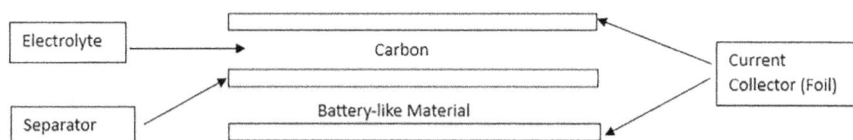

FIGURE 2.8 Schematic of a hybrid electrochemical capacitor.

TABLE 2.2
Current Density Change with Specific Capacitance

I (A)	mA/cm^2	C (F)	R (ohm)	Dry Electrode (F/g)
0.2	66	5.72	0.37	163
0.3	100	5.58	0.151	159
0.5	167	5.3	0.120	151
1.0	333	4.80	0.164	153

2.2.3.1 Activated Carbon

Electrodes are basically what? They are basically a thin layer that is being grown upon the current collector. To make the active material, we need to make a slurry by adding PVDF as a binder and carbon black as a conductive agent. The thickness is in between 100 to 400 microns; also it should stick well to the current collector, for that the contact resistance should be low. Here, activated carbon is a prime example of that as it has a higher surface area [40]. Depending upon this and pore size distribution, its specific capacitance changes gradually. The specific capacitance can be estimated by

$$(F/g)_{max} = (m^2/g) * (\mu F/cm^2) * 10^{-2}$$

Apart from the surface area, the pore size distribution also plays a huge role for specific capacitance. Current density is another aspect which significantly varies the specific capacitances. Table 2.2 shows how the current density changes are important for the change of specific capacitance.

2.2.3.2 Metal Oxides

Metal oxides are also good examples of electrode materials for capacitors. MnO_2 and RuO_2 are such metal oxides. Here, the charge transfer and storage mechanism depend upon the redox reaction and the reaction happens at the interfacial region of the electrode and electrolyte. In the previous section, we have discussed the preparation of active material's slurry making, where it is added with carbon black and binder PVDF to make a uniform, porous structure of the electrode. This active material based on metal oxides is made up of both aqueous and organic electrolytes with the cell potential ranging from 1 to 3.5 V. In recent research, different metal-organic frameworks have also been developed for the enhancement of specific capacitance [53]. Also, some new composites are made up of different metal oxides to make the different types of nanostructures so as to make it more porous, which in turn increases its energy density. These types of electrodes can also be utilized in hybrid capacitors as well as double-layer or pseudo-capacitors. The active material is not only the single constraint that is a key factor for specific capacitance; the metallic current collector also plays an important role, which will be discussed in the next section [41].

2.2.4 MATERIALS FOR CURRENT COLLECTOR

This is the most important constraint for any type of device, either it is a capacitor or a battery. The key feature associated with the current collector is that there should be zero contact resistance with the prepared active material's thin film and it should be stable with the as-prepared electrode. These are important as they are directly related to specific capacitances and cycle life. Copper and aluminium are the most used current collectors for batteries, but nowadays researchers have invented new carbon materials like carbon cloth; also Ni-foam can be used as a good current collector for different devices [56].

2.2.5 ELECTROLYTES

Electrolytes are the most important thing in any kind of electrochemical device; they ionically transport the ions from one electrode to another during the charging–discharging mechanism. Even the specific capacitance [57] and life cycle directly dependent upon this electrolyte's parameter. There are different types of electrolytes for electrochemical energy storage systems such as (i) aqueous, (ii) organic, (iii) ionic, (iv) recent trends in molten salt-based electrolytes. All of these electrolytes have different properties, which leads to the large difference in different cells. For example, activated carbon-based electrode shows OCV of 2.3–2.7 V by using organic electrolytes[58], whereas for aqueous, it shows only 0.7–1.1 V. Ionic liquid basically showed a room temperature melting point [59,60]. Some electrolytes are still of concern while used in practical implementation in vehicles, for example acetonitrile. So, researchers try to develop a low-resistive, non-toxic electrolyte that is much safer [61]. Researchers have also tried ionic liquid, as it is non-flammable and has lower toxicity [43]. But one concern about ionic liquid is its higher ionic resistivity at room temperature and cost wise it is not much feasible [46]. Some examples of different electrolytes are enlisted [62] in Table 2.3.

In the previous section, we discussed capacitors, different working mechanisms, and their charge storage mechanisms. In the case of capacitors, the charge transfer rate and amount of capacitance are lower due to lower dielectric strength. There is a need for improvement of research where supercapacitors come and play an important role, because of its higher power density, superfast charging time, and wide operating

TABLE 2.3
Different Electrolytes and Their Parameters

Electrolyte	Density (gm/cm³)	Resistivity (Ohm-cm)	Voltage (V)
KOH (aqueous)	1.29	1.9	1.0
Acetonitrile	0.78	18	2.5–3.0
Ionic liquid	1.3–1.5	126 (25°C)	4.0
		27 (100°C)	3.25

temperature. In the next section, we are going to discuss supercapacitors and their mechanisms.

2.2.6 IDEA AND INTRODUCTION OF SUPERCAPACITORS

The charge accumulation of an ideal capacitor can be characterized by equation (1)–$C = \varepsilon_0 * A/d$, where a single constant capacitance is available. The amount of energy stored by a normal capacitor is calculated by accumulating the charge between the positive and negative electrodes separated by an insulating dielectric material. In the equation (1), D represents the distance between the two plates, A represents the area of the electrode; hence, capacitance can be enhanced by increasing the specific area of the plate by decreasing the distance between the plates [30] as seen in Figure 2.9.

Supercapacitors differ from normal electrolytic capacitors; it does not involve any type of redox reaction. It yields a higher surface area electrode; the thin layer of dielectric increases its specific capacitance, higher power density, as it has lower ESR. That is why supercapacitors have an advantage over batteries, capacitors, and fuel cell in terms of power density, life cycle, and faster charging kinetics. Supercapacitors work electrostatically, where charges created by the electrolyte ion move towards the electrode surface with the opposing polarity [63]. The current collectors are coated with active material; then it is immersed into an electrolyte, being separated by an insulating layer. Due to this, when charged, opposing polarity has formed two sides of the plates, which helps to create an electric double layer. Supercapacitors are hence often called electric double-layer capacitors (EDLC) as shown in Figure 2.10. That is why supercapacitors have higher energy density than conventional capacitors. There are two types of electrochemical supercapacitors: one mechanism is based on the interfacial double layer, and the second type of supercapacitors is called 'redox capacitors' or pseudo-capacitors. There is a more

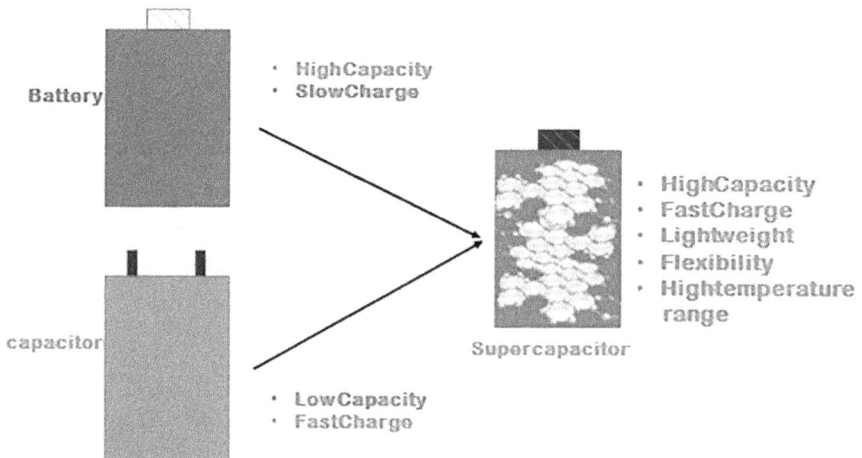

FIGURE 2.9 Supercapacitor, an advanced energy storage device.

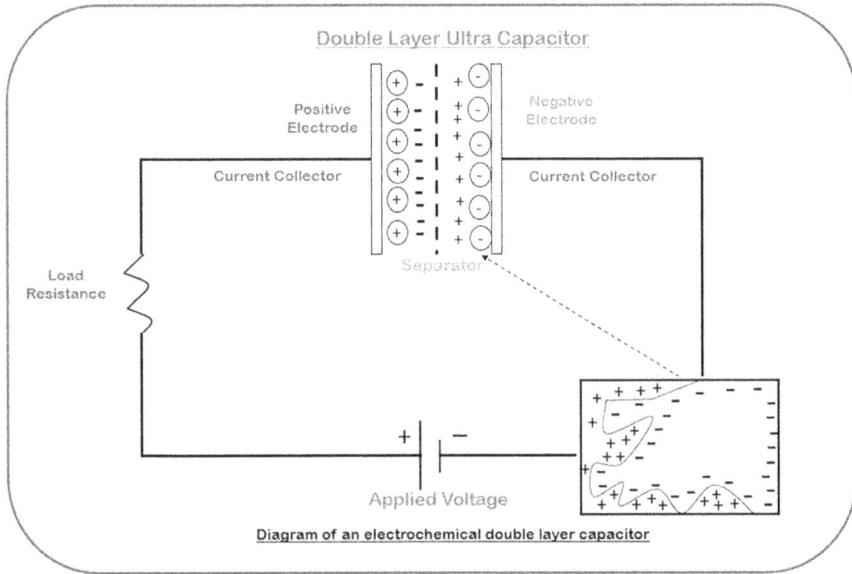

FIGURE 2.10   Schematic diagram of a double-layer supercapacitor.

prominent way to fabricate supercapacitors based on carbon-based materials. This type of electrode is prepared via pyrolysis ranging from 800 to 1100°C. This helps to enhance the specific surface area of the electrode. The specific surface area is a function of variation specific capacitance. In the case of supercapacitors, the electrolyte plays an important role in the formation of its voltage. Two types of electrolytes can be used here: (i) aqueous (ex-KOH) and (ii) Organic. The aqueous electrolyte has shown a lower resistance compared to the non-aqueous one. The aqueous one shows 10^1–10^{-2} S/cm, while the non-aqueous shows 10^{-3}–10^{-4} S/cm. In the case of organic electrolyte, the decomposition voltage is higher, i.e., 3.5 V while in the case of aqueous electrolyte, it is 1.2 V. Solid-state supercapacitors can also be fabricated, as in 1960, Gould Ionic developed a device based on silver carbon electrodes, separated by a solid electrolyte $RbAg_4I_5$, which has given a maximum decomposition potential of 0.66 V. The device was designed as a backup device for volatile memory, but it is not commercialized due to higher cost and other difficulties (Figure 2.10).

In the year 1980, Quadri Electronics developed a new improved supercapacitor device, which possesses higher peak power and current pulses higher than 10 A with rise time [43].

Some of the examples of double-layer capacitor and their characteristics are described in Table 2.4.

2.2.6.1   Redox Electrochemical Capacitors

This is the second type of supercapacitors rather than the double-layer super-capacitors, where the charge transfer mechanism is based on the non-faradaic process, termed as redox supercapacitor. This non-faradaic double-layer charging is associated with a charge transfer mechanism [56].

TABLE 2.4

Performance Parameter of Different Electrodes

Electrode Material	Electrolyte	Energy Density (Wh/kg)	Power Density (W/kg)
Carbon	Organic	2.2	400
Composite C/metal	KOH	1.2	800
Composite C/metal	Organic	7	2000
C (Synthetic)	Aqueous	1.4	1000

Generally, two types of redox electrochemical capacitors are there; this depends upon the nature of the electrode, where similar electrodes are used; they are called symmetric capacitors; when a mixture or different electrodes are used, they are called asymmetric capacitors. A few examples of metal oxide (Mos) electrodes are RuO_2 and MnO_2. Commonly used Mos is RuO_2 as it shows a higher specific surface area of around 100 m^2/g [64]. There is another class of supercapacitors that are made up of electronically conducting polymers like polyaniline (PANI) and polypyrrole [65]. These polymers have a very good specific surface area, due to which it shows higher value of specific capacitance as well as rate capability [47]. Also, these polymers can be doped or undoped based on our applications, which will prove it as a more convenient material [66], [67]. Though all of the advantages of polymer-based supercapacitors are not well commercialized yet, it is expected to explore more research in the near future. Supercapacitors have some important features and characteristics, which is enlisted below in the next section [68].

2.2.7 CHARACTERISTICS AND FEATURES OF SUPERCAPACITORS

2.2.7.1 Power Density

In comparison to conventional capacitors, the supercapacitors show higher power density. The maximum power of supercapacitors can be calculated by the maximum voltage and the ESR (electrochemical series resistance) ($P_{max} = 0.25V^2_{max} * ESR$) [69]. The power density of supercapacitors is higher as activated carbon is used as the electrode, whose surface area is higher [70]. The calculated power density is higher when the value of ESR is higher, and the value of load is the same. The typical Ragone plot [71] shows an important data about different energy storage devices (Figure 2.11).

2.2.7.2 Energy Density

Energy density is the amount of energy stored in a supercapacitor per unit mass. It is also calculated by a simple mathematical formula by integrating the area under the curve of cyclic voltammetry (CV) analysis [72]. Though it shows higher power density among all energy storage devices, but in terms of energy density, its value is lower than fuel cells and Li-ion cells [73].

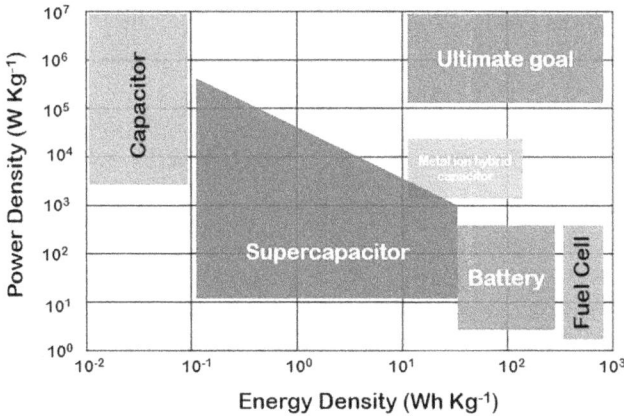

FIGURE 2.11 Ragone plot for energy storage devices [71].

2.2.7.3 Analysis by Equivalent Circuit Analysis

After the fabrication of supercapacitors, there needs to be electrochemical analysis and equivalent circuit analysis by implementing a simplified circuit model. Supercapacitors show an interesting feature; it behaves non-ideally, rather than ideally, because here the electrode is porous, which causes the distribution of capacitance and resistance in such a transmission line characteristic [74]. That is why supercapacitors have two parts of components: one is resistive and the other is capacitive. Also, the lower the ESR, the higher the specific capacitance; ESR depends upon the size of the cell and impedance. Resistance is made up of electrolyte conductivity, impurity, and several electronic parts. Thus, electrode materials, electrolyte conductivity, and size are the main factors for ESR; manufacturing capabilities as well as fabricating conditions are also important parameters [75]. That is why lab and research-level development is quite important in the development of supercapacitor devices [52].

ESR is calculated as ESR = $D/2*\Pi*f_c$ = $D*X_C$, where D is the dissipation factor, and it is the inverse of quality factor Q, when D = tanα; the more the ESR, the more the dissipation power, which indicates a capacitive loss. The capacitor impedance can be expressed as $Z = \sqrt{(ESR)^2 + (ESL - Xc)^2}$; here ESL is the series inductance, which can be expressed as ESL = $2*\Pi*f*L$. So, the value of ESL can be reduced using the design of different dielectric materials [76]. Thus, ESR plays an important role in the design of supercapacitor devices [12].

2.2.7.4 Efficiency

Supercapacitor efficiency mainly depends upon its charging-discharging cycle, which solely depends upon the electrode material used. The Coulombic efficiency of supercapacitors is often more than 100% at a very high current, and even some loss of that efficiency implies that some amount of charge is lost during charging cycling. Based on that, it is applicable where a higher amount of load fluctuation is there and needs a backup device for that also it has good applications in the field of regenerative power and faster charge kinetics. Unlike battery (100–200 and more cycles), they can withstand a huge amount of charging–discharging cycle (more than 10,000 cycles) due to their unique ability to store the charge in double layer.

2.2.8 MATERIALS FOR SUPERCAPACITOR DEVELOPMENT

We have discussed the electrode in the previous section that possesses a higher specific surface area and porous morphology. Besides this, surface functional groups like N and O can also significantly affect the capacitance properties by involving a good faradaic redox reaction. So, in the next section, we will discuss our positive and negative electrode-based novel materials for supercapacitor development [77].

2.2.8.1 Positive Electrode

Metal oxides and MOFs are the optimum examples for positive electrode materials owing to their higher specific surface area and higher specific capacitance and also a wide range of potential windows. 3D porous RuO_2/ graphene nanosheets showed a higher amount of specific capacitance = 1140 F/g as reported [78]. MnO_2 can also be used as a positive electrode material, but is not used due to its less surface area as well as lower conductivity; however incorporating with other materials like graphene, it can be widely used due to improvement in surface area, which will provide a good amount of specific capacitance [29].

2.2.8.2 Negative Electrode

Carbon materials like activated carbon, graphene, and carbon nanotubes (CNTs) have attracted a lot more attention due to their various unique features cost, abundancy, higher electronic conductivity, and specific surface area. A recently developed 3D structure of vertically aligned CNTs grown on carbon nanofibres has exhibited a energy density of 70.7 Wh/kg, and it retains 97% of its capacitance even after 20,000 and more charge–discharge cycles. Also, graphene has shown very interesting properties in terms of surface area rather than activated carbon and CNTs; so it is also a promising candidate for negative electrode materials for the development of supercapacitor devices [79].

In the near future, our goal should be to achieve more than 100 Wh/kg of energy density; apart from the design of the electrolyte and its larger potential window, we have also designed novel materials with higher surface area and pores, higher conductivity, and a superior ion diffusing channel to achieve more energy as well as power density.

2.2.9 ECONOMIC SCALE OF SUPERCAPACITORS

Price estimation is one of the important parameters for the commercialization of supercapacitors, its price depends upon the active material used as an electrode. The cost of high-purity carbon in supercapacitors is $70/kg. It has been just a few years since supercapacitors came into the global market business. This implies that it is closing the gap between the conventional capacitors and batteries. Carbon is the main material used for the development of supercapacitors, which costs around $28/kg.

With day-by-day improvement in the development of supercapacitors, the market value has increased day by day globally. In 2010, the global market estimated for supercapacitors was $470 million, in the year 2015 it went up to $1.2 billion, and it is estimated that by 2025 the global market will reach up to $5.0 billion. In the electronic industry, it is used mainly in memory protection applications; thus, its market in the electronic industry is around 1.34 billion/year [80]. The costly component of supercapacitors is the active material that is used as an electrode, separator, and electrolyte. Considering all these facts, we should design such materials, so that its performance will be enhanced and cost will be dropped, as it is bridging the gap between batteries and conventional capacitors [31]. By seeing all these trends, it is expected that supercapacitors will become a power storage device of choice in various applications [81].

2.2.10 CONCLUSION

This chapter provides a dynamic overview of capacitors and supercapacitors and their working principles. It also covers the selection of electrode materials for modern-age capacitors as well as supercapacitors. Capacitors and supercapacitors are an essential technology in today's life, as they are the future of clean energy, which indirectly helps to reduce global warming. As technology advances, supercapacitors will continue to play an important role in shaping the future of energy. Researchers and manufacturers are continually exploring new materials, designs, and processes to improve battery performance and reduce costs. Furthermore, it has a wide range of applications, from powering portable things to providing energy storage for renewable energy systems and powering medical devices. It has potential applications in defence and aerospace industries as well.

In addition, researchers use different modelling and simulation techniques to better understand the electrochemical behaviour of supercapacitors and optimize its design and different parameters. So by developing new materials and composites as next-generation electrode materials, researchers and engineers are continuing to improve the performance of next-generation supercapacitors and unlock new applications and markets for the future energy sector.

ACKNOWLEDGEMENT

Authors would like to thank Indian Institute of Technology Jodhpur for providing the resources and facilities to carry out this work.

CONFLICTS OF INTEREST

The authors declare no competing interest.

REFERENCES

1. M. S. Whittingham, "Introduction: batteries," *Chemical Reviews*, vol. 114, no. 23. American Chemical Society, p. 11413, Dec. 10, 2014. doi: 10.1021/cr500639y.

2. T. H. Nguyen, A. Fraiwan, and S. Choi, "Paper-based batteries: a review," *Biosensors and Bioelectronics*, vol. 54. Elsevier Ltd, pp. 640–649, Apr. 15, 2014. doi: 10.1016/j.bios.2013.11.007.

3. Full-Scale Battery Materials Charactereization: Solution over small (nm) to large (cm) scales by Physical Electronics. www.phi.com. "Flyer".

4. I. Dincer, and M. A. Rosen, "Exergy, environment, and sustainable development," in *Exergy*, Elsevier, 2021, pp. 61–89. doi: 10.1016/B978-0-12-824372-5.00004-X.

5. A. H. Fathima, and K. Palanisamy, "Renewable systems and energy storages for hybrid systems," in *Hybrid-Renewable Energy Systems in Microgrids: Integration, Developments and Control*, Elsevier, 2018, pp. 147–164. doi: 10.1016/B978-0-08-102493-5.00008-X.

6. R. Holze, "Self-Disxharge of Batteries: Cause, Mechanism and Remedies.Self-Disxharge of Batteries: Cause, Mechanism and Remedies." https://doi.org/10.37155/2717-526X-0402-3

7. Shichao Zhang, Siyuan Li, Yingying Lu, Designing safer lithium-based batteries with nonflammable electrolytes. *A review, eScience*, vol. 1, no. 2, 2021, pp. 163–177, ISSN 2667–1417, https://doi.org/10.1016/j.esci.2021.12.003.

8. B. Scrosati, J. Hassoun, and Y. K. Sun, "Lithium-ion batteries. A look into the future," *Energy and Environmental Science*, vol. 4, no. 9. pp. 3287–3295, Sep. 2011. doi: 10.1039/c1ee01388b.

9. S. Sengupta, A. Patra, S. Jena, K. Das, and S. Das, "A study on the effect of electrodeposition parameters on the morphology of porous nickel electrodeposits," *Metallurgical and Materials Transactions A Physical Metallurgy and Materials Science*, vol. 49, no. 3. pp. 920–937, Mar. 2018, doi: 10.1007/s11661-017-4452-8.

10. H. Cheng, J. G. Shapter, Y. Li, and G. Gao, "Recent progress of advanced anode materials of lithium-ion batteries," *Journal of Energy Chemistry*, vol. 57. Elsevier B.V., pp. 451–468, Jun. 1, 2021. doi: 10.1016/j.jechem.2020.08.056.

11. P. V. Chombo, and Y. Laoonual, "A review of safety strategies of a Li-ion battery," *Journal of Power Sources*, vol. 478. Elsevier B.V., Dec. 1, 2020. doi: 10.1016/j.jpowsour.2020.228649.

12. X. Wang, K. Jiang, and G. Shen, "Flexible fiber energy storage and integrated devices: recent progress and perspectives," *Materials Today,* vol. 18, no. 5. Elsevier B.V., pp. 265–272, Jun. 1, 2015. doi: 10.1016/j.mattod.2015.01.002.

13. D. Remler Supratim Das, A. Jayanti, P. Shearing, and J. Li, "Tech factsheets for policymakers battery technology authors," 2021. [Online]. Available: https://engineering.mit.edu/engage/ask-an-engineer/

14. C. Iclodean, B. Varga, N. Burnete, D. Cimerdean, and B. Jurchiş, "Comparison of different battery types for electric vehicles," in *IOP Conference Series: Materials Science and Engineering*, Institute of Physics Publishing, Oct. 2017. doi: 10.1088/1757-899X/252/1/012058.

15. B. Horstmann et al., "Strategies towards enabling lithium metal in batteries: Interphases and electrodes," *Energy and Environmental Science*, vol. 14, no. 10. Royal Society of Chemistry, pp. 5289–5314, Oct. 01, 2021. doi: 10.1039/d1ee00767j.

16. N. Nitta, F. Wu, J. T. Lee, and G. Yushin, "Li-ion battery materials: present and future," *Materials Today*, vol. 18, no. 5. Elsevier B.V., pp. 252–264, Jun. 1, 2015. doi: 10.1016/j.mattod.2014.10.040.

17. J. Ma et al., "The 2021 battery technology roadmap." *Journal of Physics D: Applied Physics*, vol. 54, no. 18. May 2021. doi: 10.1088/1361-6463/abd353.

18. G. Varshney, and P. Jaiswal, "A comparative study on advanced NASICON type and other effective materials for sodium ion batteries (SIBs)," in *Materials Today: Proceedings*, Elsevier Ltd, 2021, pp. 1776–1782. doi: 10.1016/j.matpr.2020.11.961.

19. Maximilian Fichtner et al., "Rechargeable Batteries of the Future—The State of the Art from a BATTERY 2030+ Perspective," *Adv. Energy Mater.* 10.1002/aenm. 202102904
20. C. H. Clark "Primary batteries," 10.1109/EE.1950.6433909.
21. Ankit Dev Singh et al., "Volume contractible antimony bromide electrode as negative electrode for long-lasting Li-ion batteries," *Journal of Alloys and Compounds*, https:// doi.org/10.1016/j.jallcom.2023.173305
22. Alvaro Masia et al., "Opportunities and Challenges of Lithium Ion Batteries in Automotive Applications," https://doi.org/10.1021/acsenergylett.0c02584
23. J. Zhu, R. Duan, S. Zhang, N. Jiang, Y. Zhang, and J. Zhu, "The application of graphene in lithium ion battery electrode materials," *Journal of the Korean Physical Society*, vol. 3, no. 1. Korean Physical Society, pp. 1–8, Sep. 24, 2014. doi: 10.1186/21 93-1801-3-585.
24. B. Chen, M. Liang, Q. Wu, S. Zhu, N. Zhao, and C. He, "Recent developments of antimony-based anodes for sodium- and potassium-ion batteries," *Transactions of Tianjin University*, vol. 28, no. 1. Tianjin University, pp. 6–32, Feb. 01, 2022. doi: 10.1007/s12209-021-00304-9.
25. D. S. Kim, J. Bae, S. H. Kwon, J. Hur, S. G. Lee, and I. T. Kim, "Synergistic effect of antimony-triselenide on addition of conductive hybrid matrix for high-performance lithium-ion batteries," *Journal of Alloys and Compounds*, vol. 828, Jul. 2020. doi: 10.1016/j.jallcom.2020.154410.
26. Y. Wang, H. Li, P. He, E. Hosono, and H. Zhou, "Nano active materials for lithium-ion batteries," *Nanoscale*, vol. 2, no. 8. pp. 1294–1305, Aug. 2010, doi: 10.1039/ c0nr00068j.
27. J. R. Rodriguez, H. J. Hamann, G. M. Mitchell, V. Ortalan, V. G. Pol, and P. V. Ramachandran, "Three-dimensional antimony nanochains for lithium-ion storage," *ACS Applied Nano Materials*, vol. 2, no. 9. pp. 5351–5355, Sep. 2019, doi: 10.1021/ acsanm.9b01316.
28. M. Mayo, and A. J. Morris, "Structure prediction of Li-Sn and Li-Sb intermetallics for Lithium-ion batteries anodes," *Chemistry of Materials*, vol. 29, no. 14. pp. 5787–5795, Jul. 2017, doi: 10.1021/acs.chemmater.6b04914.
29. Nisha Kamboj, "A closed-shell phenalenyl-based dinuclear iron(III) complex as a robust cathode for a one-compartment H2O2 fuel cell," *Dalton Transaction*, vol. 52, October, 2023. doi: https://doi.org/10.1039/D3DT02975A.
30. V. S. M. Halper, and J. C. Ellenbogen, "Supercapacitors: a brief overview MITRE MITRE MITRE MITRE," 2006. [Online]. Available: http://www.mitre.org/tech/nanotech
31. J. Sun, B. Luo, and H. Li, "A review on the conventional capacitors, supercapacitors, and emerging hybrid ion capacitors: past, present, and future," *Advanced Energy and Sustainability Research*, vol. 3, no. 6. p. 2100191, Jun. 2022, doi: 10.1002/aesr. 202100191.
32. Jiale Sun, and Bingcheng Luo, "A Review on the Conventional Capacitors, Supercapacitors, and Emerging Hybrid Ion Capacitors: Past, Present, and Future." 10.1002/aesr.202100191
33. Ayan Dey, and Sudipta Goswami, "Cu2+ at the surface / sub-surface region of CuFeO2 rhombohedral nanostructures facilitates specific capacitance (~611 F g–1): An under-standing of the solvation energy dependent charge transfer mechanism," *Physica B: Condensed Matter*, vol. 667, no. 6, August. 2023, doi: 10.1016/j.physb.2023.415207.
34. H. C. Song, J. E. Zhou, D. Maurya, Y. Yan, Y. U. Wang, and S. Priya, "Compositionally graded multilayer ceramic capacitors, 12353," *Scientific Reports*, vol. 7, no. 1. Dec. 2017, doi: 10.1038/s41598-017-12402-7.
35. K. Hong, T. H. Lee, J. M. Suh, S. H. Yoon, and H. W. Jang, "Perspectives and challenges in multilayer ceramic capacitors for next generation electronics," *Journal*

of Materials Chemistry C, vol. 7, no. 32. Royal Society of Chemistry, pp. 9782–9802, 2019. doi: 10.1039/c9tc02921d.

36. N. N. Loganathan, V. Perumal, B. R. Pandian, R. Atchudan, T. N. J. I. Edison, and M. Ovinis, "Recent studies on polymeric materials for supercapacitor development,104149, ISSN 2352-152X," *Journal of Energy Storage*, vol. 49. Elsevier Ltd, May 1, 2022. doi: 10.1016/j.est.2022.104149.

37. P. H. Patil, V. V. Kulkarni, and S. A. Jadhav, "An overview of recent advancements in conducting polymer–metal oxide nanocomposites for supercapacitor application.6(12):363," *Journal of Composites Science*, vol. 6, no. 12. MDPI, Dec. 01, 2022. doi: 10.3390/jcs6120363.

38. Awitdrus, D. A. Suwandi, Agustino, E. Taer, R. Farma, and R. F. Syahputra, "Effect of aqueous electrolyte to the supercapacitor electrode performance made from sugar palm fronds waste," in *Journal of Physics: Conference Series*, IOP Publishing Ltd, Jul. 2021. doi: 10.1088/1742-6596/1951/1/012009.

39. Kirti Sharma, "Review of supercapacitors: Materials and devices," 2019. doi: https://doi.org/10.1016/j.est.2019.01.010.

40. D. P. Chatterjee, and A. K. Nandi, "A review on the recent advances in hybrid supercapacitors," *Journal of Materials Chemistry A*, vol. 9, no. 29. Royal Society of Chemistry, pp. 15880–15918, Aug. 07, 2021. doi: 10.1039/d1ta02505h.

41. H. Xiong, E. J. Dufek, and K. L. Gering, "Batteries," in *Comprehensive Energy Systems*, Elsevier Inc., 2018, pp. 629–662. doi: 10.1016/B978-0-12-809597-3.00245-5.

42. N. Ogihara, M. Hasegawa, H. Kumagai, R. Mikita, and N. Nagasako, "Heterogeneous intercalated metal-organic framework active materials for fast-charging non-aqueous Li-ion capacitors,16;14(1):1472," *Nature Communications*, vol. 14, no. 1, Mar. 2023, doi: 10.1038/s41467-023-37120-9.

43. A. A. Ahmed, A. Alsharif, and Y. Fathi Nassar, "Recent advances in energy storage technologies: the future of renewable energy in Libya solar energy view project." [Online]. Available: https://ijees.org/index.php/ijees/index

44. C. T. Tshiani, and P. Umenne, "The impact of the electric double-layer capacitor (EDLC) in reducing stress and improving battery lifespan in a hybrid energy storage system (HESS)," *Energies (Basel)*, vol. 15, no. 22, Nov. 2022, doi: 10.3390/en15228680.

45. X. You, M. Misra, S. Gregori, and A. K. Mohanty, "Preparation of an electric double layer capacitor (EDLC) using miscanthus-derived biocarbon," *ACS Sustainable Chemistry & Engineering*, vol. 6, no. 1, pp. 318–324, Jan. 2018, doi: 10.1021/acssuschemeng.7b02563.

46. S. T. Senthilkumar, R. K. Selvan, Y. S. Lee, and J. S. Melo, "Electric double layer capacitor and its improved specific capacitance using redox additive electrolyte," *Journal of Materials Chemistry A. Materials*, vol. 1, no. 4, pp. 1086–1095, Jan. 2013, doi: 10.1039/c2ta00210h.

47. C. Chukwuka, and K. A. Folly, "Batteries and super-capacitors," in *IEEE Power and Energy Society Conference and Exposition in Africa: Intelligent Grid Integration of Renewable Energy Resources, PowerAfrica 2012*, 2012. doi: 10.1109/PowerAfrica.2012.6498634.

48. Y. J. Kim, M. Endo, T. Takeda, Y. J. Kim, K. Koshiba, and K. Ishii, "High power electric double layer capacitor (EDLC's); from operating principle to pore size control in advanced activated carbons," 2001. [Online]. Available: https://www.researchgate.net/publication/264114172

49. G. Pavaskar, K. Ramakrishnasubramanian, V. S. Kandagal, and P. Kumar, "Modeling electric double-layer capacitors using charge variation methodology in Gibbs ensemble, Article-36," *Frontiers in Energy Research*, vol. 5, no. JAN, Jan. 2018, doi: 10.3389/fenrg.2017.00036.

50. Y. Jiang, and J. Liu, "Definitions of pseudocapacitive materials: a brief review," *Energy and Environmental Materials*, vol. 2, no. 1. John Wiley and Sons Inc, pp. 30–37, Mar. 1, 2019. doi: 10.1002/eem2.12028.

51. S. Fleischmann et al., "Pseudocapacitance: from fundamental understanding to high power energy storage materials," *Chemical Reviews*, vol. 120, no. 14. American Chemical Society, pp. 6738–6782, Jul. 22, 2020. doi: 10.1021/acs.chemrev.0c00170.

52. G. Gautham Prasad, N. Shetty, S. Thakur, Rakshitha, and K. B. Bommegowda, "Supercapacitor technology and its applications: a review," in *IOP Conference Series: Materials Science and Engineering*, Institute of Physics Publishing, Nov. 2019. doi: 10.1088/1757-899X/561/1/012105.

53. J. R. Rani, R. Thangavel, S. I. Oh, Y. S. Lee, and J. H. Jang, "An ultra-high-energy density supercapacitor; fabrication based on thiol-functionalized graphene oxide scrolls," *Nanomaterials*, vol. 9, no. 2, Feb. 2019, doi: 10.3390/nano9020148.

54. A. Vlad, N. Singh, J. Rolland, S. Melinte, P. M. Ajayan, and J. F. Gohy, "Hybrid supercapacitor-battery materials for fast electrochemical charge storage," *Scientific Reports*, vol. 4, Mar. 2014, doi: 10.1038/srep04315.

55. S. M. Benoy, M. Pandey, D. Bhattacharjya, and B. K. Saikia, "Recent trends in supercapacitor-battery hybrid energy storage devices based on carbon materials," *Journal of Energy Storage*, vol. 52. Elsevier Ltd, Aug. 15, 2022. doi: 10.1016/j.est.2022.104938.

56. M. I. A. Abdel Maksoud et al., "Advanced materials and technologies for super-capacitors used in energy conversion and storage: a review," *Environmental Chemistry Letters*, vol. 19, no. 1. Springer Science and Business Media Deutschland GmbH, pp. 375–439, Feb. 1, 2021. doi: 10.1007/s10311-020-01075-w.

57. L. Zhang et al., Redox electrolytes for supercapacitors," *Frontiers in Chemistry*, vol. 8. Frontiers Media S.A., Jun. 3, 2020. doi: 10.3389/fchem.2020.00413.

58. L. Yu, and G. Z. Chen, "Ionic liquid-based electrolytes for supercapacitor and supercapattery," *Frontiers in Chemistry*, vol. 7, no. APR. Frontiers Media S.A., 2019. doi: 10.3389/fchem.2019.00272.

59. B. Pal, S. Yang, S. Ramesh, V. Thangadurai, and R. Jose, "Electrolyte selection for supercapacitive devices: a critical review," *Nanoscale Advances*, vol. 1, no. 10, pp. 3807–3835, 2019, doi: 10.1039/c9na00374f.

60. S. J. Rajasekaran, A. N. Grace, G. Jacob, A. Alodhayb, S. Pandiaraj, and V. Raghavan, "Investigation of different aqueous electrolytes for biomass-derived activated carbon-based supercapacitors," *Catalysts*, vol. 13, no. 2, Feb. 2023, doi: 10.3390/catal13020286.

61. B. Streipert et al., "Conventional and inactive electrode materials in lithium-ion batteries: determining cumulative impact of oxidative decomposition at" , *ChemSusChem*, vol. 13, no. 19, pp. 5301–5307, Oct. 2020, doi: 10.1002/cssc.202001530.

62. C. Zhong, Y. Deng, W. Hu, J. Qiao, L. Zhang, and J. Zhang, "A review of electrolyte materials and compositions for electrochemical supercapacitors," *Chemical Society Reviews*, vol. 44, no. 21. Royal Society of Chemistry, pp. 7484–7539, Nov. 7, 2015. doi: 10.1039/c5cs00303b.

63. Y. Wang et al., "Supercapacitor devices based on graphene materials," *Journal of Physical Chemistry C*, vol. 113, no. 30, pp. 13103–13107, Jul. 2009, doi: 10.1021/jp902214f.

64. N. Choudhary. "Asymmetric supercapacitor electrodes," *Advanced Materials*, vol. 29, no. 21, Jun. 2017, doi: 10.1002/adma.201605336.

65. R. Brooke et al., "Large-scale paper supercapacitors on demand," *Journal of Energy Storage*, vol. 50, Jun. 2022, doi: 10.1016/j.est.2022.104191.

66. M. Bora, J. Tamuly, S. Maria Benoy, S. Hazarika, D. Bhattacharjya, and B. K. Saikia, "Highly scalable and environment-friendly conversion of low-grade coal to activated

carbon for use as electrode material in symmetric supercapacitor," *Fuel*, vol. 329, Dec. 2022, doi: 10.1016/j.fuel.2022.125385.

67. P. Bhojane, "Recent advances and fundamentals of pseudocapacitors: materials, mechanism, and its understanding,103654," *Journal of Energy Storage*, vol. 45. Elsevier Ltd, Jan. 1, 2022. doi: 10.1016/j.est.2021.103654.

68. Y. Zhong, J. Zhang, G. Li, and A. Liu, "Research on energy efficiency of supercapacitor energy storage system," *International Conference on Power System Technology*, 2006.

69. Z. Wang et al., "Enhanced power density of a supercapacitor by introducing 3D-interfacial graphene," *New Journal of Chemistry*, vol. 44, no. 31, pp. 13377–13381, Aug. 2020, doi: 10.1039/d0nj02105a.

70. X. Fan et al., "High Power- and Energy-Density Supercapacitors through the Chlorine Respiration Mechanism" *Angewandte Chemie International Edition*, vol. 62, no. 2, Jan. 2023, doi: 10.1002/anie.202215342.

71. Y. Liu, Q. Wu, L. Liu, P. Manasa, L. Kang, and F. Ran, "Vanadium nitride for aqueous supercapacitors: a topic review," *Journal of Materials Chemistry A*, vol. 8, no. 17. Royal Society of Chemistry, pp. 8218–8233, May 07, 2020. doi: 10.1039/d0ta01490g.

72. J. J. Quintana, A. Ramos, M. Diaz, and I. Nuez, "Energy efficiency analysis as a function of the working voltages in supercapacitors," *Energy*, vol. 230, Sep. 2021, doi: 10.1016/j.energy.2021.120689.

73. Q. Luo et al., 2nd *International Conference on Power Electronics and Intelligent Transportation System (PEITS)*, 2009 19–20 Dec. 2009, Shenzhen, China.

74. P. Saha and M. Khanra, "Equivalent circuit model of supercapacitor for self-discharge analysis – a comparative study," in *International Conference on Signal Processing, Communication, Power and Embedded System, SCOPES 2016 - Proceedings*, Institute of Electrical and Electronics Engineers Inc., Jun. 2017, pp. 1381–1386. doi: 10.1109/SCOPES.2016.7955667.

75. U. S. Sani and I. Shanono, "An equivalent circuit of carbon electrode supercapacitors, Conference: 2014 Nigeria Engineering Conference At: Ahmadu Bello University, Zaria, Nigeria," 2014, doi: 10.13140/2.1.2406.6243.

76. L. E. Helseth, "Modelling supercapacitors using a dynamic equivalent circuit with a distribution of relaxation times," *Journal of Energy Storage*, vol. 25, Oct. 2019, doi: 10.1016/j.est.2019.100912.

77. X. Cai, L. Lai, Z. Shen, and J. Lin, "Graphene and graphene-based composites as Li-ion battery electrode materials and their application in full cells," *Journal of Materials Chemistry A*, vol. 5, no. 30, pp. 15423–15446, 2017, doi: 10.1039/c7ta04354f.

78. L. Hu, D. Guo, G. Feng, H. Li, and T. Zhai, "Asymmetric behavior of positive and negative electrodes in carbon/carbon supercapacitors and its underlying mechanism," *Journal of Physical Chemistry C*, vol. 120, no. 43, pp. 24675–24681, Nov. 2016, doi: 10.1021/acs.jpcc.6b09898.

79. K. K. Upadhyay, "Free-standing graphene-carbon as negative and FeCoS as positive electrode for asymmetric supercapacitor," *Journal of Energy Storage*, vol. 50, Jun. 2022, doi: 10.1016/j.est.2022.104637.

80. P. Kumar, S. Roy, H. Bora Karayaka, J. He, and Y.-H. Yu, "Economic comparison between a battery and supercapacitor for hourly dispatching wave energy converter power: preprint," 2021. [Online]. Available: https://www.nrel.gov/docs/fy21osti/77398.pdf.

81. G. Wang et al., "Electroceramics for high-energy density capacitors: current status and future perspectives," *Chemical Reviews*, vol. 121, no. 10. American Chemical Society, pp. 6124–6172, May 26, 2021. doi: 10.1021/acs.chemrev.0c01264.

3 Two-Dimensional Materials in Lithium-Ion Batteries

Sarita Yadav, Srikanth Ponnada,
Maryam Sadat Kiai, and Ravinder Pawar

3.1 INTRODUCTION

Several sources of energy and energy production methods have been utilized throughout the years. The majority of available energy to humankind is obtained from those that are hostile to the environment, and which produce definite levels of danger and peril to the well-being of humans. Such sources comprise fossil fuel-based energy sources like natural gas, coal, and crude oil among others.[1] Owing to the rising more critical environmental hazards arising from these sources of energy, an augmented research endeavor has been recorded in recent years with respect to novel and recycled sources of energy, efficient approaches of energy conversion, and procedures for energy storage. During these studies, several innovative developments and pioneering progressions have been recorded in various sources of clean energy, including wind and solar energy. Though these novel sources of environmentally friendly and renewable energy are full of challenges, the foremost one is their incapability to deliver a continuous and persistent flow of electric energy.[2]

Consequently, the prerequisite for energy storage devices with substantial storage capacity and great efficiency for the application of generated energy in storage comes to the fore. First, lead-acid batteries were utilized for energy storage as they possess an admirable stable charge/discharge state. Nevertheless, some of their drawbacks like large weight and enormous size limit their usage in portable, light electric devices.[3] The main necessities for an energy storage medium in electronic and electrical applications in current years are long life span, lightweight, cyclability, high energy density, and enhanced charging rate. Additionally, Nickel-cadmium (Ni-Cd) and Nickel-metal hydride (Ni-MH) batteries are some of the original energy storage tools that exhibit applications in conveyable electronic devices (like digital cameras and phones) and implements.[4,5] Although Ni-Cd battery unveiled prodigious capacity when exposed to high currents, it still exhibited unfavorable memory effects, which led to a noteworthy decline in battery life with hazardous toxicity produced as a consequence of its components.[6,7] However, the Ni-MH battery showed a prolonged life span and was eco-friendly; it encountered the challenge of leakage.[2,8]

DOI: 10.1201/9781003404729-3

Hence, the lithium-ion battery (LIB) was revolutionized with high forecasts. The major problem, i.e., the liquid challenge originated by the conventional secondary batteries is conveniently fixed by the introduction of solid polymer electrolyte in LIB. Further, in lightweight construction, its volume condenses to a condensed size, meeting the necessities of portable devices.[9] Nowadays, pure electric vehicles (PEVs) and hybrid electric vehicles (HEVs) have been established swiftly, but the energy density has limited their progress.[10–12] To enhance the energy density of LIBs, materials with great capacity can be exploited as anode electrodes.[13–22] Hence, it is impending to examine novel materials with high energy density and voltage platforms. Two-dimensional (2D) materials are well-defined as crystals in which electrons can move spontaneously on the nanometer scale in two dimensions.

Since the finding of graphene in 2004 by Novoselov et al.,[23] numerous 2D materials with exceptional physicochemical properties have been revealed like transition metal carbides and nitrides (MXenes),[24–28] transition metal dichalcogenides (TMDs),[29,30] Xenes[31–37] and transition metal oxides (TMOs).[38–42] The atomic-level thickness of 2D materials can offer a large number of electrochemical active sites for the adsorption of the target molecules, enhancing the catalytic property of the material and increasing the sensitivity for electrochemical ascertainment. Additionally, atoms on the 2D materials surface can simply exude from the lattice, generating vacancy defects. These vacancy defects can result in the reduction of the coordination number of atoms on the surfaces of 2D materials and can amend the structure, and the band gap that can be governed through the doping of atoms. The proliferation of electrons and ions in 2D materials is almost unrelated to other conditions, which can transport swiftly in the channel. Consequently, 2D nanomaterials have progressively become auspicious candidates for LIBs.[43–52]

Almost all the 2D nanomaterials share comparable fundamental advantages, like large surface area and admirable mechanical properties and these act as significant characteristics of LIBs anode materials. Still, there are various obstinate challenges allied with the advancement of 2D materials for LIBs like tedious fabrication approaches, predictable defective structures, and unidentified electrochemical processes. This chapter emphasizes recent advancements of 2D materials for LIBs applications along with future perspectives for the development of advanced 2D materials.

3.2 ROLE OF 2D MATERIALS IN LIBS

The utilization of 2D materials for energy storage has fascinated more and attained more attention.[53–60] Rise in the demand for battery capacity in the market and advancement of cathode materials with greater energy density of LIBs have been propagated. Hence, abundant materials have evolved for LIBs.[61–68] On the basis of elemental compositions and crystal structures, 2D materials can be coarsely grouped into the following types: graphene, TMDs, TMOs, MXenes, and Xenes as shown in Figure 3.1.

These 2D nanomaterials possess exceptional structures and properties, which can be extensively utilized in the diverse fields of electronics.[69–77] Graphene is the utmost familiar 2D material with high specific area and good conductivity which offers high

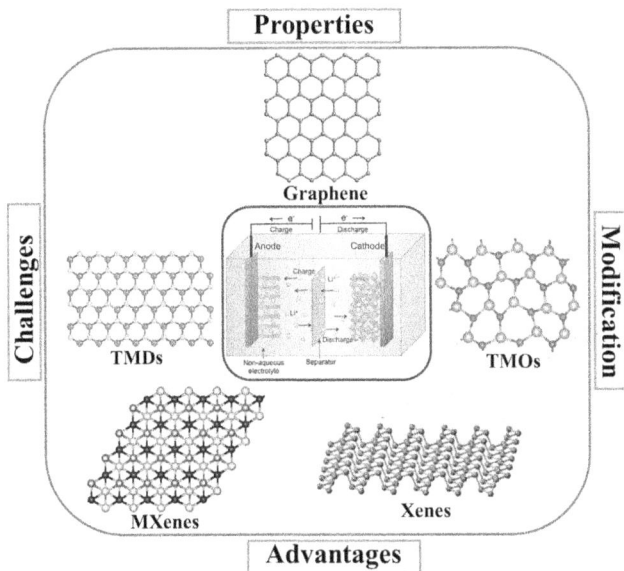

FIGURE 3.1   2D anode materials applied to LIBs.

capacity and rate characteristics in LIBs. TMOs have plentiful electrochemical reactivity sites and quickest lithium ion diffusion paths, but the single 2D nanosheet is attributed to staking and consequently, diminishing the available surface area of activity. Furthermore, metal cations in TMOs turn out to be zero-valent metals via redox reaction, which results in the alteration in morphology and crystal structure. This will utterly affect the cycle performance of LIBs, and thus, conductive carbon-based materials can be exploited to compound with TMOs to conquer these shortcomings. TMDs have a graphene-like structure that can be manufactured by using stripping methods. The single-layer structure of TMDs can encourage the transfer of lithium ions, and the loosened layered structure can also reduce the volume extension during charging and discharging. Nevertheless, TMDs have lower electrical conductivity and are usually combined by adding conductive materials to prepare the heterogeneous structure that leads in enhancing the lithium storage performance of materials. Like TMOs, pristine MXenes possess virtuous metallic conductivity and low working voltage. Additionally, it is easy to stack among layers, which is behind the collapse of the structure and degradation of the electrochemical performance. In contrast to MXenes, Xenes possess a single-layer structure consisting of single atoms, which have good electrical conductivity and flexibility. Nevertheless, the preparation of Xenes is comparatively complex, which is still an enormous challenge. The advantages and disadvantages of various available 2D materials are presented in Table 3.1.

3.2.1 GRAPHENE

Graphene is a 2D nanomaterial composed of sp^2 hybrid carbon orbitals in a hexagonal shape. It has fascinated prevalent consideration in several applications since its

TABLE 3.1
Merits and Demerits of Graphene, TMDs, TMOs, MXenes, and Xenes

	Graphene	TMDs	TMOs	MXenes	Xenes
✓	large specific surface area; great electrical conductivity	the graphene-like layered framework; less space between layers	rich in electrochemical reactive sites; short lithium-ion paths	Low working voltage; high metallic conductivity	swift electronic transference; great flexibility
✗	environment unfriendly; high cost	low electrical conductivity	monolayer easy to stack; volume extrusion	monolayer easy to stack	intricate synthesis method

discovery in 2004.[78–85] Individual carbon atoms in graphene can create a large π-bond of numerous atoms penetrating the whole layer, hence having outstanding optical and electrical properties. Graphene has attracted the scientific community as electrode materials in recent times for energy devices and primarily comprises the following characteristics: (1) graphene can rectify the intercalation/deintercalation and transference rate of lithium ions in electrode materials to accomplish the swift charging; (2) graphene can upsurge the reversible specific capacity of the battery and protract the battery life; (3) graphene is mechanical elastic, appropriate for emerging flexible LIBs.

Graphene is acknowledged for its exceptional electronic properties and prodigious potential for energy storage.[86–90] In addition, investigators have observed that the addition of heteroatoms into the graphene framework frequently can efficiently modify its chemical and electronic properties. In current years, due to virtuous electrochemical performance in LIBs, the development of doped graphene has been repetitively reported in the literature such as B-doped graphene, N-doped graphene, and P-doped graphene. Wang et al.[91] synthesized the N-doped graphene nanosheets via a heat-treating method in the presence of ammonia gas and exhibited higher specific capacity, i.e., 900 mAh/g relative at 42 mA/g and improved rate performance (250 mA/g at 2.1 A/g) as compared to pristine graphene. Accidentally, Wu et al.[92] have also performed likewise experiments by mingling graphene with nitrogen or boron, respectively. It was observed from an electrochemical performance test that doping of two elements can efficiently advance the energy density of LIBs. Difference in the electroneutrality of the doped N and B atoms results in the breakage of electrical neutrality of graphene. During the redox reaction, the charged points produced are advantageous for the adsorption of oxygen atoms; hence, the discharge performance of graphene could be enhanced. Besides, phosphorus is less electronegative than carbon and has an advanced electron donor capacity. P-doped graphene with a considerably higher specific discharge capacity, i.e., 460 mAh/g at 0.1 A/g compared to pristine graphene (460 mAh/g at 0.1 A/g). Hence, the doping of graphene can efficiently enhance the potential of graphene in LIBs.[93] Maximum studies have mainly taken into consideration graphene's outstanding electrical conductivity to boost the diffusion rate of lithium ions. Graphene like graphite can also be exploited as a carbonaceous medium when amalgamated with other materials. Idrees et al.[94] prepared silicon boron carbon nitride/nitrogen sulfur-doped graphene (SiBCN/NSGs) via adding pyrolyzed NSGs into polymer-derived SiBCN by the ball milling method (Figure 3.2(a)). The final product exhibited significantly enhanced Li-ion loading capacity accompanied by higher rate capability (Figure 3.2(b)). The nanocomposites showed a reversive capacity of 785 mAh/g even at a high current density, i.e., 450 mA/g over 800 cycles ((Figure 3.2(c))) with a high retention capacity of 780 mAh/g. Furthermore, the charge capacity of the composite was still 365 mAh/g even after 500 cycles. Song et al.[95] prepared WS_2/rGO composites by the enhanced one-step hydrothermal method as shown in Figure 3.3. Graphene plays a vital role for transporting the morphology from the nanowire microporous spheres to graphene-supported nano-honeycomb plane structures with a higher specific surface area and conductivity. This anode can distribute the excellent electrochemical performances of sodium-ion batteries/LIBs with long cycling life (more than 350/200 cycles at 1 A/g) and great specific charge capacity (953.1/522.3 mAh/g at 0.1 A/g).

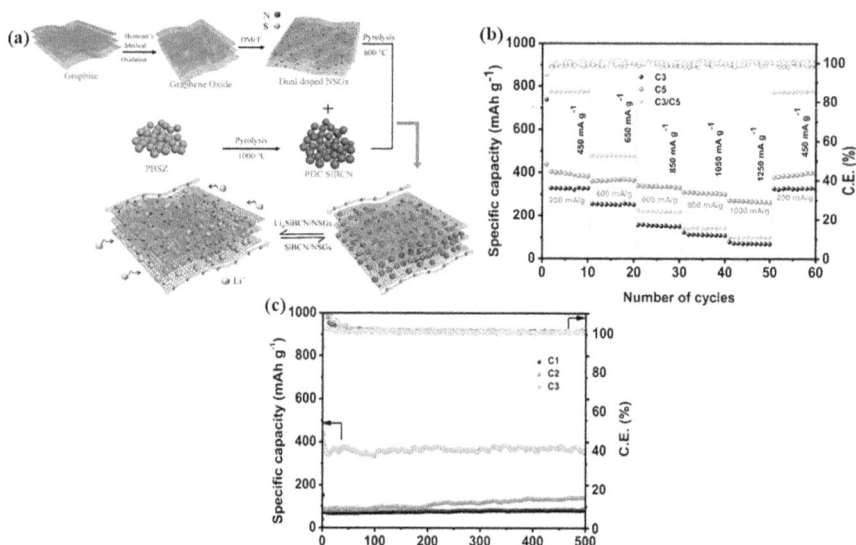

FIGURE 3.2 (a) Graphical representation of the fabrication of SiBCN/NSGs, (b) rate capabilities Li-storage performance of SiBCN/NSGs, and (c) cycling performance and Coulombic efficiency of as-prepared SiBCN/NSGs at 450 mA/g.[94]

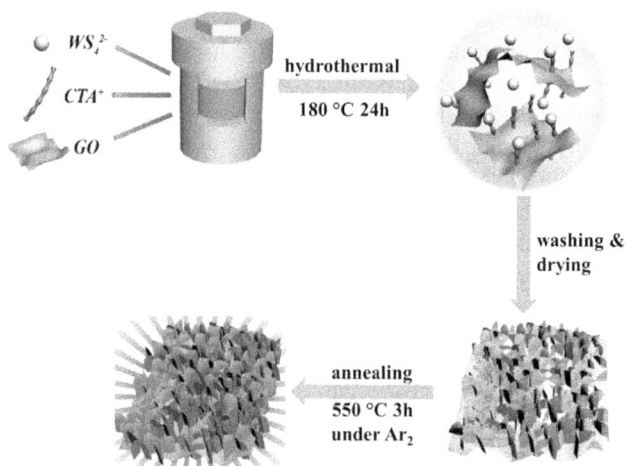

FIGURE 3.3 Schematic representation of the synthesis of WS_2/rGO nano-honeycomb.[95]

However, graphene is one of the toughest materials and it is also very difficult to bend. Hence, several investigators have evolved flexible electrodes for LIBs by compounding graphene with other anode materials.[96,97] The freestanding binder-free SnS_2 nanoplates/graphene/carbon cloth flexible anodes were synthesized via a simple two-step solvothermal method. The as-synthesized SnS_2 nanoplates/graphene/carbon cloth electrode offers great specific surface area and advanced mass loading compared to graphene/SnS$_2$/carbon cloth and facilitates the transport path and diffusion kinetics for both Li ions and electrons with the improved synergic interactions between layered graphene and SnS$_2$. The flexible electrodes offers great specific surface area and an astonishing initial capacity of 1987.4 mAh/g, and a specific capacity of up to 638.1 mAh/g over 150 cycles with around 100% efficiency.[98]

3.2.2 TMOs

TMOs are commonly implemented in LIBs owing to their brilliant electrochemical active interface and high theoretical capacity.[99–105] According to the diverse materials, TMOs have dissimilar extraction and addition mechanisms, which are primarily classified into two categories. The first type is a lithium-insertion; in that lithium ions are introduced in the crystal lattice of the materials during charging and discharging and the structure of the material has not transformed likewise MoO$_2$, TiO$_2$, and so on.[106–113] The other type of TMOs is to form active Li$_2$O while introducing lithium ions, including NiO, ZnO, and Co$_3$O$_4$.[114–120] It can also perform reversible reactions which advances the cycle reversibility of the material during the charging and discharging process.

Among them, TiO$_2$ has been commonly examined for its sustainable structure and low cost during the process of lithiation and delithiation. However, owing to its slow dispersion of lithium ions and poor electron conductivity, the utilization of TiO$_2$ in LIBs is interrupted. To resolve this issue, it is an efficient approach to modify the size and shape of materials. As can be observed from Figure 3.4(a-f), the bundle-like hierarchical structure (TiO$_2$(B)-BH),[121] the exceptional walnut-like porous shell/core structure,[122] and Ti^{3+} self-doped TiO$_2$(TiO$_{2-\delta}$)[123] nanofilm offer an efficient electron migration path and outstanding structural stability, which considerably advances the lithium storage performance. Furthermore, TMOs like V$_2$O$_5$, MoO$_2$, and Mn$_3$O$_4$ have fascinated researchers with their good energy storage potential. Xia et al.[124] used an innovative monomer-assisted reduction method to prepare high-quality 2D sheets of nonlayered MoO$_2$. The usage of these ultrathin 2D-MoO$_2$ electrodes as LIB anodes exhibited a high value of reversible capacity, i.e., amicro supercapacitors showed greater cycle stability (86% retention even after 10,000 cycles), high-rate performance (81% retention from 0.1 to 2 mA/cm^2), great areal capacitance (63.1 mF/cm^2 at 0.1 mA/cm^2). Ma et al.[125] employed a superficial template-free process to manufacture several V$_2$O$_5$ hierarchical structures by annealing various morphological and structural VO$_2$ pioneers which can be easily tailored by regulating the solvothermal reaction duration. After controlling the annealing time, yolk-shell V$_2$O$_5$ composed of nanoplates showed brilliant cycling stability at high currents and great reversible capacity at a high current density rate. As the cathode materials for LIBs, the electrode offers 119.2 and 87.3 mAh/g of reversible capacities at high current densities of 2400 and 3600 mA/g, respectively, along with 78.31% of capacity retention even after 80 cycles at 1200 mA/g. The one-pot hydrothermal method was employed for the design and preparation of highly porous Mn$_3$O$_4$@C micro/nanocuboids to improve the electrochemical performance of materials. The prepared Mn$_3$O$_4$@C micro/nanocuboids showed a high value of reversible specific, i.e., 879 mAh/g at the 100 mA/g

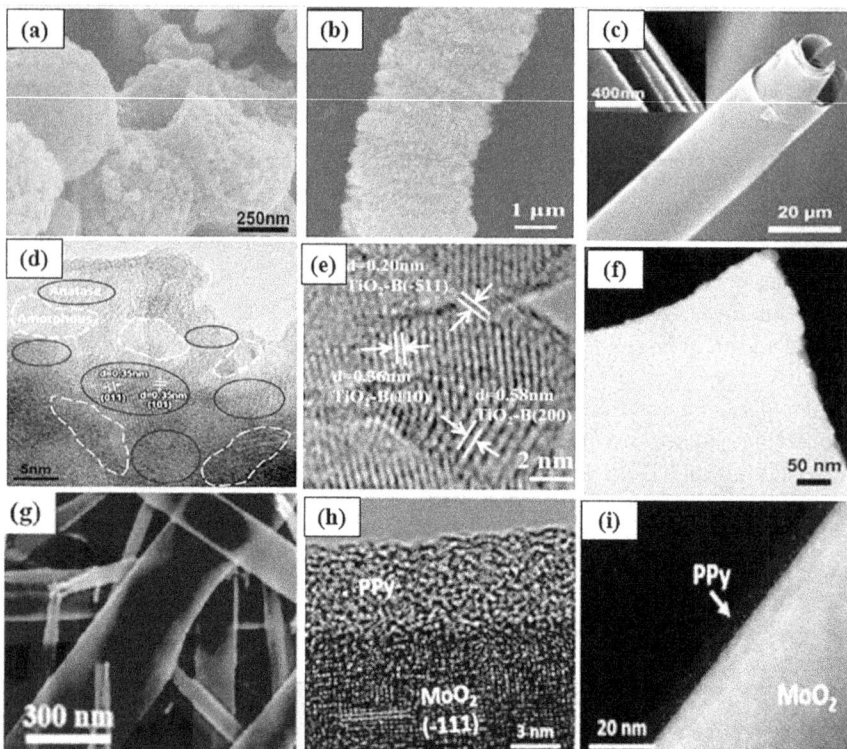

FIGURE 3.4 Some electron microscope pictures of TMOs in LIBs: (a) SEM and (d) HRTEM images of shell/core structure. (b) SEM and (e) HRTEM images of TiO₂ (B)-BH. (c) SEM and (f) TEM images of TiO$_{2-\delta}$,[121] (g) SEM, (h) TEM, and (i) STEM images of pyrrole-reduced MoO₂/PPy.[124]

current density as well as magnificent cyclic stability (i.e., 86% capacity retention even after 500 cycles), making it a suitable candidate as anode material for LIBs.[126]

The influence of volume expansion and the small conductivity are the two prime difficulties of TMO usage in LIBs. The volume alteration during the process of electrode reactions demolishes the microstructure of electrode and electrolyte drenching corrodes the material surface. It is important to mention that failure to manufacture an electronic passage network between the microstructures of TMO will result in the poor conductivity of the electrode. Hence, many researchers advance the electrochemical performance of TMOs via the introduction of supplementary conductive materials. For instance, Moon et al.[127] prepared a nanocage-like complex compound that consisted of CNTs and N-doped ZnO. CNTs could augment the mechanical stability and electrical conductivity to the complex composite (Figure 3.5 (a,b)). The internal space around the nanocage-like structure was conducive owing to the ion diffusion and electrolyte permeation and provided an assurance for the proficient charge/discharge cycles of Li-ion. V₂O₅ and graphene combined sandwich-like structures formed by accumulating them layer by layer from

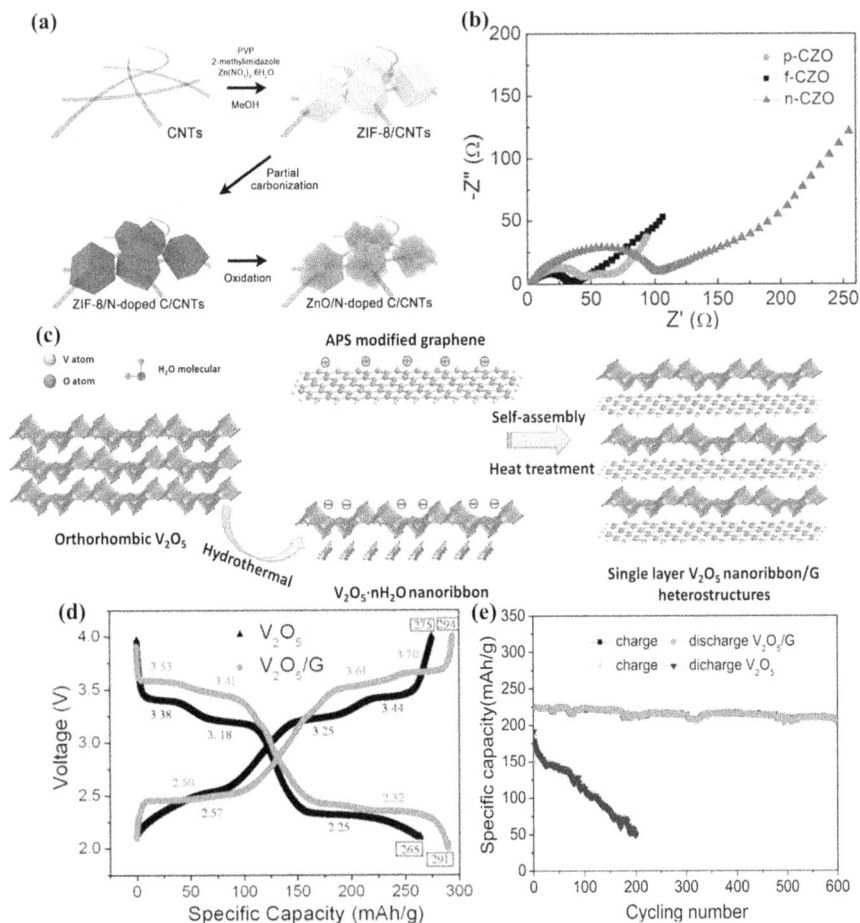

FIGURE 3.5 Experiments of TMO exploited in LIBs: (a) fabrication process and (b) EIS of ZnO/N-doped C/CNTs,[127] (c) synthesis process, (d) primary charge–discharge voltage, and (e) galvanostatic cycling properties of V_2O_5/G and V_2O_5.[128]

the bottom up (Figure 3.5 (c,d,e)). By the addition of graphene, the charging voltage platform of V_2O_5 was found to be amplified. V_2O_5/graphene material exhibits a very high value of discharge capacity, i.e., 225 mAh/g and about 99.6% of Coulombic efficiency at a high current density. In addition, V_2O_5/graphene material has prodigious cyclic stability that maintains 92.8% capacity even after 600 cycles.[128] Sulfur-doped Mn_3O_4 and rGO was fabricated and exhibits exceptional performance owing to its exceptional mesoporous structure. Mesoporous structure rendered efficient pathways, which can truncate the diffusion path of Li^+. Furthermore, Mn and S doping might also surge the conductivity of the electrode. Alternatively, the flexible mesoporous structure could prevent the degradation of materials during the cycling process and adjust the volume change of Mn_3O_4. Concurrently, the amalgamation of Mn_3O_4 and rGO could further modify the electrochemical performance.[129]

3.2.3 TMDs

TMDs with layered structures like graphene exhibit great potential to be utilized in LIBs. TMDs have the MX_2 molecular formula, where M= Ti, V, W, and Mo. (transition metal element) and X= S, Se, or Te. Most common 2D TMDs like VS_2, VSe_2, MoS_2, WS_2, and so on are extensively utilized in the literature.[130-137] Individual layer of TMDs is bonded through the van der Waals force, and the monolayer of TMDs consists of stacked triatomic layers. TMDs act as perfect candidates for lithium energy storage as they possess better cycle stability and greater specific capacity compared to traditional carbon anodes owing to their distinct lamellar structure. During the lithiation process, the volume expansion of TMDs is significantly lesser than that of Si-based materials. Furthermore, TMDs have many active sites over the surface for energy storage via reversible redox reaction. The physical properties and the characteristic color can be modified and changed, respectively, at the same time to advance the potential of TMDs in diverse fields by the introduction of doping and defects.

A consolidated and red $MoSe_2$ with tunable band gap was synthesized directly using a fast oxidation method (Figure 3.6(a)). With the addition of an oxygen gradient with the formation of mixed Mo-O bonds (Mo^{5+} and Mo^{6+}), the color of the $MoSe_2$ found to be changed from black to red; at the same time, the conductivity of the red $MoSe_2$ is also boosted compared to the black or blue MoO_3. The accurately tailored

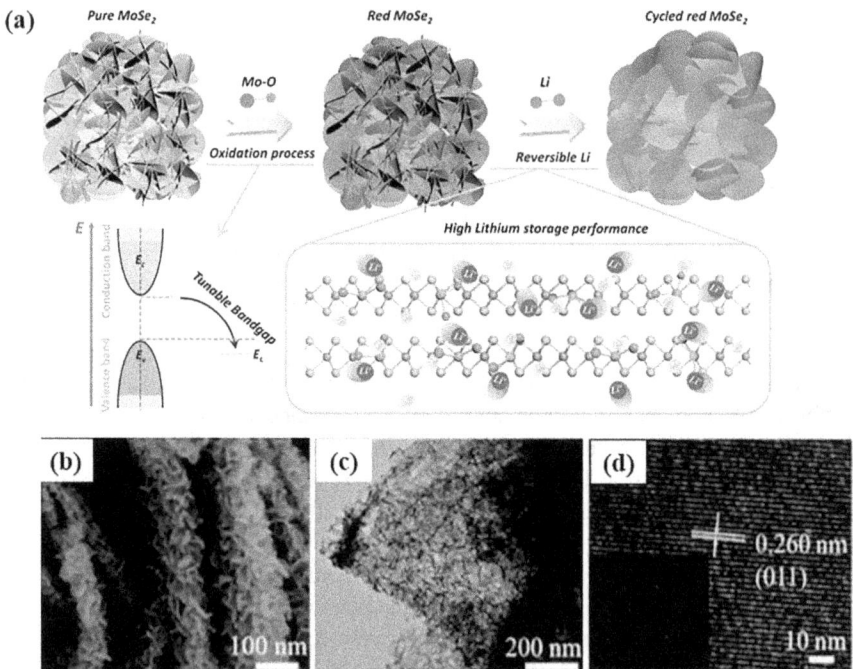

FIGURE 3.6 Experiments of TMDs utilized in LIBs (a) preparation of red $MoSe_2$.[138] (b) SEM, (c) TEM, and (d) HRTEM images of VSe_2 nanosheets.[139]

properties directly advance the Li storage capacity and electrochemical activity. Utilization of red $MoSe_2$ as anode materials of LIBs offers a great value of discharge capacity, i.e., 1125.7 mAh/g after 500 cycles at a current density of 1 A/g. After 4000 cycles, the battery can still retain a 649.5 mAh/g capacity at a current density of 5 A/g.[138] However, VSe_2 is very much different from other characteristics of TMDs. VSe_2 has a semi-metallic nature due to the overlapping of the conduction and valence band. Furthermore, an additional outstanding application of VSe_2 is the probable multielectron transfer mechanism, which can offer very high energy density. For example, VSe_2 nanosheets with selenium vacancy were prepared by a one-step solvothermal method with the utilization of appropriate starting materials and further findings reveal that the VSe_2 nanosheets with selenium vacancy could significantly increase the rate of lithium-ion diffusion (Figure 3.6(b-d)).[139]

TMDs act as perfect candidates for energy storage applications due to the presence of large intercalation space which can endorse the swift movement of lithium ions. The findings reveal that the single or multilayer TMDs have an improved spatial framework and can buffer the volume changes produced by charging and discharging. MoS_2 is the most extensively considered TMDs in recent times.[140–142] Hence, flower-like firmed MoS_2/N-doped carbon composites were synthesized and unveiled reversible capacity of about 904.7 mAh/g over 100 cycles.[143] Dong et al.[144] synthesized the MoS_2/rGO nanovesicles with the help of polystyrene nanosphere template. The structure of the special nanovesicle caused the shortage of diffusion path of lithium ions and also enhanced the percolation of electrolyte at the active sites. Additionally, the introduction of rGO nanovesicles could augment the conductivity of the materials resulting in better electrochemical performance. This complex compound could be utilized for flexible energy storage devices owing to the exceptional mechanical properties of rGO. An innovative few-atomic-layered hollow nanospheres were obtained by alternate insertion of N-doped carbon and MoS_2 monolayers. The hollow nanospheres as anode materials deliver the reversible discharge capacity of 1025 mAh/g over 80 cycles at 1000 mA/g. The admirable electrochemical performance can be accredited to the cooperative effect of alternate addition of monolayer carbon, ultrathin characteristic, and expanded interlayer spacing.[145]

Besides MoS_2, additional TMDs have been exploited in the literature. For example, N-doped mesoporous carbon (N-CMK-3) and SnS_2 with a large porosity (126.7 m^2/g) were prepared via a low-temperature thermal process. The cyclic stability and capacity of SnS_2@N-CMK-3 electrodes were found to be greater than those of pristine SnS_2 electrodes. The specific capacity of L-SnS_2@N-CMK-3 was retained at about 680 mAh/g at 1000 mA/g.[146] Owing to the unique 2D hierarchical structure, the conductivity of materials was enhanced apparently. The mesoporous network shortens the transportation distance of electrons and lithium ions between the electrolytes and electrodes. Wu et al.[147] prepared the complex compound between graphene and porous bowl-shaped VS_2 by a solvothermal methodology and displayed outstanding cycle stability and rate performance. The charge transfer impedance (R_{ct}) plays a key role in charge transfer kinetics. The composite H-VS_2@rGO electrode compared with VS_2 and H-VS_2 displayed a smaller value of R_{ct}, i.e., 140 Ω, which results in a high-rate property during charging and discharging (Figure 3.7(a-d)).

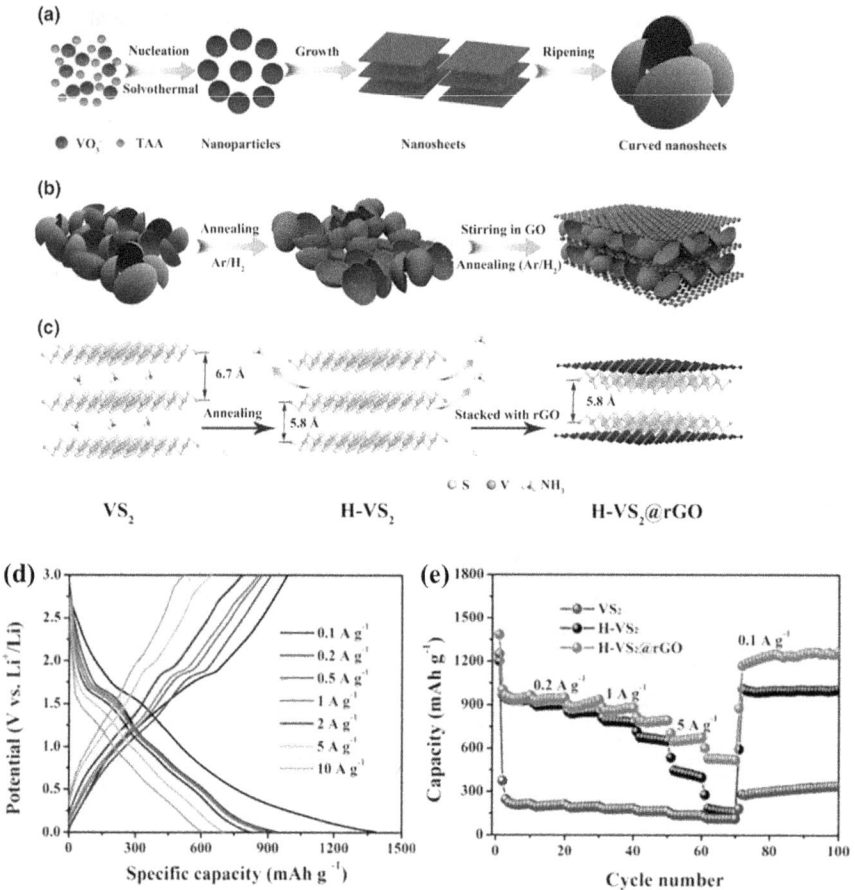

FIGURE 3.7 Schematic representation of the solvothermal growth process of (a) mono-disperse of bowl-shaped VS_2 nanosheets, (b) the production of H-VS_2 and H-VS_2@r-GO composite, (c) schematic structure of the corresponding structure, (d) discharge–charge profiles of H-VS_2@r-GO at different values of current densities, and (e) rate capability for VS_2, H-VS_2, and H-VS_2@r-GO.[147]

3.2.4 MXenes

In recent years, another category of 2D material, i.e., MXenes has drawn excessive curiosity among researchers in the field of energy storage devices.[148–154] It was discovered in 2011, and it is a collective name for a novel type of 2D materials of nitride, carbide, or carbonitride. It is primarily attained by an etching method from the MAX phase. MAX has a general formula $M_{n+1}AX_n$ (n=1, 2, or3), where M= transition metal elements (such as V, Cr, Mo, and Ni), A = group of IIIA or IVA elements (like Al, Si, and Ge), and X=C and /or N elements. MXenes have exceptional lamellar structures and plenty active sites over the surface, exhibit a great potential in the application of LIBs energy storage.[155,156] For instance, $V_4C_3T_X$ MXene ($V_4C_3T_X$-HF) was synthesized by selective etching of V_4AlC_3

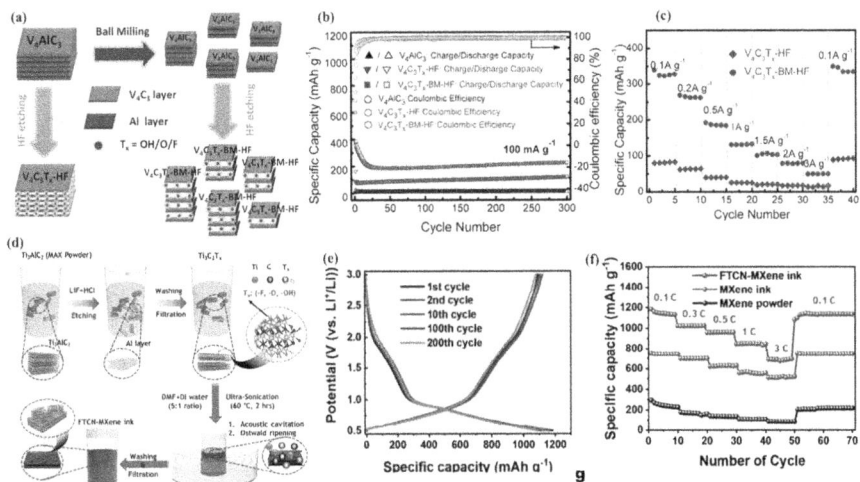

FIGURE 3.8 Examples based on MXenes applied to LIBs: (a) graphical illustration of the two different processes for the synthesis of $V_4C_3T_X$ MXene, (b) cycling performance and Coulombic efficacy for V_4AlC_3, $V_4C_3T_x$-HF, and $V_4C_3T_x$-BM-HF, and (c) rate capacity of $V_4C_3T_X$-HF, and $V_4C_3T_X$-BM-HF at various rates.[157] (d) Schematic representation of the synthesis of functionalized titanium carbide nanorods (FTCNs) over the Ti_3C_2 nanosheets, (e) charge–discharge curve at 67.57 mA/g (0.1 C), and (f) rate capability of FTCN-MXene, MXene, and MXene powder.[158]

using hydrofluoric acid solution at 55°C.[157] With a large specific area and interlayer spacing, $V_4C_3T_X$-BM-HF was obtained by performing a modest ball-milling treatment on the $V_4C_3T_X$ starting material, which is valuable for Li-ion storage as shown in Figure 3.8(a). It offers a great specific capacity of 225 mAh/g at a 0.1 A/g after 300 cycles and 125 mAh/g at 1 A/g (Figure 3.8(b) and (c)). Nam et al.[158] produced functionalized titanium carbide nanorods (FTCNs) over the Ti_3C_2 nanosheets via the liquid phase epitaxial growth and Sono-chemical approaches (Figure 3.8(d)). The intercalation of ions results in the increase of spacing between interlayers of Ti_3C_2, which could augment the reaction sites of lithium ions. Higher electronegativity of fluoride anions (F^-) that are grown over the surface of Ti_3C_2 nanosheets helps to improve the cyclic stability of FTCN. FTCN-MXene showed excellent specific capacity of 1034 mAh/g and great Coulombic efficiency, i.e., 98.78% over 250 cycles compared to pristine $Ti_3C_2T_X$ MXene (Figure 3.8(e) and (f)).

Owing to the high comparative area, narrow band gap semiconductor properties, and weak interlayer forces, MXenes demonstrate great potential in the field of LIBs. Some of the corresponding cyclic properties of MXenes are presented in Table 3.2. Among them, $Ti_3C_2T_X$ showed much lower surface energy compared to other MXenes, which will decrease the rate of mutual adsorption of $Ti_3C_2T_X$ layers and broaden the interlayer spacing for Li ions. Increasing the spacing between interlayers can efficiently provide a higher number of active sites that lead to noteworthy enhancement in the electrochemical performance of LIBs. Hence, $Ti_3C_2T_X$ is recommended as a preferable anode material for LIBs. $Ti_3C_2T_X$ can be

TABLE 3.2

Cyclic Properties of MXenes Utilized in LIBs

Electrode	Specific Capacity (mAh/g)	Current Density (A/g)	Cycle No. (R)	Ref
Nb_2CT_x	336	2	2000	163
$V_2CT_x@SnO_2$	1	768	200	164
LTP-TiO_2/Ti_3C_2	0.1	193	500	165
N, S-V_2CT_x	0.1	590	100	166
$Ti_3C_2T_x$	5	155	1000	167
SnS/ $Ti_3C_2T_x$	0.5	866	300	168
P- Ti_3C_2	1	97	3000	169
Oligolayered $Ti_3C_2T_x$	1	330	800	170
N(V)- $Ti_3C_2T_x$	3 C	92	1000	171
Co_3O_4/$Ti_3C_2T_x$	0.5	999.3	900	172
$Ni(OH)_2$/d-Ti_3C_2	1	372.0	1000	173
Sn@Ti_3C_2	0.5	666	250	174
Mo_2C	0.01	90	140	175
V_2C	0.02	230.3	480	176
$Ti_3C_2(OH)_2$	0.1	143.4	250	177

composite with carbon-based materials, oxides, and other materials to manufacture a heterogeneous assembly, reforming the functionality of lithium storage as shown in Figure 3.9. Zhang et al.[159] introduced the CNTs into $Ti_3C_2T_X$ to form a network structure and then $Ti_3C_2T_X$/CNTs are combined with red phosphorus using a ball milling scheme. The oxygen-containing functional groups present on the $Ti_3C_2T_X$ surface interact with P to create a stable Ti-O-P bond. Such $Ti_3C_2T_X$/CNTs@P nanohybrids showed exceptional reversible capacity of 2598 mAh/g at 0.05°C, better rate capability (454 mAh/g at 30°C), and admirable cycling stability (2078 mAh/g over 500 cycles). Zhang et al.[160] used the vapor deposition method to synthesize N-doped $Ti_3C_2T_X$ and red phosphorus composites. The red phosphorus adsorbed over the $Ti_3C_2T_X$ matrix facilitates the transportation of electrons and lithium ions. The capacity of $Ti_3C_2T_X$ was found to upsurge meaningfully with the increase in the interlayer distance at a high current. The bottom-up method was utilized to affix Co_3O_4 nanocrystals with $Ti_3C_2T_X$ to significantly enhance the cycle life of the battery.[161] Further, it was validated by the first-principles computations that the introduced Co_3O_4 nanocrystals were advantageous in improving the movement of electrons at the material interface. He et al.[162] congregated the cationic RDS over the surface of anionic $Ti_3C_2T_X$ nanosheets via electrostatic interaction. The electrochemical test displayed that the charge–discharge finding of RDS/d-$Ti_3C_2T_X$ offered the first discharge capacity of 458 mAh/g at a current density of 50 mA/g and a reversible capacity of 160 mAh/g under a current density of 1000 mA/g, which was meaningfully higher than that of pristine $Ti_3C_2T_X$ and laponite RDS.

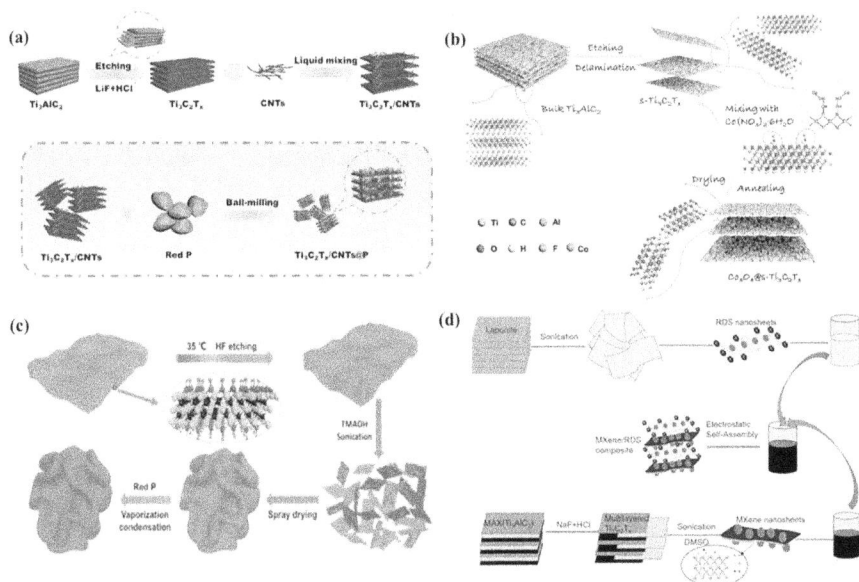

FIGURE 3.9 Fabrication processes of (a) $Ti_3C_2T_x$/CNTs@P,[159] (b) Co_3O_4@s-$Ti_3C_2T_x$,[161] (c) N- $Ti_3C_2T_x$/P,[160] and (d) RDS/d- $Ti_3C_2T_x$.[162]

3.2.5 XENES

Graphite as first commercial anode materials suffers from low theoretical capacity, i.e., 372 mAh/g and poor Li adsorption strength leads to the establishment of other 2D materials with monoatomic layers of B, Si, Ge, P, As, and other elements contiguous to carbon in successive years.[178–184] This series of materials was termed as Xenes (X= B, Si, Ge, P, As). The atoms present in these single layers are almost entirely uncovered compared to bulk materials that result in the improvement of atom utilization rate. The electrical properties and band structures can be more simply controlled by element doping and thickness control. Xenes can increase the electron movement, which is beneficial to the performance of electronic devices. Furthermore, Xenes are transparent and elastic, with great potential and can be exploited in wearable devices. Now, most Xenes have been discovered by first-principles calculations to have great theoretical specific capacity and a small movement barrier of lithium ions, which is advantageous for anode materials and the findings are represented in Table 3.3.

Black phosphorene, or phosphorene,[192] is a 2D material consisting of ordered P atoms uncovered from black phosphorus along with a direct band gap. Phosphene has wide-ranging applications in gas sensors, field effect transistors, optoelectronic devices, and LIBs.[193–195] Additionally, C_3N is a new 2D semiconductor material with a nonporous and graphene-like assembly, which has admirable electronic and mechanical properties. Thus, Guo et al.[196] amalgamated the two materials to diminish each other's limitations. The first-principles calculation reported that the

TABLE 3.3

Computational Calculation of Xenes Used in LIBs

Material	Diffusion Barrier (eV)	Theoretical Specific Capacity (mAh/g)	Adsorption Energy of Li (eV)	Ref
Silicene/BN	0.17	1015	−2.07	185
χ3 borophene	0.21	1396	−1.75	186
Graphene/antimonene	0.59	369.03	−2.92	187
Arsenene	0.16	1430	−2.55	188
Blue phosphorene/ borophene	0.30	1019	−2.83	189
Borophene	0.325	1860	−1.12	190
Black phosphorus	0.08	2596	−1.9	191

stiffness of C_3N/phosphene complex was higher than that of graphene with 463.84 mAh/g theoretical capacity.

Borophene is another 2D planar structure consisting of boron atoms with an exceptional hexagonal honeycomb structure.[190] Borophene is not found directly in nature but it can only be formed by artificial methods. Owing to the unusually complex structure, no method was reported in the literature for the synthesis until 2015.[197] Theoretical findings over the years exhibit that cellular borophene cannot be found stable in the free state owing to the absence of electrons.[198] In addition, it has an insignificant energy barrier and open circuit voltage but still its theoretical capacity is comparatively large, which was appropriate for LIBs. Boraphene showed good electrochemical properties by the embedment of lithium ions.

Arsenene is a 2D monoatomic layer material with a honeycomb framework. Theoretical calculations anticipated that arsenene had a low energy barrier (0.16 eV), low lithium-atom adsorption energy (−2.55 eV) but high theoretical capacity, i.e., 1430 mAh/g.[188]

Progressively, scientists have successfully synthesized 2D Xenes like black phosphorus and silicene via the stripping and chemical method[199–202] and have been exploited in LIBs, exhibiting outstanding lithium storage performance. It may take the place of graphite and emerge as a dominating anode material with great energy density in the upcoming years.

3.3 CONCLUSION AND FUTURE ASPECTS

With the growing energy crisis of society, LIBs are progressively emerging toward quicker charging rates and greater storage capacity. Nevertheless, the recent viable carbon rot cannot meet this prerequisite. The development of 2D materials can offer a solution to this rising crisis. The 2D materials have a robust interlayer adsorption energy and high specific surface area, which are appropriate to produce the anode of LIBs. This chapter encompasses various distinctive 2D materials (graphene, TMOs, TMDs, MXenes, and Xenes), and briefly describes the research implemented in LIBs.

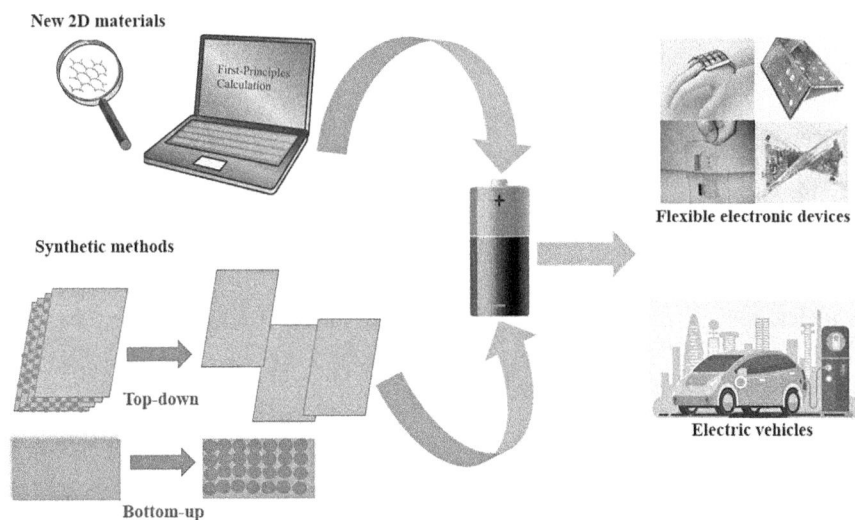

New 2D materials

First-Principles Calculation

Flexible electronic devices

Synthetic methods

Top-down

Bottom-up

Electric vehicles

FIGURE 3.10 Applications and challenges of promising 2D materials.

Though numerous triumphs on 2D materials for LIBs have been conveyed, there are still several prospects and challenges to be explored, as represented in Figure 3.10.

Foremost, the fabrication of 2D materials is the main aspect to resolve industrialization. The preparation methods of 2D materials mainly include the top-down and bottom-up methods, with chemical vapor deposition, micromechanical exfoliation, and physical vapor deposition. Most 2D layered materials can be prepared through the mechanical stripping method, but this method is ineffective and cannot attain large area production. Though the type of 2D materials is inadequate less to vapor deposition, the number of layers, the temperature of the materials, and the phase transition will affect the ultimate quality of the obtained 2D materials. Subsequently, the upsurge of flexible wearable electronics governs the forthcoming growth in the direction of LIBs. Nevertheless, the structural properties of 2D materials will alter capriciously after twisting and folding. Presently, carbonaceous materials are usually utilized to advance the flexibility of electrodes, but there is still a lot of research to be done in this field. Investigating the mechanism of internal reactions in 2D materials is the most challenging matter of concern. Owing to the enormous reaction swiftness, usual instruments are incapable of performing real-time monitoring. This entails scientists to develop testing instruments with higher precision and collaborate with theoretical examination to overcome this issue. The investigation of novel 2D nanomaterials is enduring, and the main goal is to perceive new 2D materials and to reconnoiter their exciting properties. The involvement with 2D materials displays that computational calculation is an essential criterion to understand the properties of materials. Based on first-principles calculations, 2D nanomaterials have totally diverse electrical properties and energy bands, covering material types from metals, semimetals, superconductors, and semiconductors to insulators. Hence, the enormous demands of society validate the persistence of appropriate scientific research despite prevailing several technological problems. The amalgamation of experimental experience and theoretical scrutiny will aid overcome the difficulties in the way of flexible electrode LIBs in the forthcoming time.

ACKNOWLEDGMENTS

SY thanks the National Institute of Technology Warangal (NITW) for the Postdoctoral grant and for providing the facility to work. Financial support from DST-SERB (ECR/2018/002346 and EEQ/2019/000656) is gratefully acknowledged. We are also grateful to the Director of the National Institute of Technology Warangal for providing the facilities. SP would like to thank Colorado School of Mines, USA, and the Indian Institute of Technology Jodhpur, India for resources and technical support.

DECLARATION OF COMPETING INTEREST

The authors declare no conflicts of interest.

REFERENCES

1. Smith, C. J.; Forster, P. M.; Allen, M.; Fuglestvedt, J.; Millar, R. J.; Rogelj, J.; Zickfeld, K. Current Fossil Fuel Infrastructure Does Not Yet Commit Us to 1.5 C Warming. *Nat. Commun.* 2019, *10* (1), 101.
2. Nzereogu, P. U.; Omah, A. D.; Ezema, F. I.; Iwuoha, E. I.; Nwanya, A. C. Anode Materials for Lithium-Ion Batteries: A Review. *Appl. Surf. Sci. Adv.* 2022, *9*, 100233.
3. Manwell, J. F.; McGowan, J. G. Lead Acid Battery Storage Model for Hybrid Energy Systems. *Sol. Energy* 1993, *50* (5), 399–405.
4. Bernard, P.; Lippert, M. Nickel–Cadmium and Nickel–Metal Hydride Battery Energy Storage. In *Electrochemical energy storage for renewable sources and grid balancing*; Elsevier, 2015; pp 223–251.
5. Zelinsky, M. A.; Koch, J. M.; Young, K.-H. Performance Comparison of Rechargeable Batteries for Stationary Applications (Ni/MH vs. Ni–Cd and VRLA). *Batteries* 2018, *4* (1), 1. 10.3390/batteries4010001.
6. Chen, H.; Cong, T. N.; Yang, W.; Tan, C.; Li, Y.; Ding, Y. Progress in Electrical Energy Storage System: A Critical Review. *Prog. Nat. Sci.* 2009, *19* (3), 291–312. 10.1016/j.pnsc.2008.07.014.
7. Abdin, Z.; Khalilpour, K. R. Single and Polystorage Technologies for Renewable-Based Hybrid Energy Systems. In *Polygeneration with polystorage for chemical and energy hubs*; Elsevier, 2019; pp 77–131.
8. Ouyang, L.; Huang, J.; Wang, H.; Liu, J.; Zhu, M. Progress of Hydrogen Storage Alloys for Ni-MH Rechargeable Power Batteries in Electric Vehicles: A Review. *Mater. Chem. Phys.* 2017, *200*, 164–178.
9. Roy, P.; Srivastava, S. K. Nanostructured Anode Materials for Lithium Ion Batteries. *J. Mater. Chem. A* 2015, *3* (6), 2454–2484.
10. Zhu, J.; Huang, B.; Zhao, C.; Xu, H.; Wang, S.; Chen, Y.; Xie, L.; Chen, L. Benzoic Acid-Assisted Substrate-Free Synthesis of Ultrathin Nanosheets Assembled Two-Dimensional Porous Co3O4 Thin Sheets with 3D Hierarchical Micro-/Nano-Structures and Enhanced Performance as Battery-Type Materials for Supercapacitors. *Electrochim. Acta* 2019, *313*, 194–204.

11. Xiang, J.; Liu, Z.; Song, T. Bi@ C Nanoplates Derived from (BiO)2CO3 as an Enhanced Electrode Material for Lithium/Sodium-Ion Batteries. *ChemistrySelect* 2018, *3* (31), 8973–8979.

12. Xu, K.; Ma, L.; Shen, X.; Ji, Z.; Yuan, A.; Kong, L.; Zhu, G.; Zhu, J. Bimetallic Metal-Organic Framework Derived Sn-Based Nanocomposites for High-Performance Lithium Storage. *Electrochim. Acta* 2019, *323*, 134855.

13. Idrees, M.; Batool, S.; Zhuang, Q.; Kong, J.; Seok, I.; Zhang, J.; Liu, H.; Murugadoss, V.; Gao, Q.; Guo, Z. Achieving Carbon-Rich Silicon-Containing Ceramic Anode for Advanced Lithium Ion Battery. *Ceram. Int.* 2019, *45* (8), 10572–10580.

14. Balogun, M.-S.; Yang, H.; Luo, Y.; Qiu, W.; Huang, Y.; Liu, Z.-Q.; Tong, Y. Achieving High Gravimetric Energy Density for Flexible Lithium-Ion Batteries Facilitated by Core–Double-Shell Electrodes. *Energy Environ. Sci.* 2018, *11* (7), 1859–1869.

15. Sun, Y.; Li, S.; Zhuang, Y.; Liu, G.; Xing, W.; Jing, W. Adjustable Interlayer Spacing of Ultrathin MXene-Derived Membranes for Ion Rejection. *J. Membr. Sci.* 2019, *591*, 117350.

16. Liu, C.; Li, F.; Ma, L.-P.; Cheng, H.-M. Advanced Materials for Energy Storage. *Adv. Mater.* 2010, *22* (8), E28–E62.

17. Jiang, N.; Li, B.; Ning, F.; Xia, D. All Boron-Based 2D Material as Anode Material in Li-Ion Batteries. *J. Energy Chem.* 2018, *27* (6), 1651–1654.

18. Goriparti, S.; Miele, E.; De Angelis, F.; Di Fabrizio, E.; Zaccaria, R. P.; Capiglia, C. Review on Recent Progress of Nanostructured Anode Materials for Li-Ion Batteries. *J. Power Sources* 2014, *257*, 421–443.

19. Wei, W.; Wan, L.; Du, C.; Zhang, Y.; Xie, M.; Tian, Z.; Chen, J. Redox-Active Mesoporous Carbon Nanosheet with Rich Cracks for High-Performance Electrochemical Energy Storage. *J. Alloys Compd.* 2019, *794*, 247–254.

20. Xia, H.; Xu, Q.; Zhang, J. Recent Progress on Two-Dimensional Nanoflake Ensembles for Energy Storage Applications. *Nano-Micro Lett.* 2018, *10*, 1–30.

21. Xu, B.; Qian, D.; Wang, Z.; Meng, Y. S. Recent Progress in Cathode Materials Research for Advanced Lithium Ion Batteries. *Mater. Sci. Eng. R Rep.* 2012, *73* (5–6), 51–65.

22. Hu, H.; Zavabeti, A.; Quan, H.; Zhu, W.; Wei, H.; Chen, D.; Ou, J. Z. Recent Advances in Two-Dimensional Transition Metal Dichalcogenides for Biological Sensing. *Biosens. Bioelectron.* 2019, *142*, 111573.

23. Novoselov, K. S.; Geim, A. K.; Morozov, S. V.; Jiang, D.; Katsnelson, M. I.; Grigorieva, I. V.; Dubonos, Sv.; Firsov, A. A. Two-Dimensional Gas of Massless Dirac Fermions in Graphene. *Nature* 2005, *438* (7065), 197–200.

24. Zhao, X.; Liu, H.; Feng, Y.; Pang, L.; Ding, M.; Deng, L.; Zhu, J. In-Situ Constructing of Hierarchical Li4Ti5O12-TiO2 Microspheres Assembled by Nanosheets for Lithium-Ion Batteries. *Mater. Lett.* 2018, *231*, 130–133.

25. Wang, C.-H.; Kurra, N.; Alhabeb, M.; Chang, J.-K.; Alshareef, H. N.; Gogotsi, Y. Titanium Carbide (MXene) as a Current Collector for Lithium-Ion Batteries. *ACS Omega* 2018, *3* (10), 12489–12494.

26. Tang, X.; Guo, X.; Wu, W.; Wang, G. 2D Metal Carbides and Nitrides (MXenes) as High-Performance Electrode Materials for Lithium-Based Batteries. *Adv. Energy Mater.* 2018, *8* (33), 1801897.

27. Naguib, M.; Mochalin, V. N.; Barsoum, M. W.; Gogotsi, Y. 25th Anniversary Article: MXenes: A New Family of Two-Dimensional Materials. *Adv. Mater.* 2014, *26* (7), 992–1005.

28. Wang, L.; Zhang, K.; Pan, H.; Wang, L.; Wang, D.; Dai, W.; Qin, H.; Li, G.; Zhang, J. 2D Molybdenum Nitride Nanosheets as Anode Materials for Improved Lithium Storage. *Nanoscale* 2018, *10* (40), 18936–18941.

29. Yun, Q.; Li, L.; Hu, Z.; Lu, Q.; Chen, B.; Zhang, H. Layered Transition Metal Dichalcogenide-Based Nanomaterials for Electrochemical Energy Storage. *Adv. Mater.* 2020, *32* (1), 1903826.

30. Wang, H.; Jiang, H.; Hu, Y.; Deng, Z.; Li, C. Interface Engineering of Few-Layered MoS2 Nanosheets with Ultrafine TiO2 Nanoparticles for Ultrastable Li-Ion Batteries. *Chem. Eng. J.* 2018, *345*, 320–326.

31. Lin, H.; Jin, R.; Zhu, S.; Huang, Y. C3N/Blue Phosphorene Heterostructure as a High Rate-Capacity and Stable Anode Material for Lithium Ion Batteries: Insight from First Principles Calculations. *Appl. Surf. Sci.* 2020, *505*, 144518.

32. Javadian, S.; Atashzar, S. M.; Gharibi, H.; Vafaee, M. Phosphorene and Graphene Flakes under the Effect of External Electric Field as an Anode Material for High-Performance Lithium-Ion Batteries: A First-Principles Study. *Comput. Mater. Sci.* 2019, *165*, 144–153.

33. Barik, G.; Pal, S. Energy Gap-Modulated Blue Phosphorene as Flexible Anodes for Lithium-and Sodium-Ion Batteries. *J. Phys. Chem. C* 2019, *123* (5), 2808–2819.

34. Shojaei, F.; Kang, H. S. Electronic Structures and Li-Diffusion Properties of Group IV–V Layered Materials: Hexagonal Germanium Phosphide and Germanium Arsenide. *J. Phys. Chem. C* 2016, *120* (41), 23842–23850.

35. Hu, J.; Zhong, C.; Wu, W.; Liu, N.; Liu, Y.; Yang, S. A.; Ouyang, C. 2D Honeycomb Borophene Oxide: A Promising Anode Material Offering Super High Capacity for Li/Na-Ion Batteries. *J. Phys. Condens. Matter* 2019, *32* (6), 065001.

36. Su, J.; Duan, T.; Li, W.; Xiao, B.; Zhou, G.; Pei, Y.; Wang, X. A First-Principles Study of 2D Antimonene Electrodes for Li Ion Storage. *Appl. Surf. Sci.* 2018, *462*, 270–275.

37. Zhu, J.; Chroneos, A.; Schwingenschlögl, U. Silicene/Germanene on MgX 2 (X= Cl, Br, and I) for Li-Ion Battery Applications. *Nanoscale* 2016, *8* (13), 7272–7277.

38. Li, Y.; Huang, H.; Chen, S.; Yu, X.; Wang, C.; Ma, T. 2D Nanoplate Assembled Nitrogen Doped Hollow Carbon Sphere Decorated with Fe 3 O 4 as an Efficient Electrocatalyst for Oxygen Reduction Reaction and Zn-Air Batteries. *Nano Res.* 2019, *12*, 2774–2780.

39. Liu, Y.; Wang, J.; Wu, J.; Ding, Z.; Yao, P.; Zhang, S.; Chen, Y. 3D Cube-Maze-like Li-Rich Layered Cathodes Assembled from 2D Porous Nanosheets for Enhanced Cycle Stability and Rate Capability of Lithium-Ion Batteries. *Adv. Energy Mater.* 2020, *10* (5), 1903139.

40. Wang, Y.; Wu, Z.; Jiang, L.; Tian, W.; Zhang, C.; Cai, C.; Hu, L. A Long-Lifespan, Flexible Zinc-Ion Secondary Battery Using a Paper-Like Cathode from Single-Atomic Layer MnO2 Nanosheets. *Nanoscale Adv.* 2019, *1* (11), 4365–4372.

41. Nithya, C.; Vishnuprakash, P.; Gopukumar, S. A Mn 3 O 4 Nanospheres@ RGO Architecture with Capacitive Effects on High Potassium Storage Capability. *Nanoscale Adv.* 2019, *1* (11), 4347–4358.

42. Ao, L.; Wu, C.; Xu, Y.; Wang, X.; Jiang, K.; Shang, L.; Li, Y.; Zhang, J.; Hu, Z.; Chu, J. A Novel Sn Particles Coated Composite of SnOx/ZnO and N-Doped Carbon Nanofibers as High-Capacity and Cycle-Stable Anode for Lithium-Ion Batteries. *J. Alloys Compd.* 2020, *819*, 153036.

43. Wang, L.; Guo, W.; Lu, P.; Zhang, T.; Hou, F.; Liang, J. A Flexible and Boron-Doped Carbon Nanotube Film for High-Performance Li Storage. *Front. Chem.* 2019, *7*, 832.

44. Li, S.; Zhao, X.; Feng, Y.; Yang, L.; Shi, X.; Xu, P.; Zhang, J.; Wang, P.; Wang, M.; Che, R. A Flexible Film toward High-Performance Lithium Storage: Designing Nanosheet-Assembled Hollow Single-Hole Ni–Co–Mn–O Spheres with Oxygen Vacancy Embedded in 3D Carbon Nanotube/Graphene Network. *Small* 2019, *15* (27), 1901343.

45. Zhang, X.; Li, S.; Li, J.; Ye, M.; Song, Z.; Jin, S.; Shi, B.; Pan, Y.; Yan, J.; Wang, Y. Absorption and Diffusion of Lithium on Layered InSe. *Comput. Condens. Matter* 2019, *21*, e00404.

46. Han, X.; Sun, L.; Wang, F.; Sun, D. MOF-Derived Honeycomb-like N-Doped Carbon Structures Assembled from Mesoporous Nanosheets with Superior Performance in Lithium-Ion Batteries. *J. Mater. Chem. A* 2018, *6* (39), 18891–18897.

47. Gan, Q.; He, H.; Zhao, K.; He, Z.; Liu, S. Morphology-Dependent Electrochemical Performance of Ni-1, 3, 5-Benzenetricarboxylate Metal-Organic Frameworks as an Anode Material for Li-Ion Batteries. *J. Colloid Interface Sci.* 2018, *530*, 127–136.

48. Jia, J.; Li, B.; Duan, S.; Cui, Z.; Gao, H. Monolayer MBenes: Prediction of Anode Materials for High-Performance Lithium/Sodium Ion Batteries. *Nanoscale* 2019, *11* (42), 20307–20314.

49. Sharma, D. K.; Kumar, S.; Laref, A.; Auluck, S. Mono and Bi-Layer Germanene as Prospective Anode Material for Li-Ion Batteries: A First-Principles Study. *Comput. Condens. Matter* 2018, *16*, e00314.

50. Wang, H.; Wu, Q.; Wang, Y.; Wang, X.; Wu, L.; Song, S.; Zhang, H. Molecular Engineering of Monodisperse SnO2 Nanocrystals Anchored on Doped Graphene with High-Performance Lithium/Sodium-Storage Properties in Half/Full Cells. *Adv. Energy Mater.* 2019, *9* (3), 1802993.

51. Jing, Y.; Zhou, Z.; Cabrera, C. R.; Chen, Z. Metallic VS2 Monolayer: A Promising 2D Anode Material for Lithium Ion Batteries. *J. Phys. Chem. C* 2013, *117* (48), 25409–25413.

52. Jing, Y.; Liu, J.; Zhou, Z.; Zhang, J.; Li, Y. Metallic Nb2S2C Monolayer: A Promising Two-Dimensional Anode Material for Metal-Ion Batteries. *J. Phys. Chem. C* 2019, *123* (44), 26803–26811.

53. Li, X.; Wang, N.; He, J.; Yang, Z.; Tu, Z.; Zhao, F.; Wang, K.; Yi, Y.; Huang, C. Designing the Efficient Lithium Diffusion and Storage Channels Based on Graphdiyne. *Carbon* 2020, *162*, 579–585.

54. Jiang, M.; Zhang, F.; Zhu, G.; Ma, Y.; Luo, W.; Zhou, T.; Yang, J. Interface-Amorphized Ti3C2@ Si/SiO X@ TiO2 Anodes with Sandwiched Structures and Stable Lithium Storage. *ACS Appl. Mater. Interfaces* 2020, *12* (22), 24796–24805.

55. Zhang, M.; Xie, H.; Fan, H.; Zeng, T.; Yang, W.; Zheng, W.; Liang, H.; Liu, Z. Two-Dimensional Carbon-Coated CoS2 Nanoplatelets Issued from a Novel Co (OH) (OCH3) Precursor as Anode Materials for Lithium Ion Batteries. *Appl. Surf. Sci.* 2020, *516*, 146133.

56. Ji, H.; Hu, S.; Jiang, Z.; Shi, S.; Hou, W.; Yang, G. Directly Scalable Preparation of Sandwiched MoS2/Graphene Nanocomposites via Ball-Milling with Excellent Electrochemical Energy Storage Performance. *Electrochim. Acta* 2019, *299*, 143–151.

57. Li, C.; Liu, S.; Shi, C.; Liang, G.; Lu, Z.; Fu, R.; Wu, D. Two-Dimensional Molecular Brush-Functionalized Porous Bilayer Composite Separators toward Ultrastable High-Current Density Lithium Metal Anodes. *Nat. Commun.* 2019, *10* (1), 1363.

58. Yao, W.; Wu, S.; Zhan, L.; Wang, Y. Two-Dimensional Porous Carbon-Coated Sandwich-like Mesoporous SnO2/Graphene/Mesoporous SnO2 Nanosheets towards High-Rate and Long Cycle Life Lithium-Ion Batteries. *Chem. Eng. J.* 2019, *361*, 329–341.

59. Mei, J.; He, T.; Zhang, Q.; Liao, T.; Du, A.; Ayoko, G. A.; Sun, Z. Carbon–Phosphorus Bonds-Enriched 3D Graphene by Self-Sacrificing Black Phosphorus Nanosheets for Elevating Capacitive Lithium Storage. *ACS Appl. Mater. Interfaces* 2020, *12* (19), 21720–21729.

60. Qin, G.; Liu, Y.; Liu, F.; Sun, X.; Hou, L.; Liu, B.; Yuan, C. Magnetic Field Assisted Construction of Hollow Red P Nanospheres Confined in Hierarchical N-Doped Carbon Nanosheets/Nanotubes 3D Framework for Efficient Potassium Storage. *Adv. Energy Mater.* 2021, *11* (4), 2003429.

61. Sannyal, A.; Zhang, Z.; Gao, X.; Jang, J. Two-Dimensional Sheet of Germanium Selenide as an Anode Material for Sodium and Potassium Ion Batteries: First-Principles Simulation Study. *Comput. Mater. Sci.* 2018, *154*, 204–211.

62. Xiong, P.; Peng, L.; Chen, D.; Zhao, Y.; Wang, X.; Yu, G. Two-Dimensional Nanosheets Based Li-Ion Full Batteries with High Rate Capability and Flexibility. *Nano Energy* 2015, *12*, 816–823.

63. Eickhoff, H.; Sedlmeier, C.; Klein, W.; Raudaschl-Sieber, G.; Gasteiger, H. A.; Fässler, T. F. Polyanionic Frameworks in the Lithium Phosphidogermanates Li2GeP2 and LiGe3P3 – Synthesis, Structure, and Lithium Ion Mobility. *Z. Für Anorg. Allg. Chem.* 2020, *646* (3), 95–102.

64. Guo, Z.; Zhou, J.; Sun, Z. New Two-Dimensional Transition Metal Borides for Li Ion Batteries and Electrocatalysis. *J. Mater. Chem. A* 2017, *5* (45), 23530–23535.

65. Zhang, H.; Xin, X.; Liu, H.; Huang, H.; Chen, N.; Xie, Y.; Deng, W.; Guo, C.; Yang, W. Enhancing Lithium Adsorption and Diffusion toward Extraordinary Lithium Storage Capability of Freestanding Ti3C2T x MXene. *J. Phys. Chem. C* 2019, *123* (5), 2792–2800.

66. Shi, H.; Qin, J.; Huang, K.; Lu, P.; Zhang, C.; Dong, Y.; Ye, M.; Liu, Z.; Wu, Z.-S. A Two-Dimensional Mesoporous Polypyrrole–Graphene Oxide Heterostructure as a Dual-Functional Ion Redistributor for Dendrite-Free Lithium Metal Anodes. *Angew. Chem.* 2020, *132* (29), 12245–12251.

67. Wu, Y.-Y.; Bo, T.; Zhu, X.; Wang, Z.; Wu, J.; Li, Y.; Wang, B.-T. Two-Dimensional Tetragonal Ti2BN: A Novel Potential Anode Material for Li-Ion Batteries. *Appl. Surf. Sci.* 2020, *513*, 145821.

68. Guo, G.; Wang, R.; Luo, S.; Ming, B.; Wang, C.; Zhang, M.; Zhang, Y.; Yan, H. Metallic Two-Dimensional C3N Allotropes with Electron and Ion Channels for High-Performance Li-Ion Battery Anode Materials. *Appl. Surf. Sci.* 2020, *518*, 146254.

69. A Review on Synthesis of Graphene, h-BN and MoS2 for Energy Storage Applications: Recent Progress and - Search. https://www.bing.com/search?q= A+Review+on+Synthesis+of+Graphene%2C+h-BN+and+MoS2+for+Energy +Storage+Applications%3A+Recent+Progress+and&cvid=cc0bac0f4b8840af-b356abdddb3746d0&aqs=edge..69i57.740j0j1&pglt=43&FORM=ANNTA1&PC= LCTS (accessed 2023-02-13).

70. Akinwande, D.; Huyghebaert, C.; Wang, C.-H.; Serna, M. I.; Goossens, S.; Li, L.-J.; Wong, H.-S. P.; Koppens, F. H. Graphene and Two-Dimensional Materials for Silicon Technology. *Nature* 2019, *573* (7775), 507–518.

71. Chen, H.; Deng, L.; Luo, S.; Ren, X.; Li, Y.; Sun, L.; Zhang, P.; Chen, G.; Gao, Y. Flexible Three-Dimensional Heterostructured ZnO-Co3O4 on Carbon Cloth as Free-Standing Anode with Outstanding Li/Na Storage Performance. *J. Electrochem. Soc.* 2018, *165* (16), A3932.

72. Dong, X.; Liu, W.; Chen, X.; Yan, J.; Li, N.; Shi, S.; Zhang, S.; Yang, X. Novel Three Dimensional Hierarchical Porous Sn-Ni Alloys as Anode for Lithium Ion Batteries with Long Cycle Life by Pulse Electrodeposition. *Chem. Eng. J.* 2018, *350*, 791–798. 10.1016/j.cej.2018.06.031.

73. Fang, C.; Wan, C. H.; Guo, C. Y.; Feng, C.; Wang, X.; Xing, Y. W.; Zhao, M. K.; Dong, J.; Yu, G. Q.; Zhao, Y. G.; Han, X. F. Observation of Large Anomalous Nernst Effect in 2D Layered Materials Fe3GeTe2. *Appl. Phys. Lett.* 2019, *115* (21), 212402. 10.1063/1.5129370.

74. Shi, N.; Xi, B.; Huang, M.; Tian, F.; Chen, W.; Li, H.; Feng, J.; Xiong, S. One-Step Construction of MoS0. 74Se1. 26/N-Doped Carbon Flower-like Hierarchical Microspheres with Enhanced Sodium Storage. *ACS Appl. Mater. Interfaces* 2019, *11* (47), 44342–44351.

75. Jiang, B.; Yang, Z.; Liu, X.; Liu, Y.; Liao, L. Interface Engineering for Two-Dimensional Semiconductor Transistors. *Nano Today* 2019, *25*, 122–134.

76. Zhang, Q.; Dai, M.; Shao, H.; Tian, Z.; Lin, Y.; Chen, L.; Zeng, X. C. Insights into High Conductivity of the Two-Dimensional Iodine-Oxidized Sp2-c-COF. *ACS Appl. Mater. Interfaces* 2018, *10* (50), 43595–43602.

77. Qi, H.; Wang, L.; Zuo, T.; Deng, S.; Li, Q.; Liu, Z.-H.; Hu, P.; He, X. Hollow Structure VS2@ Reduced Graphene Oxide (RGO) Architecture for Enhanced Sodium-Ion Battery Performance. *ChemElectroChem* 2020, *7* (1), 78–85.

78. Zhao, X.; E, J.; Wu, G.; Deng, Y.; Han, D.; Zhang, B.; Zhang, Z. A Review of Studies Using Graphenes in Energy Conversion, Energy Storage and Heat Transfer Development. *Energy Convers. Manag.* 2019, *184*, 581–599. 10.1016/j.enconman.2019.01.092.

79. Wu, K.; Yang, H.; Jia, L.; Pan, Y.; Hao, Y.; Zhang, K.; Du, K.; Hu, G. Smart Construction of 3D N-Doped Graphene Honeycombs with (NH_4) $_2SO_4$ as a Multifunctional Template for Li-Ion Battery Anode: "A Choice That Serves Three Purposes." *Green Chem.* 2019, *21*(6), 1472–1483.

80. Wu, Z.-S.; Zhou, G.; Yin, L.-C.; Ren, W.; Li, F.; Cheng, H.-M. Graphene/Metal Oxide Composite Electrode Materials for Energy Storage. *Nano Energy* 2012, *1* (1), 107–131.

81. Li, N.; Chen, Z.; Ren, W.; Li, F.; Cheng, H.-M. Flexible Graphene-Based Lithium Ion Batteries with Ultrafast Charge and Discharge Rates. *Proc. Natl. Acad. Sci.* 2012, *109* (43), 17360–17365.

82. Shao, J.; Feng, J.; Zhou, H.; Yuan, A. Graphene Aerogel Encapsulated Fe-Co Oxide Nanocubes Derived from Prussian Blue Analogue as Integrated Anode with Enhanced Li-Ion Storage Properties. *Appl. Surf. Sci.* 2019, *471*, 745–752.

83. Antonatos, N.; Ghodrati, H.; Sofer, Z. Elements beyond Graphene: Current State and Perspectives of Elemental Monolayer Deposition by Bottom-up Approach. *Appl. Mater. Today* 2020, *18*, 100502.

84. Ma, X.; Ning, G.; Sun, Y.; Pu, Y.; Gao, J. High Capacity Li Storage in Sulfur and Nitrogen Dual-Doped Graphene Networks. *Carbon* 2014, *79*, 310–320.

85. Mo, R.; Tan, X.; Li, F.; Tao, R.; Xu, J.; Kong, D.; Wang, Z.; Xu, B.; Wang, X.; Wang, C. Tin-Graphene Tubes as Anodes for Lithium-Ion Batteries with High Volumetric and Gravimetric Energy Densities. *Nat. Commun.* 2020, *11* (1), 1374.

86. Raccichini, R.; Varzi, A.; Passerini, S.; Scrosati, B. The Role of Graphene for Electrochemical Energy Storage. *Nat. Mater.* 2015, *14* (3), 271–279.

87. Yoo, E.; Kim, J.; Hosono, E.; Zhou, H.; Kudo, T.; Honma, I. Large Reversible Li Storage of Graphene Nanosheet Families for Use in Rechargeable Lithium Ion Batteries. *Nano Lett.* 2008, *8* (8), 2277–2282.

88. Huang, X.; Qi, X.; Boey, F.; Zhang, H. Graphene-Based Composites. *Chem. Soc. Rev.* 2012, *41* (2), 666–686.

89. Wang, K.; Li, Z. Synthesis of Nitrogen and Phosphorus Dual-Doped Graphene Oxide as High-Performance Anode Material for Lithium-Ion Batteries. *J. Nanosci. Nanotechnol.* 2020, *20* (12), 7673–7679.

90. Pomerantseva, E.; Bonaccorso, F.; Feng, X.; Cui, Y.; Gogotsi, Y. Energy Storage: The Future Enabled by Nanomaterials. *Science* 2019, *366* (6468), eaan8285.

91. Wang, H.; Zhang, C.; Liu, Z.; Wang, L.; Han, P.; Xu, H.; Zhang, K.; Dong, S.; Yao, J.; Cui, G. Nitrogen-Doped Graphene Nanosheets with Excellent Lithium Storage Properties. *J. Mater. Chem.* 2011, *21* (14), 5430–5434.

92. Wu, Z.-S.; Ren, W.; Xu, L.; Li, F.; Cheng, H.-M. Doped Graphene Sheets as Anode Materials with Superhigh Rate and Large Capacity for Lithium Ion Batteries. *ACS Nano* 2011, *5* (7), 5463–5471.

93. Zhang, C.; Mahmood, N.; Yin, H.; Liu, F.; Hou, Y. Synthesis of Phosphorus-Doped Graphene and Its Multifunctional Applications for Oxygen Reduction Reaction and Lithium Ion Batteries. *Adv. Mater.* 2013, *25* (35), 4932–4937.

94. Idrees, M.; Batool, S.; Kong, J.; Zhuang, Q.; Liu, H.; Shao, Q.; Lu, N.; Feng, Y.; Wujcik, E. K.; Gao, Q. Polyborosilazane Derived Ceramics-Nitrogen Sulfur Dual Doped Graphene Nanocomposite Anode for Enhanced Lithium Ion Batteries. *Electrochim. Acta* 2019, *296*, 925–937.

95. Song, Y.; Liao, J.; Chen, C.; Yang, J.; Chen, J.; Gong, F.; Wang, S.; Xu, Z.; Wu, M. Controllable Morphologies and Electrochemical Performances of Self-Assembled Nano-Honeycomb WS2 Anodes Modified by Graphene Doping for Lithium and Sodium Ion Batteries. *Carbon* 2019, *142*, 697–706.

96. Hou, W.; He, J.; Yu, B.; Lu, Y.; Zhang, W.; Chen, Y. One-Pot Synthesis of Graphene-Wrapped NiSe2-Ni0. 85Se Hollow Microspheres as Superior and Stable Electrocatalyst for Hydrogen Evolution Reaction. *Electrochim. Acta* 2018, *291*, 242–248.

97. Liang, L.; Sun, X.; Zhang, J.; Hou, L.; Sun, J.; Liu, Y.; Wang, S.; Yuan, C. In Situ Synthesis of Hierarchical Core Double-Shell Ti-Doped LiMnPO4@ NaTi2 (PO4) 3@ C/3D Graphene Cathode with High-Rate Capability and Long Cycle Life for Lithium-Ion Batteries. *Adv. Energy Mater.* 2019, *9* (11), 1802847.

98. Wang, M.; Huang, Y.; Zhu, Y.; Wu, X.; Zhang, N.; Zhang, H. Binder-Free Flower-like SnS2 Nanoplates Decorated on the Graphene as a Flexible Anode for High-Performance Lithium-Ion Batteries. *J. Alloys Compd.* 2019, *774*, 601–609.

99. Zheng, H.; Zhang, Q.; Gao, H.; Sun, W.; Zhao, H.; Feng, C.; Mao, J.; Guo, Z. Synthesis of Porous MoV2O8 Nanosheets as Anode Material for Superior Lithium Storage. *Energy Storage Mater.* 2019, *22*, 128–137.

100. Fang, S.; Bresser, D.; Passerini, S. Transition Metal Oxide Anodes for Electrochemical Energy Storage in Lithium-and Sodium-Ion Batteries. *Transit. Met. Oxides Electrochem. Energy Storage* 2022, 55–99.

101. Liu, J.; Wu, J.; Zhou, C.; Zhang, P.; Guo, S.; Li, S.; Yang, Y.; Li, K.; Chen, L.; Wang, M. Single-Phase ZnCo2O4 Derived ZnO–CoO Mesoporous Microspheres Encapsulated by Nitrogen-Doped Carbon Shell as Anode for High-Performance Lithium-Ion Batteries. *J. Alloys Compd.* 2020, *825*, 153951.

102. Feng, Q.; Du, Y.; Liang, S.; Li, H. Reduced Graphene Oxide Supported Quasi-Two-Dimensional ZnCo2O4 Nanosheets for Lithium Ion Batteries with High Electrochemical Stability. *Nanotechnology* 2019, *31* (4), 045402.

103. Zhao, N.; Chen, J.; Liu, Z.-Q.; Ban, K.-J.; Duan, W.-J. Porous LiNi1/3Co1/3Mn1/3O2 Microsheets Assembled with Single Crystal Nanoparticles as Cathode Materials for Lithium Ion Batteries. *J. Alloys Compd.* 2018, *768*, 782–788.

104. Ning, W.-W.; Chen, L.-B.; Wei, W.-F.; Chen, Y.-J.; Zhang, X.-Y. NiCoO 2/ NiCoP@ Ni Nanowire Arrays: Tunable Composition and Unique Structure Design for High-Performance Winding Asymmetric Hybrid Supercapacitors. *Rare Met.* 2020, *39*, 1034–1044.

105. Le, S.; Li, Y.; Xiao, S.; Sun, T.; Yao, J.; Zou, Z. Enhanced Reversible Lithium Storage Property of Sn0. 1V2O5 in the Voltage Window of 1.5–4.0 V. *Solid State Ion.* 2019, *341*, 115028.

106. Yu, J.; Dai, G.; Xiang, Q.; Jaroniec, M. Fabrication and Enhanced Visible-Light Photocatalytic Activity of Carbon Self-Doped TiO 2 Sheets with Exposed ${\{\$001\}} $ Facets. *J. Mater. Chem.* 2011, *21* (4), 1049–1057.

107. Shi, Y.; Guo, B.; Corr, S. A.; Shi, Q.; Hu, Y.-S.; Heier, K. R.; Chen, L.; Seshadri, R.; Stucky, G. D. Ordered Mesoporous Metallic MoO2 Materials with Highly Reversible Lithium Storage Capacity. *Nano Lett.* 2009, *9* (12), 4215–4220.

108. Xu, G.; Liu, P.; Ren, Y.; Huang, X.; Peng, Z.; Tang, Y.; Wang, H. Three-Dimensional MoO2 Nanotextiles Assembled from Elongated Nanowires as Advanced Anode for Li Ion Batteries. *J. Power Sources* 2017, *361*, 1–8.

109. Ni, J.; Zhao, Y.; Li, L.; Mai, L. Ultrathin MoO2 Nanosheets for Superior Lithium Storage. *Nano Energy* 2015, *11*, 129–135.

110. Zhong, W.; Huang, J.; Liang, S.; Liu, J.; Li, Y.; Cai, G.; Jiang, Y.; Liu, J. New Prelithiated V2O5 Superstructure for Lithium-Ion Batteries with Long Cycle Life and High Power. *ACS Energy Lett.* 2019, *5* (1), 31–38.

111. Kim, U.-H.; Jun, D.-W.; Park, K.-J.; Zhang, Q.; Kaghazchi, P.; Aurbach, D.; Major, D. T.; Goobes, G.; Dixit, M.; Leifer, N. Pushing the Limit of Layered Transition Metal Oxide Cathodes for High-Energy Density Rechargeable Li Ion Batteries. *Energy Environ. Sci.* 2018, *11* (5), 1271–1279.

112. Li, Y.; Wang, S.; He, Y.-B.; Tang, L.; Kaneti, Y. V.; Lv, W.; Lin, Z.; Li, B.; Yang, Q.-H.; Kang, F. Li-Ion and Na-Ion Transportation and Storage Properties in Various Sized TiO2 Spheres with Hierarchical Pores and High Tap Density. *J. Mater. Chem. A* 2017, *5* (9), 4359–4367.

113. Grosjean, R.; Fehse, M.; Pigeot-Remy, S.; Stievano, L.; Monconduit, L.; Cassaignon, S. Facile Synthetic Route towards Nanostructured Fe–TiO2 (B), Used as Negative Electrode for Li-Ion Batteries. *J. Power Sources* 2015, *278*, 1–8.

114. Pan, Y.; Xu, M.; Yang, L.; Yu, M.; Liu, H.; Zeng, F. Porous Architectures Assembled with Ultrathin Cu2O–Mn3O4 Hetero-Nanosheets Vertically Anchoring on Graphene for High-Rate Lithium-Ion Batteries. *J. Alloys Compd.* 2020, *819*, 152969.

115. Chen, T.; Zhou, Y.; Zhang, J.; Cao, Y. Two-Dimensional MnO2/Reduced Graphene Oxide Nanosheet as a High-Capacity and High-Rate Cathode for Lithium-Ion Batteries. *Int. J. Electrochem. Sci.* 2018, *13* (9), 8575–8588.

116. Yoon, C. S.; Choi, M.-J.; Jun, D.-W.; Zhang, Q.; Kaghazchi, P.; Kim, K.-H.; Sun, Y.-K. Cation Ordering of Zr-Doped LiNiO2 Cathode for Lithium-Ion Batteries. *Chem. Mater.* 2018, *30* (5), 1808–1814.

117. Chu, Y.; Guo, L.; Xi, B.; Feng, Z.; Wu, F.; Lin, Y.; Liu, J.; Sun, D.; Feng, J.; Qian, Y. Embedding MnO@ Mn3O4 Nanoparticles in an N-Doped-Carbon Framework Derived from Mn-Organic Clusters for Efficient Lithium Storage. *Adv. Mater.* 2018, *30* (6), 1704244.

118. Wu, G.; Jia, Z.; Cheng, Y.; Zhang, H.; Zhou, X.; Wu, H. Easy Synthesis of Multi-Shelled ZnO Hollow Spheres and Their Conversion into Hedgehog-like ZnO Hollow Spheres with Superior Rate Performance for Lithium Ion Batteries. *Appl. Surf. Sci.* 2019, *464*, 472–478.

119. Kundu, S.; Sain, S.; Yoshio, M.; Kar, T.; Gunawardhana, N.; Pradhan, S. K. Structural Interpretation of Chemically Synthesized ZnO Nanorod and Its Application in Lithium Ion Battery. *Appl. Surf. Sci.* 2015, *329*, 206–211.

120. Yue, Y.; Liang, H. Micro-and Nano-Structured Vanadium Pentoxide (V2O5) for Electrodes of Lithium-Ion Batteries. *Adv. Energy Mater.* 2017, *7* (17), 1602545.

121. Li, X.; Wu, G.; Liu, X.; Li, W.; Li, M. Orderly Integration of Porous TiO2 (B) Nanosheets into Bunchy Hierarchical Structure for High-Rate and Ultralong-Lifespan Lithium-Ion Batteries. *Nano Energy* 2017, *31*, 1–8.

122. Cai, Y.; Wang, H.-E.; Zhao, X.; Huang, F.; Wang, C.; Deng, Z.; Li, Y.; Cao, G.; Su, B.-L. Walnut-like Porous Core/Shell TiO2 with Hybridized Phases Enabling Fast and Stable Lithium Storage. *ACS Appl. Mater. Interfaces* 2017, *9* (12), 10652–10663.

123. Huang, S.; Zhang, L.; Lu, X.; Liu, L.; Liu, L.; Sun, X.; Yin, Y.; Oswald, S.; Zou, Z.; Ding, F. Tunable Pseudocapacitance in 3D TiO2- δ Nanomembranes Enabling Superior Lithium Storage Performance. *ACS Nano* 2017, *11* (1), 821–830.

124. Xia, C.; Zhou, Y.; Velusamy, D. B.; Farah, A. A.; Li, P.; Jiang, Q.; Odeh, I. N.; Wang, Z.; Zhang, X.; Alshareef, H. N. Anomalous Li Storage Capability in Atomically Thin Two-Dimensional Sheets of Nonlayered MoO2. *Nano Lett.* 2018, *18* (2), 1506–1515.

125. Ma, Y.; Huang, A.; Zhou, H.; Ji, S.; Zhang, S.; Li, R.; Yao, H.; Cao, X.; Jin, P. Template-Free Formation of Various V 2 O 5 Hierarchical Structures as Cathode Materials for Lithium-Ion Batteries. *J. Mater. Chem. A* 2017, *5* (14), 6522–6531.

126. Jiang, Y.; Yue, J.-L.; Guo, Q.; Xia, Q.; Zhou, C.; Feng, T.; Xu, J.; Xia, H. Highly Porous Mn3O4 Micro/Nanocuboids with in Situ Coated Carbon as Advanced Anode Material for Lithium-Ion Batteries. *Small* 2018, *14* (19), 1704296.

127. Moon, J. H.; Oh, M. J.; Nam, M. G.; Lee, J. H.; Min, G. D.; Park, J.; Kim, W.-J.; Yoo, P. J. Carbonization/Oxidation-Mediated Synthesis of MOF-Derived Hollow Nanocages of ZnO/N-Doped Carbon Interwoven by Carbon Nanotubes for Lithium-Ion Battery Anodes. *Dalton Trans.* 2019, *48* (31), 11941–11950.

128. Zeng, J.; Huang, J.; Liu, J.; Xie, T.; Peng, C.; Lu, Y.; Lu, P.; Zhang, R.; Min, J. Self-Assembly of Single Layer V2O5 Nanoribbon/Graphene Heterostructures as Ultrahigh-Performance Cathode Materials for Lithium-Ion Batteries. Carbon 2019, *154*, 24–32.

129. Yao, W.; Qiu, W.; Xu, Z.; Xu, J.; Luo, J.; Wen, Y. Two-Dimensional Sulfur-Doped Mn3O4 Quantum Dots/Reduced Graphene Oxide Nanosheets as High-Rate Anode Materials for Lithium Storage. *Ceram. Int.* 2018, *44* (17), 21734–21741.

130. Liu, H.; Huang, Z.; Wu, G.; Wu, Y.; Yuan, G.; He, C.; Qi, X.; Zhong, J. A Novel WS 2/NbSe 2 VdW Heterostructure as an Ultrafast Charging and Discharging Anode Material for Lithium-Ion Batteries. *J. Mater. Chem. A* 2018, *6* (35), 17040–17048.

131. Li, Z.; Zhang, L. Y.; Zhang, L.; Huang, J.; Liu, H. ZIF-67-Derived CoSe/NC Composites as Anode Materials for Lithium-Ion Batteries. *Nanoscale Res. Lett.* 2019, *14*, 1–11.

132. Chaturvedi, A.; Hu, P.; Ray, A.; Kloc, C.; Madhavi, S.; Aravindan, V. Unusual Li-Storage Behaviour of Two-Dimensional ReS2 Single Crystals. *Batter. Supercaps* 2018, *1* (2), 69–74.

133. Gao, Y.-P.; Wu, X.; Huang, K.-J.; Xing, L.-L.; Zhang, Y.-Y.; Liu, L. Two-Dimensional Transition Metal Diseleniums for Energy Storage Application: A Review of Recent Developments. *CrystEngComm* 2017, *19* (3), 404–418.

134. Radisavljevic, B.; Radenovic, A.; Brivio, J.; Giacometti, V.; Kis, A. Single-Layer MoS2 Transistors. *Nat. Nanotechnol.* 2011, *6* (3), 147–150.

135. Yang, W.; Lu, H.; Cao, Y.; Jing, P. Single-/Few-Layered Ultrasmall WS2 Nanoplates Embedded in Nitrogen-Doped Carbon Nanofibers as a Cathode for Rechargeable Aluminum Batteries. *J. Power Sources* 2019, *441*, 227173.

136. Yin, H.; Liu, Y.; Yu, N.; Qu, H.-Q.; Liu, Z.; Jiang, R.; Li, C.; Zhu, M.-Q. Graphene-like MoS2 Nanosheets on Carbon Fabrics as High-Performance Binder-Free Electrodes for Supercapacitors and Li-Ion Batteries. *ACS Omega* 2018, *3* (12), 17466–17473.

137. Zhang, Y.; Wang, P.; Yin, Y.; Zhang, X.; Fan, L.; Zhang, N.; Sun, K. Heterostructured SnS-ZnS@ C Hollow Nanoboxes Embedded in Graphene for High Performance Lithium and Sodium Ion Batteries. *Chem. Eng. J.* 2019, *356*, 1042–1051.

138. Zhang, S.; Wang, G.; Jin, J.; Zhang, L.; Wen, Z.; Yang, J. Robust and Conductive Red MoSe2 for Stable and Fast Lithium Storage. *ACS Nano* 2018, *12* (4), 4010–4018.

139. Jiang, Q.; Wang, J.; Jiang, Y.; Li, L.; Cao, X.; Cao, M. Selenium Vacancy-Rich and Carbon-Free VSe 2 Nanosheets for High-Performance Lithium Storage. *Nanoscale* 2020, *12* (16), 8858–8866.

140. Ette, P. M.; Chithambararaj, A.; Prakash, A. S.; Ramesha, K. MoS2 Nanoflower-Derived Interconnected CoMoO4 Nanoarchitectures as a Stable and High Rate Performing Anode for Lithium-Ion Battery Applications. *ACS Appl. Mater. Interfaces* 2020, *12* (10), 11511–11521.

141. Zhang, Z.; Wu, S.; Cheng, J.; Zhang, W. MoS2 Nanobelts with (002) Plane Edges-Enriched Flat Surfaces for High-Rate Sodium and Lithium Storage. *Energy Storage Mater.* 2018, *15*, 65–74.

142. Wu, H.; Hou, C.; Shen, G.; Liu, T.; Shao, Y.; Xiao, R.; Wang, H. MoS 2/C/C Nanofiber with Double-Layer Carbon Coating for High Cycling Stability and Rate Capability in Lithium-Ion Batteries. *Nano Res.* 2018, *11*, 5866–5878.

143. Wang, X.; Tian, J.; Cheng, X.; Na, R.; Wang, D.; Shan, Z. Chitosan-Induced Synthesis of Hierarchical Flower Ridge-like MoS2/N-Doped Carbon Composites with Enhanced Lithium Storage. *ACS Appl. Mater. Interfaces* 2018, *10* (42), 35953–35962.

144. Dong, Y.; Jiang, H.; Deng, Z.; Hu, Y.; Li, C. Synthesis and Assembly of Three-Dimensional MoS2/RGO Nanovesicles for High-Performance Lithium Storage. *Chem. Eng. J.* 2018, *350*, 1066–1072.

145. Jing, L.; Lian, G.; Niu, F.; Yang, J.; Wang, Q.; Cui, D.; Wong, C.-P.; Liu, X. Few-Atomic-Layered Hollow Nanospheres Constructed from Alternate Intercalation of Carbon and MoS2 Monolayers for Sodium and Lithium Storage. *Nano Energy* 2018, *51*, 546–555.

146. Liang, J.; Li, H.; Sun, Z.; Jia, D. N-Doped CMK-3 Anchored with SnS2 Nanosheets as Anode of Lithium Ion Batteries with Superior Cyclic Performance and Enhanced Reversible Capacity. *J. Solid State Chem.* 2018, *265*, 424–430.

147. Wu, D.; Wang, C.; Wu, M.; Chao, Y.; He, P.; Ma, J. Porous Bowl-Shaped VS2 Nanosheets/Graphene Composite for High-Rate Lithium-Ion Storage. *J. Energy Chem.* 2020, *43*, 24–32.

148. Mojtabavi, M.; VahidMohammadi, A.; Liang, W.; Beidaghi, M.; Wanunu, M. Single-Molecule Sensing Using Nanopores in Two-Dimensional Transition Metal Carbide (MXene) Membranes. *ACS Nano* 2019, *13* (3), 3042–3053.

149. Li, D.; Chen, X.; Xiang, P.; Du, H.; Xiao, B. Chalcogenated-Ti3C2X2 MXene (X= O, S, Se and Te) as a High-Performance Anode Material for Li-Ion Batteries. *Appl. Surf. Sci.* 2020, *501*, 144221.

150. Liu, R.; Cao, W.; Han, D.; Mo, Y.; Zeng, H.; Yang, H.; Li, W. Nitrogen-Doped Nb2CTx MXene as Anode Materials for Lithium Ion Batteries. *J. Alloys Compd.* 2019, *793*, 505–511.

151. Ding, W.; Wang, S.; Wu, X.; Wang, Y.; Li, Y.; Zhou, P.; Zhou, T.; Zhou, J.; Zhuo, S. Co0. 85Se@ C/Ti3C2Tx MXene Hybrids as Anode Materials for Lithium-Ion Batteries.*J. Alloys Compd.* 2020, *816*, 152566.

152. Liu, Y.; Yu, J.; Guo, D.; Li, Z.; Su, Y. Ti3C2Tx MXene/Graphene Nanocomposites: Synthesis and Application in Electrochemical Energy Storage. *J. Alloys Compd.* 2020, *815*, 152403.

153. Chen, X.; Zhu, Y.; Zhu, X.; Peng, W.; Li, Y.; Zhang, G.; Zhang, F.; Fan, X. Partially Etched Ti3AlC2 as a Promising High-Capacity Lithium-Ion Battery Anode. *ChemSusChem* 2018, *11* (16), 2677–2680.

154. Melchior, S. A.; Palaniyandy, N.; Sigalas, I.; Iyuke, S. E.; Ozoemena, K. I. Probing the Electrochemistry of MXene (Ti2CTx)/Electrolytic Manganese Dioxide (EMD) Composites as Anode Materials for Lithium-Ion Batteries. *Electrochim. Acta* 2019, *297*, 961–973.

155. Wang Hu, J. M.; Zhang, H.; Li, Z. J.; Hu, M. M.; Wang, X. H. Vibrational Properties of Ti3C2 and Ti3C2T2 (T= O, F, OH) Monosheets by First-Principles Calculations: A Comparative Study. 2015, *17*(15), 9997–10003

156. Cheng, R.; Hu, T.; Zhang, H.; Wang, C.; Hu, M.; Yang, J.; Cui, C.; Guang, T.; Li, C.; Shi, C. Understanding the Lithium Storage Mechanism of Ti3C2T x MXene. *J. Phys. Chem. C* 2018, *123* (2), 1099–1109.

157. Zhou, J.; Lin, S.; Huang, Y.; Tong, P.; Zhao, B.; Zhu, X.; Sun, Y. Synthesis and Lithium Ion Storage Performance of Two-Dimensional V4C3 MXene. *Chem. Eng. J.* 2019, *373*, 203–212.

158. Nam, S.; Umrao, S.; Oh, S.; Shin, K. H.; Park, H. S.; Oh, I.-K. Sonochemical Self-Growth of Functionalized Titanium Carbide Nanorods on Ti3C2 Nanosheets for High Capacity Anode for Lithium-Ion Batteries. *Compos. Part B Eng.* 2020, *181*, 107583.

159. Zhang, S.; Liu, H.; Cao, B.; Zhu, Q.; Zhang, P.; Zhang, X.; Chen, R.; Wu, F.; Xu, B. An MXene/CNTs@ P Nanohybrid with Stable Ti–O–P Bonds for Enhanced Lithium Ion Storage.*J. Mater. Chem. A* 2019, *7* (38), 21766–21773.

160. Zhang, S.; Ying, H.; Guo, R.; Yang, W.; Han, W.-Q. Vapor Deposition Red Phosphorus to Prepare Nitrogen-Doped Ti3C2T x MXenes Composites for Lithium-Ion Batteries. *J. Phys. Chem. Lett.* 2019, *10* (21), 6446–6454.

161. Sun, X.; Tan, K.; Liu, Y.; Zhang, J.; Denis, D. K.; uz Zaman, F.; Hou, L.; Yuan, C. A Two-Dimensional Assembly of Ultrafine Cobalt Oxide Nanocrystallites Anchored on Single-Layer Ti3C2Tx Nanosheets with Enhanced Lithium Storage for Li-Ion Batteries. *Nanoscale* 2019, *11* (36), 16755–16766.

162. He, Y.; Zhou, A.; Liu, D.; Hu, Q.; Liu, X.; Wang, L. Self-Assemble and In-Situ Formation of Laponite RDS-Decorated d-Ti3C2Tx Hybrids for Application in Lithium-Ion Battery. *ChemistrySelect* 2019, *4* (36), 10694–10700.

163. Zhao, J.; Wen, J.; Xiao, J.; Ma, X.; Gao, J.; Bai, L.; Gao, H.; Zhang, X.; Zhang, Z. Nb2CTx MXene: High Capacity and Ultra-Long Cycle Capability for Lithium-Ion Battery by Regulation of Functional Groups. *J. Energy Chem.* 2021, *53*, 387–395.

164. Liu, D.; Wang, L.; He, Y.; Liu, L.; Yang, Z.; Wang, B.; Xia, Q.; Hu, Q.; Zhou, A. Enhanced Reversible Capacity and Cyclic Performance of Lithium-Ion Batteries Using SnO2 Interpenetrated MXene V2C Architecture as Anode Materials. *Energy Technol.* 2021, *9* (2), 2000753.

165. Niu, R.; Han, R.; Wang, Y.; Zhang, L.; Qiao, Q.; Jiang, L.; Sun, Y.; Tang, S.; Zhu, J. MXene-Based Porous and Robust 2D/2D Hybrid Architectures with Dispersed Li3Ti2 (PO4) 3 as Superior Anodes for Lithium-Ion Battery. *Chem. Eng. J.* 2021, *405*, 127049.

166. Zhang, Y.; Li, J.; Gong, Z.; Xie, J.; Lu, T.; Pan, L. Nitrogen and Sulfur Co-Doped Vanadium Carbide MXene for Highly Reversible Lithium-Ion Storage. *J. Colloid Interface Sci.* 2021, *587*, 489–498.

167. Barmann, P.; Nölle, R.; Siozios, V.; Ruttert, M.; Guillon, O.; Winter, M.; Gonzalez-Julian, J.; Placke, T. Solvent Co-Intercalation into Few-Layered Ti3C2T x MXenes in Lithium Ion Batteries Induced by Acidic or Basic Post-Treatment. *ACS Nano* 2021, *15* (2), 3295–3308.

168. Zhang, S.; Ying, H.; Huang, P.; Wang, J.; Zhang, Z.; Yang, T.; Han, W.-Q. Rational Design of Pillared SnS/Ti3C2T x MXene for Superior Lithium-Ion Storage. *ACS Nano* 2020, *14* (12), 17665–17674.

169. Xue, N.; Li, X.; Zhang, M.; Han, L.; Liu, Y.; Tao, X. Chemical-Combined Ball-Milling Synthesis of Fluorine-Free Porous MXene for High-Performance Lithium Ion Batteries. *ACS Appl. Energy Mater.* 2020, *3* (10), 10234–10241.

170. Song, X.; Wang, H.; Jin, S.; Lv, M.; Zhang, Y.; Kong, X.; Xu, H.; Ma, T.; Luo, X.; Tan, H. Oligolayered Ti3C2Tx MXene towards High Performance Lithium/Sodium Storage. *Nano Res.* 2020, *13*, 1659–1667.

171. Cheng, R.; Wang, Z.; Cui, C.; Hu, T.; Fan, B.; Wang, H.; Liang, Y.; Zhang, C.; Zhang, H.; Wang, X. One-Step Incorporation of Nitrogen and Vanadium between Ti3C2Tx MXene Interlayers Enhances Lithium Ion Storage Capability. *J. Phys. Chem. C* 2020, *124* (11), 6012–6021.

172. Zhang, Z.; Guo, H.; Li, W.; Liu, G.; Zhang, Y.; Wang, Y. Sandwich-like Co 3 O 4/ MXene Composites as High Capacity Electrodes for Lithium-Ion Batteries. *New J. Chem.* 2020, *44* (15), 5913–5920.

173. Li, C.; Xue, Z.; Qin, J.; Sawangphruk, M.; Yu, P.; Zhang, X.; Liu, R. Synthesis of Nickel Hydroxide/Delaminated-Ti3C2 MXene Nanosheets as Promising Anode Material for High Performance Lithium Ion Battery. *J. Alloys Compd.* 2020, *842*, 155812.

174. Wang, Z.; Bai, J.; Xu, H.; Chen, G.; Kang, S.; Li, X. Synthesis of Three-Dimensional Sn@ Ti3C2 by Layer-by-Layer Self-Assembly for High-Performance Lithium-Ion Storage. *J. Colloid Interface Sci.* 2020, *577*, 329–336.

175. Mei, J.; Ayoko, G. A.; Hu, C.; Bell, J. M.; Sun, Z. Two-Dimensional Fluorine-Free Mesoporous Mo2C MXene via UV-Induced Selective Etching of Mo2Ga2C for Energy Storage. *Sustain. Mater. Technol.* 2020, *25*, e00156.

176. Liu, F.; Liu, Y.; Zhao, X.; Liu, K.; Yin, H.; Fan, L.-Z. Prelithiated V2C MXene: A High-Performance Electrode for Hybrid Magnesium/Lithium-Ion Batteries by Ion Cointercalation. Small 2020, *16* (8), 1906076.

177. Zhang, B.; Zhu, J.; Shi, P.; Wu, W.; Wang, F. Fluoride-Free Synthesis and Microstructure Evolution of Novel Two-Dimensional Ti3C2 (OH) 2 Nanoribbons as High-Performance Anode Materials for Lithium-Ion Batteries. *Ceram. Int.* 2019, *45* (7), 8395–8405.

178. Lee, Y.; Jo, J. H.; Park, H.; Ko, W.; Kang, J.; Myung, S.-T.; Sun, Y.-K.; Kim, J. Development of Novel Cathode with Large Lithium Storage Mechanism Based on Pyrophosphate-Based Conversion Reaction for Rechargeable Lithium Batteries. *Small Methods* 2020, *4* (3), 1900847.

179. Chittari, B. L.; Park, Y.; Lee, D.; Han, M.; MacDonald, A. H.; Hwang, E.; Jung, J. Electronic and Magnetic Properties of Single-Layer MPX3 Metal Phosphorous Trichalcogenides. *Phys. Rev. B* 2016, *94* (18), 184428.

180. Hu, J.; Ouyang, C.; Yang, S. A.; Yang, H. Y. Germagraphene as a Promising Anode Material for Lithium-Ion Batteries Predicted from First-Principles Calculations. *Nanoscale Horiz.* 2019, *4* (2), 457–463.

181. Bao, J.; Li, H.; Duan, Q.; Jiang, D.; Liu, W.; Guo, X.; Hou, J.; Tian, J. Graphene-like C3N/Blue Phosphorene Heterostructure as a Potential Anode Material for Li/Na-Ion Batteries: A First Principles Study. *Solid State Ion.* 2020, *345*, 115160.

182. Rao, D.; Liu, X.; Yang, H.; Zhang, L.; Qiao, G.; Shen, X.; Xiaohong, Y.; Wang, G.; Lu, R. Interfacial Competition between a Borophene-Based Cathode and Electrolyte for the Multiple-Sulfide Immobilization of a Lithium Sulfur Battery. *J. Mater. Chem. A* 2019, *7* (12), 7092–7098.

183. Grazianetti, C.; Martella, C.; Molle, A. The Xenes Generations: A Taxonomy of Epitaxial Single-Element 2D Materials. *Phys. Status Solidi RRL–Rapid Res. Lett.* 2020, *14* (2), 1900439.

184. Wang, X.; Tang, C.; Zhou, X.; Zhu, W.; Fu, L. The Good Performance of Bilayer β-Antimoneneas an Anode Material for the Li-Ion Battery Study. *Appl. Surf. Sci.* 2019, *495*, 143549.

185. Wang, T.; Zhang, S.; Yin, L.; Li, C.; Xia, C.; An, Y.; Wei, S. Silicene/Boron Nitride Heterostructure for the Design of Highly Efficient Anode Materials in Lithium-Ion Battery. *J. Phys. Condens. Matter* 2020, *32* (35), 355502.

186. Zergani, F.; Tavangar, Z. A Thorough Study on the F-Decoration of X3 Borophene and Enhancement of Anodic Performance of Lithium-Ion Batteries. *J. Mol. Liq.* 2020, *319*, 114343.

187. Wu, P.; Li, P.; Huang, M. Potential Application of Graphene/Antimonene Herterostructure as an Anode for Li-Ion Batteries: A First-Principles Study. *Nanomaterials* 2019, *9* (10), 1430.

188. Benzidi, H.; Lakhal, M.; Garara, M.; Abdellaoui, M.; Benyoussef, A.; Mounkachi, O. Arsenene Monolayer as an Outstanding Anode Material for (Li/Na/Mg)-Ion Batteries: Density Functional Theory. *Phys. Chem. Chem. Phys.* 2019, *21* (36), 19951–19962.

189. Li, Q.; Yang, J.; Zhang, L. Theoretical Prediction of Blue Phosphorene/Borophene Heterostructure as a Promising Anode Material for Lithium-Ion Batteries. *J. Phys. Chem. C* 2018, *122* (32), 18294–18303.

190. Jiang, H. R.; Lu, Z.; Wu, M. C.; Ciucci, F.; Zhao, T. S. Borophene: A Promising Anode Material Offering High Specific Capacity and High Rate Capability for Lithium-Ion Batteries. *Nano Energy* 2016, *23*, 97–104.

191. Li, W.; Yang, Y.; Zhang, G.; Zhang, Y.-W. Ultrafast and Directional Diffusion of Lithium in Phosphorene for High-Performance Lithium-Ion Battery. *Nano Lett.* 2015, *15* (3), 1691–1697.

192. Jin, H.; Xin, S.; Chuang, C.; Li, W.; Wang, H.; Zhu, J.; Xie, H.; Zhang, T.; Wan, Y.; Qi, Z. Black Phosphorus Composites with Engineered Interfaces for High-Rate High-Capacity Lithium Storage. *Science* 2020, *370* (6513), 192–197.

193. Li, L.; Yu, Y.; Ye, G. J.; Ge, Q.; Ou, X.; Wu, H.; Feng, D.; Chen, X. H.; Zhang, Y. Black Phosphorus Field-Effect Transistors. *Nat. Nanotechnol.* 2014, *9* (5), 372–377.

194. Rajapakse, M.; Anderson, G.; Zhang, C.; Musa, R.; Walter, J.; Yu, M.; Sumanasekera, G.; Jasinski, J. B. Gas Adsorption and Light Interaction Mechanism in Phosphorene-Based Field-Effect Transistors. *Phys. Chem. Chem. Phys.* 2020, *22* (10), 5949–5958.

195. Xia, Y.; Li, G.; Jiang, B.; Yang, Z.; Liu, X.; Xiao, X.; Flandre, D.; Wang, C.; Liu, Y.; Liao, L. Exploring and Suppressing the Kink Effect of Black Phosphorus Field-Effect Transistors Operating in the Saturation Regime. *Nanoscale* 2019, *11* (21), 10420–10428.

196. Guo, G.-C.; Wang, R.-Z.; Ming, B.-M.; Wang, C.; Luo, S.-W.; Zhang, M.; Yan, H. C 3 N/Phosphorene Heterostructure: A Promising Anode Material in Lithium-Ion Batteries. *J. Mater. Chem. A* 2019, *7* (5), 2106–2113.

197. Mannix, A. J.; Zhou, X.-F.; Kiraly, B.; Wood, J. D.; Alducin, D.; Myers, B. D.; Liu, X.; Fisher, B. L.; Santiago, U.; Guest, J. R. Synthesis of Borophenes: Anisotropic, Two-Dimensional Boron Polymorphs. *Science* 2015, *350* (6267), 1513–1516.

198. Huang, T.; Tian, B.; Guo, J.; Shu, H.; Wang, Y.; Dai, J. Semiconducting Borophene as a Promising Anode Material for Li-Ion and Na-Ion Batteries. *Mater. Sci. Semicond. Process.* 2019, *89*, 250–255.

199. Zhu, J.; Xiao, G.; Zuo, X. Two-Dimensional Black Phosphorus: An Emerging Anode Material for Lithium-Ion Batteries. *Nano-Micro Lett.* 2020, *12*, 1–25.

200. Zhang, X.; Qiu, X.; Kong, D.; Zhou, L.; Li, Z.; Li, X.; Zhi, L. Silicene Flowers: A Dual Stabilized Silicon Building Block for High-Performance Lithium Battery Anodes. *ACS Nano* 2017, *11* (7), 7476–7484.

201. Liu, J.; Yang, Y.; Lyu, P.; Nachtigall, P.; Xu, Y. Few-Layer Silicene Nanosheets with Superior Lithium-Storage Properties. *Adv. Mater.* 2018, *30* (26), 1800838.

202. Wang, R.; Dai, X.; Qian, Z.; Zhong, S.; Chen, S.; Fan, S.; Zhang, H.; Wu, F. Boosting Lithium Storage in Free-Standing Black Phosphorus Anode via Multifunction of Nanocellulose. *ACS Appl. Mater. Interfaces* 2020, *12* (28), 31628–31636.

4 Two-Dimensional Materials in Lithium-Sulfur Batteries

Rimpa Jaiswal, Srikanth Ponnada, and Susmita Naskar

4.1 INTRODUCTION

The electrochemical devices which could be used to store and provide high power and energy are called batteries. It is a collection of many galvanic cells comprising electrodes (anode and cathode), a current collector, a separator, and an electrolyte. There are two types of batteries found: one single-used primary and another rechargeable secondary battery [1].

In the recent past, rechargeable, eco-friendly lithium-ion batteries (LIB) have been widely explored as portable electronic devices owing to their promising characteristics like profitable energy storage capacity, lowest self-discharge, steady cycle performance, huge speculative capacity, and specific energy density. These characteristics depend on the material's composition employed in LIB and directly affect the design and development of energy storage technologies [2–4]. It gives superior energy density over the other rechargeable systems [5,6].

The massive demand for energy in daily life at the domestic and industrial level, the development of rechargeable batteries is the main center of attraction of research to fulfill essential needs such as electric vehicles and energy storage.

Moreover, lithium-sulfur (Li-S) batteries are advanced, cost-effective, and excellent new-generation rechargeable batteries that possess a high theoretical capacity of 1675 mAh g^{-1} (as a result of molecular S$_8$ reduction by taking 16 electrons during the discharge process at the cathode) and specific energy of 2600 Wh Kg^{-1}, respectively, greater than commercial Li-ion batteries.

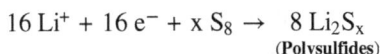

$$16\,Li^+ + 16\,e^- + x\,S_8 \rightarrow 8\,Li_2S_x$$
$$\text{(Polysulfides)}$$

In general, Li-S battery encompasses lithium anode, sulfur cathode separated by separator using electrolyte along with carbon-based porous and functional material [7,8].

The sulfur (S) of cathode gets dissolves into the electrolyte where it forms octasulfur (S$_8$) and subsequently undergoes electrochemical reduction at cathode to

DOI: 10.1201/9781003404729-4

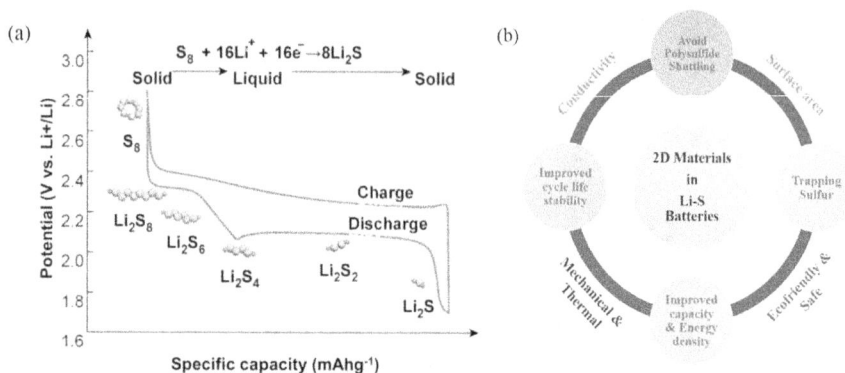

FIGURE 4.1 (a) The mechanism involves redox reaction in Li-S batteries. Represented with permission [9], and (b) improvement of Li-S batteries using 2D material.

produce lithium polysulfides (LPS) intermediates along with oxidation of Li-metal to Li^+ ion at anode. As the discharge process continues, the polysulfide (PS) chain length decreases and affects the electrolyte concentration, mobility, viscosity, and solubility of PS [9]. After complete discharge, S_8 is converted to LiS_2 (reduced form S_2^-) and causes deficiency of Li-metal at anode. However, while charging, the reverse process takes place. The PS gets oxidized from S_2^- to S_8 and Li^+ ion reduced to Li-metal at the anode as presented in Figure 4.1a [9].

Unfortunately, the commercialization of Li-S batteries still faces challenges due to the shuttle effect of PS and Li-dendrite formation. The LPS are soluble intermediates formed at the cathode and undergo dissolution into the electrolyte via separator and reach the anode and interact with Li metal, causing the loss of active sulfur and passivation of the Li-metal anode, called shuttle effect of PS. Thus, soluble LPS (Li_2S_x, where x is more than two and equal or less than 8) move from cathode to anode, consequently hindering the charge transfer and getting corroded. This results in abrupt and dramatic loss of capacity with very less Coulombic efficiency [9,10].

In the present scenario, there are several strategies that have been introduced to the Li-S battery to overcome the challenges associated with the Li-S battery which limits its efficiency, for example, fast capacity decline, the dissolution of active species, charge–discharge characteristics, and the supreme notorious and obstinate shuttle effect of PS [11–13].

Fundamentally, two main techniques are adopted to conquer the shuttle effect. One is by employing a functional, polar, and anchoring material to inhibit the movement of PS to the anode and the other is by suppressing the diffusion of soluble PS into the electrolytes [14]. The first method incorporates a material to prevent the diffusion pathway of PS and the second technique involves the diminishing of PS solubility via diverted PS toward cathode or electrolyte [15].

The diffusion process of PS would be controlled when the binding of LPS assemblages to the host material is more efficient than that to the electrolytes. Figure 4.1b displays the improvements in Li-S batteries by employing two-dimensional (2D) materials.

4.2 THE SIGNIFICANCE OF SULFUR IN LI-S BATTERIES AND THEIR LIMITATIONS

The sulfur is employed as cathode in Li-S batteries owing to the availability of its liberal resources, inexpensive, non-toxic, and non-polluting [9].

Li-S batteries have the following advantages over lithium-ion batteries (LIB):

i. The lower cost of cell of Li-S batteries (~100 USD/ kWh) than LIBs.
ii. It is a lightweight rechargeable battery than LIB, making it a potential candidate for submarine and aeronautical applications.
iii. It does not need to be successively recharged for long-term storage after complete discharge, i.e., up to 0% charge condition, without any significant loss of capacity while LIB essentially needs to be recharged previously for long-term storage to avoid irreversible degradation of capacity.

The complex redox mechanism and perceptible lesser practical value of energy density are the major hurdles to explore Li-S batteries on commercial level in spite of its enormously high theoretical energy density. The rate of reaction involved in electrochemical redox reaction of Li-S batteries is greatly affected by chemical and physical properties of electrode and electrolyte such as molecular structure, affinity, and reactivity of PS to the Li-metal, concentration, and dielectric constant [9]. The penetration ability of electrolyte decreases as the viscosity of electrolyte increases, through the whole depth of the cathode film, thus hindering the efficient species mass transport in the solution. This is the major problem in the case of Li-S batteries compared to LIB. However, a high capacity would be predictable in the case of a compact cathode possessing a large surface area along with a low-viscosity electrolyte.

Furthermore, Li metal used for anode having good gravimetric and theoretical energy capacity could also have some limitations owing to its high reduction potential (−3.04 V) and enable it to react spontaneously with most of the organic solvent causing electrolyte degradation [15] and resulting short cycle life of Li-S batteries. During charging and discharging, the stripping and plating of Li-metal anode causes roughness of the anode surface which also affects the cathode and consequently non-uniform distribution of current density across the Li-S battery and ultimately leads to a short circuit [16,17].

Therefore, the material selection (electrode and electrolyte) is very complicating and a difficult task for researchers to make a Li-S battery possessing fast mass transfer with redox reaction.

To overcome the challenges such as electrochemically induced phase transformations, deprived mechanical and electrical properties, repeated volume extension and stress concentrations at interfaces, dendritic growth of Li, low ionic conductivity, and PS shuttling, which limit the efficient performance of Li-S batteries, researchers are trying to choose an appropriate anchoring material which possesses high surface area along with porous structure and outstanding conductivity, high binding energy affinity to the PS [18,19], and avoiding dissolution of PS into the electrolytes.

Various 2D materials have been suggested as anchoring materials and are used to overwhelm the shuttling effect, for example, graphene, functional group/element

(OH, F, and S) terminated MXene (Ti$_2$C) [18], boro-phosphorene [20], graphene decorated with porous vanadium nitride nanoribbon [21], transition metal hydrides (TMH)/oxides (TMO) [22,23], transition metal dichalcogenide (TMD) [24–26], and biphenylene [27]. Some of the 2D materials are shown in Figure 4.2a.

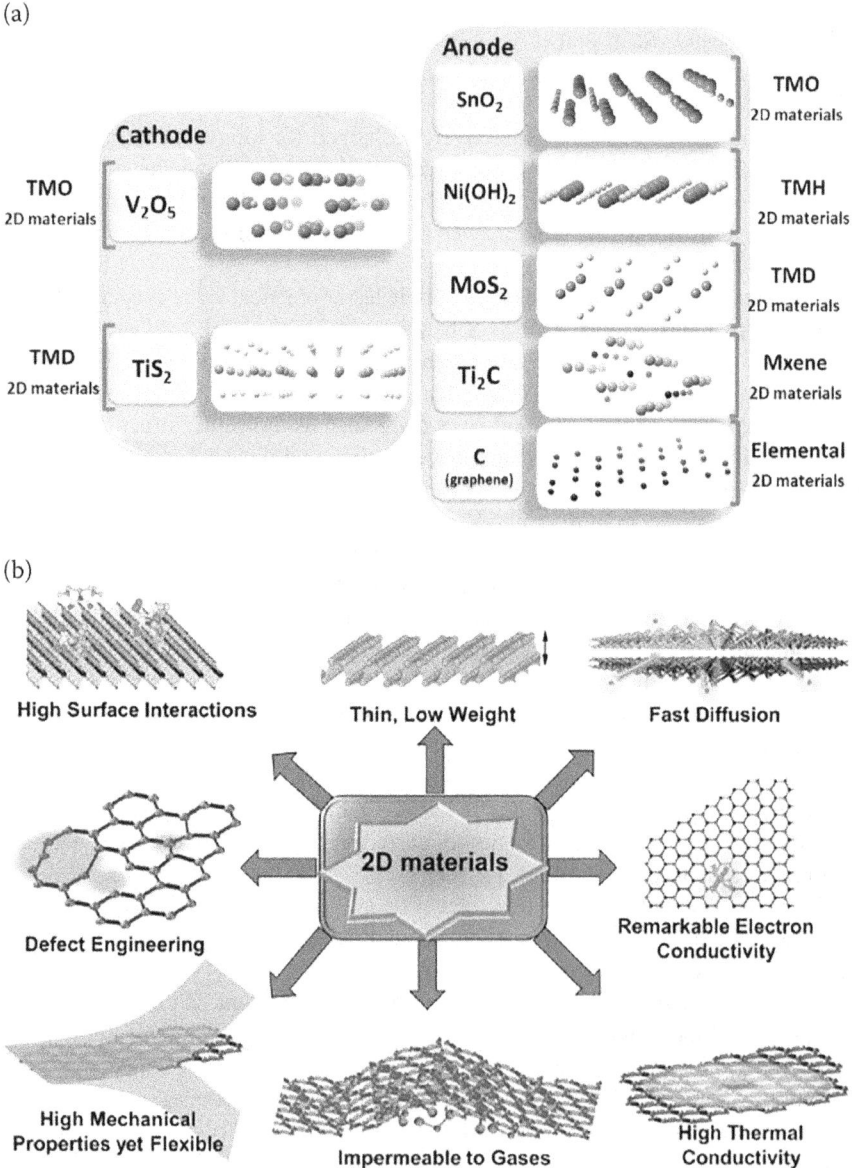

FIGURE 4.2   (a) Schematic of some 2D materials for electrode (cathode/anode) in lithium batteries, and (b) the advantages and unique qualities of 2D materials. Represented with permission [1].

The planar, puckered, and buckling structures of 2D materials play a significant role in diminishing ion diffusion distance via providing facile ultrafast ion transport channels, enhancing ion conductivity [28], and favor fast redox reactions at electrode surfaces owing to their high surface area and unique structural properties.

These 2D materials have attracted more attention owing to the conduction properties of superionic [29] and allow the facility of high current densities without compromising the chemical and mechanical stabilities.

Furthermore, these materials possess excellent mechanical properties including high Young's modulus and flexibility compared to their bulk counterparts [30]. The capability to undergo high strain and stress without structural collapse permits the utilization of 2D materials as mechanical supportive systems for electrode materials having high capacity [31–33].

Hence, 2D material coverings on electrodes and separators can block the formation of Li-metal dendrite and proliferation very effectively [34,35]. Moreover, the band gap and other physical and chemical properties can be tuned and manipulated by doping/ functionalizing with elements or groups as per the interest of applications [36–38].

2D materials have also great intrinsic thermal conductivity that allows good thermal management applications [39]. The graphene and BN nanosheets have been explored as heat dissipaters in LIB to mitigate the produced heat owing to their anisotropic thermal conductive structure [40]. Some benefits and qualities of 2D materials are presented in Figure 4.2b.

Therefore, the energy storage community utilized the intrinsic properties of 2D materials to execute intensive research on it to address the current challenges of electrodes and electrolytes in rechargeable Li-S batteries.

4.3 2D MATERIALS TO SUPPRESS THE SHUTTLING OF POLYSULFIDES

4.3.1 GRAPHENE

Graphene (G) is the most widely studied and explored 2D material in the energy storage field due to its outstanding physical, chemical, and electrical properties [41,42]. Large surface area (2630 m^2g^{-1}) [43], high charge carrier mobility (2×10^5 $cm^2/v^{-1} s^{-1}$), and high electrical conductivity (10^6 S/cm) [44] make it a promising candidate as an anode active material. It has also been employed in cathode as additives and plays a vital role in the diffusion of Li^+ ions through electrodes, and it augments the ion migration [45].

Encapsulation, wrapping, mixing, and functionalizing the active electrode materials with 2D materials are utmost promising methods in dropping electrical resistance and enhancing Li^+ ion diffusion paths via the electroactive materials (Figure 4.3) [46]. It is worth mentioning that the incorporation of graphene-based frameworks has been exposed to improve electrical contact and thus decline the interfacial impedance of active electrode material and graphene additives [47,48].

To mitigate the Li-S battery issues, graphene-based composites were widely explored and studied for their prospective applications like host electrodes,

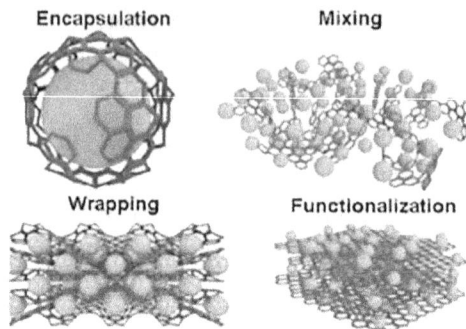

FIGURE 4.3 Enhancement of battery performance via decorating active electrode materials with graphene in different modes. Represented with permission from [1].

electrolytes, and separators [49–51]. The negatively charged reduced graphene oxide (rGO)-lignosulfonate (LS) layering on a polypropylene (PP) separator developed by Lei et al. [52] presented in Figure 4.4a. This covering layer inhibits the transport of the negatively charged PS ions by introducing a repulsive force and hence results in excellent cycling stability and retained 74% capacity over 1000 cycles at 5 mA/cm^2 current densities [52]. Similarly, Kim et al. [53] designed the interphases consisting of positively charged polyethylenimine conductive nanoparticles decorated with rGO on the cathode fronting side of the separator. These ionically protecting nanocomposites have been presented to stabilize Li surfaces and intercept the PS shuttling [53].

In general, the presence of functional groups such as C-H, C-OH, >C=O, -COOH, and >C=C<, on the surface of carbonaceous materials has displayed a high affinity to anchor with sulfur. Thus, the surface alteration of 2D materials raises the chemical adsorption of long $-S_n-$ chains and creates insoluble short-chain PS. However, the formation of both short and long-chain PS is unfavorable to the electrochemical properties of Li–S batteries. It has been observed that fragmentation of PS to shorter chains can mitigate the shuttling effect and augment electrochemical properties [54–58]. Wu et al. [59] prepared a layer-by-layer (LBL) self-assembly covering comprising multilayers of polyelectrolyte (polyethylene imine (PEI)), functionalized graphene sheets (GS), and polystyrene sulfonate (PSS) on the surface of composite particles made of hollow carbon spheres/sulfur (HCSs/S) as presented in Figure 4.4b. This study demonstrated that hydroxyl groups functionalized graphene sheets improve the development of LBL film on the HCSs/S composite, along with acting as a hindering shield to confine the S_n^{2-} anions inside the hollow carbon by strong electrostatic interactions. This strategy diminished the parasitic reactions at the cathode (sulfur) with excellent reversibility. Figure 4.4b indicates the cyclic voltammetry (CV) plots of the HCSs/S and LBL-HCSs/S composites at a scan rate of 0.1 mV/s. Noticeably, only the typical redox reaction (oxidation and reduction) peaks of the sulfur cathode were detected for LBL-HCSs/S composite with the first anodic (oxidation) peak at 2.35 V vs Li/Li$^+$. While the redox peaks of HCSs/S composite showed much broader

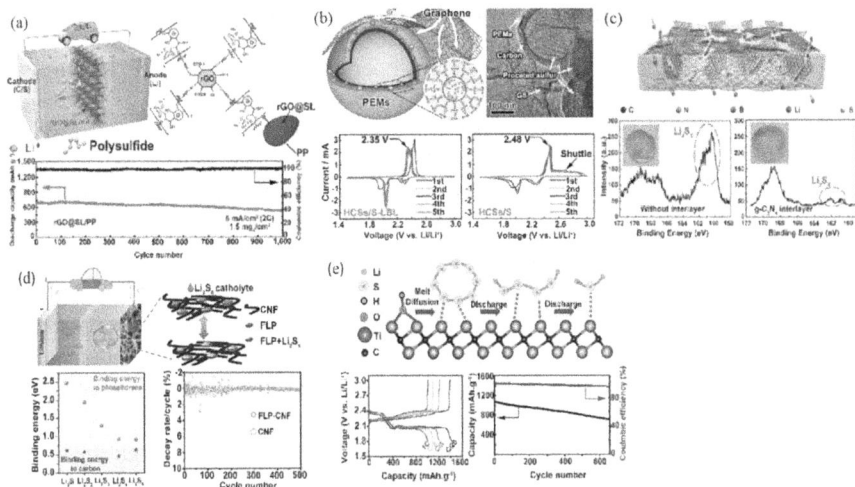

FIGURE 4.4 Polysulfide shuttling prevented by using 2D materials. (a) Schematic presentation shows the rGO functionalized with LS using polypropylene separators for inhibiting the shuttling effect of polysulfide in Li–S batteries and discharge capacity vs cycle number illustrates its cycling stability. Represented with permission from [1], (b) schematic diagram (upper-left) and TEM image (upper-right) of HCSs/S-LBL composite decorated with GS coating. The CV curves of the HCSs/S-LBL (lower-left) and HCSs/S (lowest-right) composites. Represented with permission from [1]. (c) Schematic showing the multi-functional ion sieve formed by combining three 2D materials, i.e., graphene (top), g-C3N4, and BN. After 100 cycles, S 2p of the Li metal anode display in the XPS plot with and without the interlayer coating of g-C3N4 on the separator, indicating the capability of this coating to prevent the shuttle of polysulfides (bottom). Represented with permission from [1]. (d) Schematic demonstration displays the FLP-CNF composite cathode to accommodate LiPS (top). Binding energy among lithium polysulfides and carbon hexa-atomic ring network (bottom left), and phosphorene, decay rate vs cycle number for Li–S batteries having the FLP-CNF electrode and CNF (pure) electrode (bottom right). Represented with permission from [1]. (e) Schematic showing the Ti–OH bond replacement on the MXene surface with an S–Ti–C upon interaction with polysulfides (top). Voltage vs capacity profiles show 70 wt % sulfur_d-Ti2C at several rates ranging from 0.05 to 1 C (lower left). Cycling stability for long term, 70 wt % S_d-Ti2C at 0.5 C (lower-right). Represented with permission from [1].

behavior and moved to a higher voltage of 2.48 V vs Li/Li$^+$, demonstrating a higher polarization and increased internal resistance. Furthermore, the CV graph of the HCSs/S composite displayed some activities with a high anodic current between 2.5 and 3 V vs Li/Li$^+$, indicating the PS shuttle reactions in this unprotected C/S structure [59].

Assimilation of defective 2D materials into sulfur cathodes can inhibit the polysulfides from decomposing and has been extensively investigated [60–63]. Density functional theory (DFT) calculations and experimental studies showed that the defect sites on 2D materials for example nitrogen-doping could inhibit Li_2S_n (LPS) dissolution and diffusion by incorporating a strong synergistic effect on their adsorption [64–66].

4.3.2   Graphitic Carbon Nitride and Boron Nitride

To overcome the pitiable electrochemical properties of Li–S batteries, Deng et al. [67] designed a multifunctional interlayer covering comprising graphitic 2D materials, i.e., graphitic carbon nitrides (g-C_3N_4), graphene nanosheets, and boron nitride (BN) on the separator, as depicted in Figure 4.4c. They have revealed that g-C_3N_4 crystalline nanosheets with ca. 3 Å network size occlude the PS and only permit the Li^+ ions to move freely.

The XPS plot of the separators with and without the g-C_3N_4 nanosheet interlayer covering after 100 cycles. Figure 4.4c displays the usefulness of this interlayer in hampering the PS shuttling. BN nanosheets offer outstanding catalytic performance for sulfur redox kinetics. In conclusion, the graphene additive endorses the electrical as well as ionic conductivity of the electrodes. The consequence of this synergistic effect is responsible for a high specific capacity of ~600 mAh/g after 500 cycles at 1 C and capacity declining of <0.01% per cycle at a high loading of sulfur (6 mg/cm^2) [67].

4.3.3   Phosphorene

Phosphorene is the most stable 2D nanomaterial with crystalline, puckered structure, semiconducting, and allotrope of phosphorus (black phosphorus). It is a single layer of black-layered phosphorus synthesized artificially and a monolayer of phosphorene consists of two atomic planes and has a tetragonal lattice. The unit cell of phosphorene is made of four P-atoms. It has a high mobility of ~1000 cm^2/Vs, making it a promising electrode (anode) material for rechargeable batteries. It shows high theoretical capacity (2596 mA h/g) and possesses ultrahigh Li^+ ion mobility that makes it applicable for Li^+ ion storage [1]. The Li-storage and transfer occurred in interlayer spaces of the phosphorene. The stability and capacity of anode depend on the number of layers and lateral size of phosphorene. The composite material having few layers of phosphorene (FLP) attached to carbon nanofiber (CNF) has been proposed by Li et al. [68]. They reveal the usefulness of this material as a cathode host to address the depraved performance of Li–S batteries as presented in Figure 4.4d. The LPS has a strong affinity to the phosphorene nanosheets, allowing it to design electrode material. The binding energy of several LPS to phosphorene lies in the range of 1–2.5 eV that is significantly greater than a carbon network of about 0.5 eV. Hence, the exploration of sulfur is improved by phosphorene and subsequently decreases the polarization as a sign of superior electrochemical reaction dynamics in Li-S batteries. Besides this, phosphorene is established to have a catalysis effect on redox reactions of sulfur and Li_2S and enables conversion reactions. The cycling of the Li–S batteries modified with phosphorene electrode host materials has displayed the retention of specific capacity of the battery >660 mA h/g over 500 cycles with a very less capacity attenuation of ≈0.053%, revealing a much better performance of the battery in the presence of phosphorene. However, in the absence of phosphorene, the battery shows a capacity decay of ≈0.25% in 200 cycles [68].

4.3.4 TRANSITION METAL DICHALCOGENIDES

2D TMDs are atomically thin sheets of monolayers. It is a semiconductor consisting of MX_2-type layers where M is a transition metal atom and X a chalcogen atom (i.e., S, Se, Te). A single layer of M-atom lies in between two layers of X-atoms. It has a honeycomb hexagonal lattice with three-fold symmetry. For example, MoS_2 and WS_2 have been employed to enhance the electro-catalytic activities for fast sulfur conversion reactions in batteries. In the presence of these 2D materials, the gathering of soluble polysulfides diminishes on the sulfur cathode owing to a precise chemical and electronic structure, abundant edges, and ample defects [49,69,70].

4.3.5 MXENE

2D transition metal carbides (M_2C) and nitrides (M_2N) are known as MXene (where M = scandium (Sc), Titanium (Ti), vanadium (V), and chromium (Cr)). It has been explored owing to its promising electrochemical performance in battery applications based on lithium. A large family of MXene have displayed a high discharge capacity (>400 mA h/g) [71,72]. This type of 2D materials can be selected and utilized for the choice of cathodes or anodes because it can offer a wide range of working potentials [71].

The anchoring of PS onto the 2D material's surface having functional groups along with defect sites is another approach to mitigate the shuttle effect of PS and their irreversible dissolution [49,73]. Among 2D materials, MXene with functional group on their surfaces has been profusely studied and found to be an efficient 2D material for trapping PS via physical and chemical adsorption [49,74]. Liang et al. [74] demonstrated that the surface of 2D nanosheets of Ti_2C (MXene) functionalized with hydroxyl group (-OH) has a great affinity for sulfur forming a strong bond between S and Ti_2C. Figure 4.4e shows reinforcement of Ti_2C matrix with 70 weight % of sulfur via melt-diffusion and revealed an excellent initial specific capacity of 1440 mAh/g at 83 mA/g (0.05 C) to ~1000 mAh/g at 1675 mA/g (1 C). Moreover, this strategy leads to a stable retention of capacity 723 mA h/g at a high current density of 837 mA/g (0.5 C) after 650 cycles with noticeably low deterioration rate of 0.05% per cycle and Coulombic efficiency ~94% [74].

4.3.6 BIPHENYLENE

Recently, biphenylene (BPN) is synthesized [75] and explored in many applications such as energy storage devices. It is a planar entirely sp^2-hybridized allotrope of carbon. In BPN, carbon atoms are organized in square, hexagonal, and octagonal rings. It is metallic in nature with a Dirac cone beyond the Fermi level, proposing high carrier mobility and worthy conductivity [76]. The planar lattice of biphenylene shows its one key factor for designing batteries with a fast charge/discharge rate, so it is predicted to exhibit better performance as a potential anchoring material [27]. Thus, the first-principles calculations have been carried out to estimate the potential of 2D BPN sheet as an attaching material for Li–S batteries. The binding energy (E_b) of S_8 clusters and LPS with the BPN sheet is calculated using the following equation:

$$E_b = E_{total} - (E_{biphenylene} + E_{LPSs/S8})$$

where E_{total} is the total energy of biphenylene adsorbed with Li_2S_x/S_8, $E_{biphenylene}$ refer to energy of pristine biphenylene, and $E_{LPSs/S8}$ denote the isolated Li_2S_x/S_8 clusters.

Jayyousi et al. [27] developed biphenylene as anchoring 2D material to trapped LPS and effectively inhibit the shuttle effect and augment the cycling stability of Li–S batteries. The DFT evaluations reveal that LPS bind with pristine biphenylene weakly with binding energy ranging from −0.21 to −1.22 eV. However, defect engineering via a single C atom vacancy considerably improves the binding strength (i.e., binding energy in the range of −1.07 to −4.11 eV). The Bader study divulges that LPS and S_8 clusters give the charge (ranging from −0.05 to −1.12 e) to the biphenylene sheet. The binding energy of LPS with electrolytes is lesser than those with the defective biphenylene 2D sheet, which offers its potential as an anchoring material. On comparing with other reported 2D materials such as graphene, phosphorene, and MXenes, the biphenylene sheet shows higher binding energies with polysulfides.

4.4 CONCLUSION

There are many 2D materials studied as an electrode material for the improvement of Li-S battery performance. The physical and chemical properties of 2D materials are tuned as per desirability and requirements to develop a Li-S battery. The large specific surface area and the presence of selective functional group on to the surface of 2D material prevent the shuttle effect of PS formed at sulfur cathode by absorbing/adsorbing the PS, preventing the dissolution of intermediates into the electrolytes, and hindering the movement of PS to the Li anode. The Li-dendrites formation is also controlled by employing 2D materials onto electrode via wrapping, mixing, or encapsulating. Besides these, 2D materials play a very significant role in electrochemical performances by imparting in the charge /discharge activities owing to their high mobility and conductivity. Therefore, life cycle and sustainability of Li-S batteries are enhanced by incorporating appropriate 2D materials to the cathode.

ACKNOWLEDGMENTS

All authors would like to thank Babha Atomic Research Centre, Mumbai, India, Colorado School of Mines, USA, the Indian Institute of Technology Jodhpur, India, and the University of Southampton for Technical and Resource support. The authors would also like to thank the Department of Science and Technology, India and University Grants Commission India for financial support.

CONFLICTS OF INTEREST

The authors declare no competing interest.

REFERENCES

1. R. Rojaee, R. S. Yassar, Two-dimensional materials to address the lithium battery challenges. *ACS Nano, 14* (2020), 2628–2658.
2. L. Zhou, D. L. Danilov, R. A. Eichel, P. H. L. Notten, Host materials anchoring polysulfides in Li–S batteries reviewed. *Adv. Energy Mater., 11* (2021), 2001304.
3. P. G. Bruce, S. A. Freunberger, L. J. Hardwick, J. M. Tarascon, Li–O2 and Li–S batteries with high energy storage. *Nat. Mater., 11* (2012), 19–29.
4. W. J. Chung, et al., The use of elemental sulfur as an alternative feedstock for polymeric materials. *Nat. Chem., 5* (2013), 518–524.
5. G. Crabtree, Perspective: The energy-storage revolution. *Nature, 526* (2015), S92.
6. H. Budde-Meiwes, J. Drillkens, B. Lunz, J. Muennix, S. Rothgang, J. Kowal, D. U. Sauer, A review of current automotive battery technology and future prospects. *Proc. Inst. Mech. Eng. Part D J. Automobile Eng., 227* (2013), 761–776.
7. A. Manthiram, S. Chung, C. Zu, Lithium-sulfur batteries: progress and prospects. *Adv. Mater., 27* (2015), 1980–2006.
8. S. Ponnada, M. S. Kiai, D. B. Gorle, A. Nowduri, History and recent developments in divergent electrolytes towards high-efficiency lithium–sulfur batteries – a review. *Mater. Adv., 2* (2021), 4115–4139.
9. R. Mori, Cathode materials for lithium sulfur battery: a review. *J. Solid State Electrochem., 27* (2023), 813–839.
10. H. J. Peng, Q. Zhang, Designing host materials for sulfur cathodes: from physical confinement to surface chemistry. *Angew. Chem. Int. Ed., 54* (2015), 11018–11020.
11. M. Wild, et al., Lithium sulfur batteries, a mechanistic review. *Energy Environ. Sci., 8* (2015), 3477–3494.
12. Y. Diao, K. Xie, S. Xiong, X. Hong, Shuttle phenomenon-the irreversible oxidation mechanism of sulfur active material in Li-S battery. *J. Power Sources, 235* (2013), 181–186.
13. R. Fang, S. Zhao, Z. Sun, D. W. Wang, H. M. Cheng, F. Li, More reliable lithium-sulfur batteries: status, solutions and prospects. *Adv. Mater., 29* (2017), 1606823.
14. W. Ren, W. Ma, S. Zhang, B. Tang, Recent advances in shuttle effect inhibition for lithium sulfur batteries. *Energy Storage Mater., 23* (2019), 707–732.
15. B. L. D Rinkel, D. S. Hall, I. Temprano, C. P. Grey, Electrolyte oxidation pathways in lithium-ion batteries. *J. Am. Chem. Soc., 142* (2020), 15058–15074.
16. J. Luo, R. Lee, J. Jin, Y. T. Weng, C. C. Fang, N. L. Wu, A dual functional polymer coating on a lithium anode for suppressing dendrite growth and polysulfide shuttling in Li–S batteries. *Chem. Commun., 53* (2017), 963–966.
17. B. D. McCloskey, Attainable gravimetric and volumetric energy density of Li–S and Li ion battery cells with solid separator-protected Li metal anodes. *J. Phys. Chem. Lett., 6* (2015), 4581–4588.
18. X. Liu, X. Shao, F. Li, M. Zhao, Anchoring effects of S-terminated Ti2C MXene for lithium–sulfur batteries: a first-principles study. *Appl. Surf. Sci., 455* (2018), 522–526.
19. Z. Li, J. Zhang, X. W. Lou, Hollow carbon nanofibers filled with MnO_2 nanosheets as efficient sulfur hosts for lithium-sulfur batteries. *Angew. Chem. Int. Ed., 54* (2015), 12886–12890.
20. H. Zhang, S. Wang, Y. Wang, B. Huang, Y. Dai, W. Wei, Borophosphene: A potential anchoring material for lithium-sulfur batteries. *Appl. Surf. Sci., 562* (2021), 150157.
21. Z. Sun, J. Zhang, L. Yin, G. Hu, R. Fang, H. M. Cheng, F Li, Conductive porous vanadium nitride/graphene composite as chemical anchor of polysulfides for lithium-sulfur batteries. *Nat. Commun., 8* (2017), 14627.
22. Y. Xue, Q. Zhang, W. Wang, H. Cao, Q. Yang, L. Fu, Opening two-dimensional materials for energy conversion and storage: a concept. *Adv. Energy Mater., 7* (2017), 1602684.

23. J. Mei, T. Liao, L. Kou, Z. Sun, Two-dimensional metal oxide nanomaterials for next-generation rechargeable batteries. *Adv. Mater.*, *29* (2017), 1700176.

24. Q. Yun, Q. Lu, X. Zhang, C. Tan, H. Zhang, Three-dimensional architectures constructed from transition metal dichalcogenide nanomaterials for electrochemical energy storage and conversion. *Angew. Chem. Int. Ed. Engl.*, *57* (2018), 626–646.

25. Y. Liu, J. Wu, K. P. Hackenberg, J. Zhang, Y. M. Wang, Y. Yang, K. Keyshar, J. Gu, T. Ogitsu, R. Vajtai, J. Lou, P. M. Ajayan, B. C. Wood, B. I. Yakobson, Self-optimizing, highly surface active layered metal dichalcogenide catalysts for hydrogen evolution. *Nat. Energy*, *2* (2017), 17127.

26. Y. Gao, X. Wu, K. Huang, L. Xing, Y. Zhang, L. Liu, Two-dimensional transition metal diseleniums for energy storage application: a review of recent developments. *CrystEngComm*, *19* (2017), 404–418.

27. H. K. A. Jayyousi, M. Sajjad, K. Liao, N. Singh, Two-dimensional biphenylene: a promising anchoring material for lithium-sulfur batteries. *Sci Rep.*, *12* (2022), 4653.

28. H. Tian, Z. W. Seh, K. Yan, Z. Fu, P. Tang, Y. Lu, R. Zhang, D. Legut, Y. Cui, Q. Zhang, Theoretical investigation of 2D layered materials as protective films for lithium and sodium metal anodes. *Adv. Energy Mater.*, *7* (2017), 1602528.

29. C. Chowdhury, A. Datta, Exotic physics and chemistry of two-dimensional phosphorus: phosphorene. *J. Phys. Chem. Lett.*, *8* (2017), 2909–2916.

30. J. H. Kim, J. H. Jeong, N. Kim, R. Joshi, G. H. Lee, Mechanical properties of two-dimensional materials and their applications. *J. Phys. D: Appl. Phys.*, *52* (2019), 083001.

31. C. Zhang, S.-H. Park, A. Seral-Ascaso, S. Barwich, N. McEvoy, C. S. Boland, J. N. Coleman, Y. Gogotsi, V. Nicolosi, High capacity silicon anodes enabled by MXene viscous aqueous ink. *Nat. Commun.*, *10* (2019), 849.

32. X. Yao, Y. Zhao, Three-dimensional porous graphene networks and hybrids for lithium-ion batteries and supercapacitors. *Chemistry*, *2* (2017), 171–200.

33. C. Wang, X. Wang, Y. Yang, A. Kushima, J. Chen, Y. Huang, J. Li, Slurryless Li2S/reduced graphene oxide cathode paper for high-performance lithium sulfur battery. *Nano Lett.*, *15* (2015), 1796–1802.

34. E. Cha, M. D. Patel, J. Park, J. Hwang, V. Prasad, K. Cho, W. Choi, 2D MoS2 as an efficient protective layer for lithium metal anodes in high-performance Li-S batteries. *Nat. Nanotechnol.*, *13* (2018), 337–343.

35. C. Zhang, A. Wang, J. Zhang, X. Guan, W. Tang, J. Luo, 2D materials for lithium/sodium metal anodes. *Adv. Energy Mater.*, *8* (2018), 1802833.

36. R. Ribeiro-Palau, C. R. Dean, C. Zhang, J. Hone, K. Watanabe, T. Taniguchi, Twistable electronics with dynamically rotatable heterostructures. *Science*, *361* (2018), 690–693.

37. A. Raja, A. Chaves, J. Yu, G. Arefe, H. M. Hill, A. F. Rigosi, T. C. Berkelbach, P. Nagler, C. Schüller, T. Korn, C. Nuckolls, J. Hone, L. E. Brus, T. F. Heinz, D. R. Reichman, A. Chernikov, Coulomb engineering of the bandgap and excitons in two dimensional materials. *Nat. Commun.*, *8* (2017), 1–7.

38. C. Zhang, H. Huang, X. Ni, Y. Zhou, L. Kang, W. Jiang, H. Chen, J. Zhong, F. Liu, Band gap reduction in van der Waals layered 2D materials via a de-charge transfer mechanism. *Nanoscale*, *10* (2018), 16759–16764.

39. H. Song, J. Liu, B. Liu, J. Wu, H. M. Cheng, F. Kang, Two dimensional materials for thermal management applications. *Joule*, *2* (2018), 442–463.

40. B. Mortazavi, H. Yang, F. Mohebbi, G. Cuniberti, T. Rabczuk, Graphene or H-BN paraffin composite structures for the thermal management of Li-Ion batteries: a multiscale investigation. *Appl. Energy*, *202* (2017), 323–334.

41. G. Kucinskis, G. Bajars, J. Kleperis, Graphene in lithium ion battery cathode materials: a review. *J. Power Sources*, *240* (2013), 66–79.

42. J. Han, W. Wei, C. Zhang, Y. Tao, W. Lv, G. Ling, F. Kang, Q. H. Yang, Engineering graphenes from the nano- to the macroscale for electrochemical energy storage. *Electrochem. Energy Rev.*, *1* (2018), 139–168.

43. R. Jaiswal, U. Saha, T. H. Goswami, A. Srivastava, N. E. Prasad, Pillar effect' of chemically bonded fullerene in enhancing supercapacitance performances of partially reduced fullerenol graphene oxide hybrid electrode material. *Electrochim. Acta*, *283* (2018), 269–290.

44. U. Saha, R. Jaiswal, T. H. Goswami, A facile bulk production of processable partially reduced graphene oxide as superior supercapacitor electrode material. *Electrochim. Acta*, *196* (2016), 386–404.

45. C. Wang, D. Li, C. O. Too, G. G. Wallace, Electrochemical properties of graphene paper electrodes used in lithium batteries. *Chem. Mater.*, *21* (2009), 2604–2606.

46. X. Cai, L. Lai, Z. Shen, J. Lin, Graphene and graphene based composites as Li-ion battery electrode materials and their application in full cells. *J. Mater. Chem. A*, *5* (2017), 15423–15446.

47. B. P. N. Nguyen, N. A. Kumar, J. Gaubicher, F. Duclairoir, T. Brousse, O. Crosnier, L. Dubois, G. Bidan, D. Guyomard, B. Lestriez, Nanosilicon-based thick negative composite electrodes for lithium batteries with graphene as conductive additive. *Adv. Energy Mater.*, *3* (2013), 1351–1357.

48. F. Fathollahi, M. Javanbakht, H. Omidvar, M. Ghaemi, Improved electrochemical properties of LiFePO4/graphene cathode nanocomposite prepared by one-step hydrothermal method. *J. Alloys Compd.*, *627* (2015), 146–152.

49. B. Li, H. Xu, Y. Ma, S. Yang, Harnessing the unique properties of 2D materials for advanced lithium–sulfur batteries. *Nanoscale Horizons*, *4* (2019), 77–98.

50. W. Lin, Y. Chen, P. Li, J. He, Y. Zhao, Z. Wang, J. Liu, F. Qi, B. Zheng, J. Zhou, C. Xu, F. Fu, Enhanced performance of lithium sulfur battery with a reduced graphene oxide coating separator. *J. Electrochem. Soc.*, *162* (2015), A1624–A1629.

51. W. Bao, X. Xie, J. Xu, X. Guo, J. Song, W. Wu, D. Su, G. Wang, Confined sulfur in 3D MXene/reduced graphene oxide hybrid nanosheets for lithium–sulfur battery. *Chem. – Eur. J.*, *23* (2017), 12613–12619.

52. T. Lei, W. Chen, W. Lv, J. Huang, J. Zhu, J. Chu, C. Yan, C. Wu, Y. Yan, W. He, J. Xiong, Y. Li, C. Yan, J. B. Goodenough, X. Duan, Inhibiting polysulfide shuttling with a graphene composite separator for highly robust lithium-sulfur batteries. *Joule*, *2* (2018), 2091–2104.

53. M. S. Kim, M. S. Kim, V. Do, Y. R. Lim, I. W. Nah, L. A. Archer, W. Il. Cho, Designing solid-electrolyte interphases for lithium sulfur electrodes using ionic shields. *Nano Energy*, *41* (2017), 573–582.

54. Z. Zeng, X. Liu, Sulfur immobilization by "chemical anchor" to suppress the diffusion of polysulfides in lithium–sulfur batteries. *Adv. Mater. Interfaces*, *5* (2018), 1701274.

55. X. Fan, W. Sun, F. Meng, A. Xing, J. Liu, Advanced chemical strategies for lithium–sulfur batteries: a review. *Green Energy Environ.*, *3* (2018), 2–19.

56. W. Ren, W. Ma, S. Zhang, B. Tang, Recent advances in shuttle effect inhibition for lithium sulfur batteries. *Energy Storage Mater.*, *23* (2019), 707–732.

57. E. P. Kamphaus, P. B. Balbuena, Long-chain polysulfide retention at the cathode of Li-S batteries. *J. Phys. Chem. C*, *120* (2016), 4296–4305.

58. T. G. Jeong, D. S. Choi, H. Song, J. Choi, S. A. Park, S. H. Oh, H. Kim, Y. Jung, Y. T. Kim, Heterogeneous catalysis for lithium-sulfur batteries: enhanced rate performance by promoting polysulfide fragmentations. *ACS Energy Lett.*, *2* (2017), 327–333.

59. F. Wu, J. Li, Y. Su, J. Wang, W. Yang, N. Li, L. Chen, S. Chen, R. Chen, L. Bao, Layer-by-layer assembled architecture of polyelectrolyte multilayers and graphene sheets on hollow carbon spheres/sulfur composite for high-performance lithium-sulfur batteries. *Nano Lett.*, *16* (2016), 5488–5494.

60. C. Hu, H. Chen, Y. Shen, D. Lu, Y. Zhao, A. H. Lu, X. Wu, W. Lu, L. Chen, In situ wrapping of the cathode material in lithium-sulfur batteries. *Nat. Commun.*, *8* (2017), 1–9.

61. H. Lin, D.-D. Yang, N. Lou, A.-L. Wang, S.-G. Zhu, H.-Z. Li, Defect engineering of black phosphorene towards an enhanced polysulfide host and catalyst for lithium-sulfur batteries: a first principles study. *J. Appl. Phys.*, *125* (2019), No. 094303.

62. Z. Liang, X. Fan, D. J. Singh, W. T. Zheng, Adsorption and diffusion of Li with S on pristine and defected graphene. *Phys. Chem. Chem. Phys.*, *18* (2016), 31268–31276.

63. C. Deng, Z. Wang, S. Wang, J. Yu, Inhibition of polysulfide diffusion in lithium-sulfur batteries: mechanism and improvement strategies. *J. Mater. Chem. A*, *7* (2019), 12381–12413.

64. W. Bao, L. Liu, C. Wang, S. Choi, D. Wang, G. Wang, Facile synthesis of crumpled nitrogen-doped MXene nanosheets as a new sulfur host for lithium–sulfur batteries. *Adv. Energy Mater.*, *8* (2018), 1702485.

65. D. Rao, Y. Wang, L. Zhang, S. Yao, X. Qian, X. Xi, K. Xiao, K. Deng, X. Shen, R. Lu, Mechanism of polysulfide immobilization on defective graphene sheets with N-substitution. *Carbon*, *110* (2016), 207–214.

66. P. Velez, M. L. Para, G. L. Luque, D. Barraco, E. P. M. Leiva, Modeling of substitutionally modified graphene structures to prevent the shuttle mechanism in lithium-sulfur batteries. *Electrochim. Acta*, *309* (2019), 402–414.

67. D. R. Deng, C. D. Bai, F. Xue, J. Lei, P. Xu, M.-S. Zheng, Q. F. Dong, Multifunctional ion-sieve constructed by 2D materials as an interlayer for Li-S batteries. *ACS Appl. Mater. Interfaces*, *11* (2019), 11474–11480.

68. L. Li, L. Chen, S. Mukherjee, J. Gao, H. Sun, Z. Liu, X. Ma, T. Gupta, C. V. Singh, W. Ren, H.-M. Cheng, N. Koratkar, Phosphorene as a polysulfide immobilizer and catalyst in high performance lithium-sulfur batteries. *Adv. Mater.*, *29* (2017), 1602734.

69. H. Lin, L. Yang, X. Jiang, G. Li, T. Zhang, Q. Yao, G. W. Zheng, J. Y. Lee, Electrocatalysis of polysulfide conversion by sulfur deficient MoS2 nanoflakes for lithium-sulfur batteries. *Energy Environ. Sci.*, *10* (2017), 1476–1486.

70. G. Babu, N. Masurkar, H. Al Salem, L. M. R. Arava, Transition metal dichalcogenide atomic layers for lithium polysulfides electrocatalysis. *J. Am. Chem. Soc.*, *139* (2017), 171–178.

71. B. Anasori, M. R. Lukatskaya, Y. Gogotsi, 2D metal carbides and nitrides (MXenes) for energy storage. *Nat. Rev. Mater.*, *2* (2017), 16098.

72. C. Eames, M. S. Islam, Ion intercalation into two dimensional transition-metal carbides: global screening for new high-capacity battery materials. *J. Am. Chem. Soc.*, *136* (2014), 16270–16276.

73. J. Ren, Y. Zhou, L. Xia, Q. Zheng, J. Liao, E. Long, F. Xie, C. Xu, D. Lin, Rational design of a multidimensional N-doped porous carbon/MoS$_2$/CNT nano-architecture hybrid for high performance lithium-sulfur batteries. *J. Mater. Chem. A*, *6* (2018), 13835–13847.

74. X. Liang, A. Garsuch, L. F. Nazar, Sulfur cathodes based on conductive MXene nanosheets for high-performance lithium-sulfur batteries. *Angew. Chem. Int. Ed.*, *54* (2015), 3907–3911.

75. Q. Fan, L. Yan, M. W. Tripp, O. Krejci, S. Dimosthenous, S. R. Kachel, M. Chen, A. S. Foster, U. Koert, P. Lilijeroth, J. M. Gottfried, Biphenylene network: a non-benzenoid carbon allotrope. *Science*, *372* (2021), 852–856.

76. A. Bafekry, M. Faraji, M. M. Fadlallah, H. R. Jappor, S. Karbasizadeh, M. Ghergherehchi, D. Gogova, Biphenylene monolayer as a two-dimensional nonbenzenoid carbon allotrope: a first-principles study. *J. Phys. Condens. Matter*, *34* (2021), 1–5.

5 Two-Dimensional Materials in Lead-Tin Alloys in Li Batteries

Cyril A Andrews and Srijan Sengupta

5.1 INTRODUCTION

Batteries are rapidly becoming an important component for a variety of energy storage applications. They are now incorporated into future advanced systems, appliances, Internet of Things, and electric vehicles, all of which are revolutionizing our lives. The intermittency of renewable energy sources such as sun, wind, and waves is becoming easier to deal with thanks to developments in large-scale battery-based energy storage. However, most of today's Li-ion batteries cannot be recharged quickly and efficiently; they deteriorate rapidly and must be replaced after a few hundred charging cycles. There is room for improvement in their energy density [1–3]. Rechargeable LIBs have experienced explosive growth due to their high specific energy. Graphite has a relatively high Coulombic efficiency, allowing for reversible insertion/extraction of lithium ions at intercalation potentials with reasonable specific capacity. Therefore, most commercially available LIBs use a graphite anode and $LiCoO_2$ cathode. Researchers prioritize the design and synthesis of new electrode materials with high lithium-storage capacity and quick rate capability because the electrode materials significantly contribute to determining the electrochemical performance of LIBs [4].

In general, the safety of lithium-ion batteries is strongly reliant on the efficacy of the electrodes (especially the anodes) for lithium storage. The four kinds of materials that may be used to store lithium, according to the lithium insertion technique, are surface storing, low or zero-strain insertion, phase-conversion insertion, and synergic impact insertion. Surface storing is the most common method. In point of fact, the nanostructured electrodes' structural features are primarily responsible for the method in which they store lithium as well as their capacities in this regard [5–7]. Figure 5.1 illustrates how various structures display their one-of-a-kind performance depending on their surface qualities and structural properties. 0D structures, also known as nanospheres and microspheres, have the least amount of surface area possible and high thermal stability. The ability to create micro-devices is made possible by the formation of 1D structures, such as nanotubes and nanowires, by directed growth. 2D structures, also known as nanosheets, often have unique facets and vast exposed surfaces. The majority of three-dimensional (3D)

DOI: 10.1201/9781003404729-5

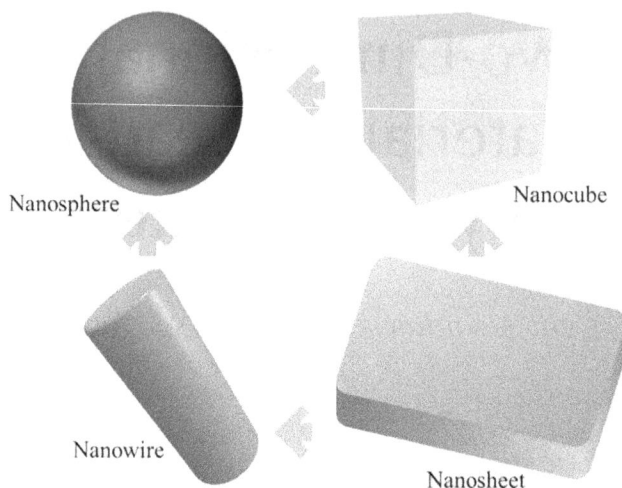

FIGURE 5.1 Surface and morphology of nanospheres, nanocubes, nanowires, and nanosheets [8].

structures may be generated with the assistance of SDAs (structure-directing agents) or formed from zero-dimensional, one-dimensional, and two-dimensional (2D) structures, all of which are capable of performing the functions of secondary structures [8,9]. Therefore, reactivity of materials under the same reaction conditions depends heavily on their dimensionality, which is why decreasing layered materials' thickness to single-atom nanosheets results in fascinating physical properties. Fast ion and electron transport are conspicuous benefits of two-dimensionality, leading researchers to exhibit increased interest in this field due to the profound scarcity of experimentally known 2D materials. Such materials' unique structural qualities and enormous specific surface area minimize ion transport distance, allowing rapid surface redox reactions. To achieve freestanding monolayer or few-layer materials from stacked 2D layered structures like graphite, metal dichalcogenides, boron, black phosphorus, and metallic antimony which possess intense chemical bonds in the plane but the weakest Van der Waals (vdWs) [10–13].

Many electrode materials for high-performance Li-ion batteries have recently been described, and they are emerging as rivals to Li-ion energy storage devices. Yet, certain intrinsic difficulties persist. A high density of power is only possible in systems with fast ion and electron transport. However, ion flow in high-volume electrodes is often inhibited owing to reduced lattice space, leading to slow charge–discharge cycles and early electrode breakdown. The performance rate is further restricted by the limited electrical conduct of popular cathode materials (oxides). The high energy density necessitates optimization of charge storage capability; yet typical materials have a limited count of ion intercalation sites and, in order to significantly improve the charge storage capability, their surface is not to be exploited. Numerous electrode materials undergo phase changes, which result in creating redox idle phases and capacity reduction. Electrode enlargement and diminution is another issue that occurs during inoculation of reversible-ion. The

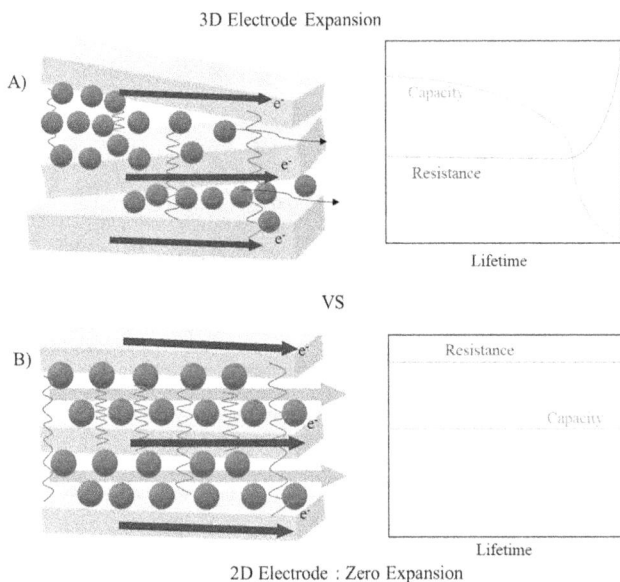

FIGURE 5.2 Using 2D materials to overcome the limitations of current batteries. (A) Illustration of a three-dimensional intercalation electrode having increased resistance and reduced ability as a consequence of mechanical and/or structural deterioration of electrode materials. (B) Near-zero expansion two-dimensional electrode with stable electrochemical performance, a longer lifetime, and improved kinetics of electrochemical processes because of the ease of electron and ion transport [14].

variations in volume cause grain splitting and delamination from the current collector, resulting in the reduction of capacity (Figure 5.2A). Consequently, the battery lifespan is fundamentally reduced by phase transitions and mechanical deterioration of the electrode materials. As a consequence of the chemical variability of 2D materials (Figure 5.2B), it is possible to rationally create intercalation electrodes with near-zero volumetric expansion by assembling elements with both positive and negative volume changes in multilayered structure. Layered heterostructures are difficult to manufacture; yet they facilitate electrodes to have exceptional mechanical and electrochemical durability [14].

LIBs efficacy and its relationship to and influence from electrode materials and electrolytes have been the subject of extensive research. As nanostructured anodes possess singular physical and electrochemical features, like short lithium ion diffusion route span, their manufacture has gained considerable observation in several investigations recently. As a result, research has been conducted on a wide range of anode materials, including those with 0D, 1D, and 2D nanostructures of varying geometries. Among carbon-based materials, graphite is widely utilized despite its low specific capacity (372 mAhg^{-1}) for high diffusion, whereas its 2D equivalent, graphene, has not only a high specific capacity (955 mAhg^{-1}) but also a barrier and lithium nucleation. High ion conductivity may be achieved by the introduction of simple ion transport channels, which may be facilitated by the structural framework of

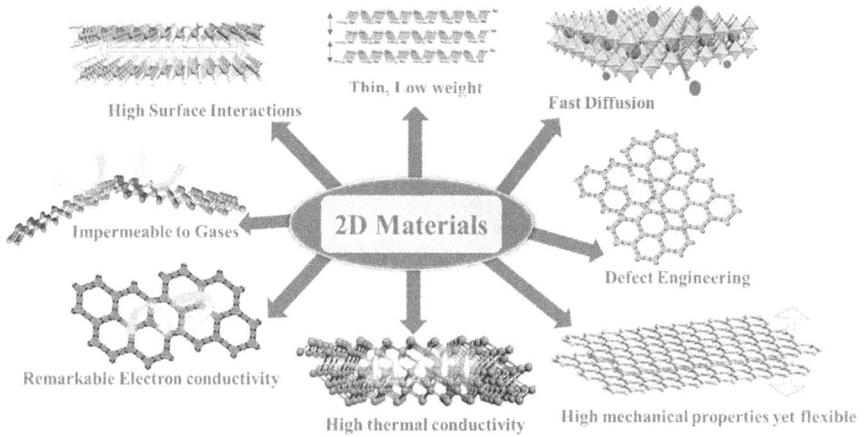

FIGURE 5.3 Merits of 2D materials. Using 2D materials in energy storage systems may bring novel features [12].

2D materials (such as planar, buckling, and puckered structures). Since it would enable large current densities to be supported without chemical and mechanical instabilities, this ability might be of tremendous importance. As can be anticipated, the superionic conduction qualities of 2D substances promote assisted diffusion features while reducing the barriers in ion mobility for ultrafast ion movement. The mechanical characteristics of these materials have been shown to be superior to those of their bulk analogues, namely a high Young's modulus and a high degree of flexibility. Some of the benefits of using 2D materials are shown in Figure 5.3 [12,13].

Among the group of 2D materials, IVA is confirmed by both theoretical and experimental results, making it a candidate for future nanodevices. Particularly, Group IVA elements of the periodic table (Sn, Pb) have significantly higher specific capacities than commercially available carbon-deployed anodes (372 mAh/g). These are regarded as more feasible anode applicants for the subsequent iteration of lithium-ion batteries. The utilization of bulk Sn and Pb is hampered by the significant volumetric increase of 230–260% (257%, 230% correspondingly) over lithiation and delithiation cycles, which results in particle pulverization and SEI film instability. These problems cause fast capacity degradation and poor Coulombic efficiency [13,15–18]. There are several effective approaches to address the challenges associated with using Group IVA elements as anodes in LIBs. The most promising strategies primarily include (1) decreasing the size of the particles to the nanoscale to diminish mechanical strain; (2) creating a porous hierarchy structure with a suitable interior permeability to enable the growth of group IV elements (Si, Ge, and Sn) and a stable SEI layer; (3) to facilitate dimensional changes and preserve mechanical integration, the group IV elements (Si, Ge, and Sn) are dispersed in nanosized form in a conductive matrix (such as carbon-based materials). (4) Reducing the intensity threshold and altering the amount of lithiation. (5) Employing intermetallic alloys in an active or inactive host matrix in composite structures. These advancements could lead the path for the next iteration of high-potential LIBs [16].

As an analysis of the latest advancements in the development of Group IVA 2D materials, the numerous electrochemical interaction mechanisms and techniques are briefly discussed. Eventually, a viewpoint on Sn and Pb 2D materials anodes, including mono- and multi-elemental materials, will be presented for impending advancements toward the development of improved rechargeable batteries.

5.2 GENERAL CHARACTERISTICS (PROPERTIES) OF TIN AND LEAD AS ANODE MATERIALS

Sn and Pb-based materials (Figure 5.4) have garnered much interest as potential candidates for use as anodes in lithium-ion batteries (LIBs) due to their inexpensive, strong theoretical capacities and increased energy density [17–19].

For next-generation LIBs, neoteric anode materials with increased Li-storage capacity and functional safeness are indeed required. Currently, commercialized graphite is not one of these materials. In search of such neoteric anode materials, researchers suggested group IV A elements, like Tin (Sn), Silicon (Si), Lead (Pb), and Germanium (Ge) that can be considered, the main reasons being high specific capacity, high energy density, low operating potentials, and high power density.

FIGURE 5.4   Sn and Pb as battery materials.

However, Sn was a potential candidate for anode materials, which has a higher gravimetric capacity (993 mAh/g) (Figure 5.5) and volumetric capacity (7214 mAh/c) than graphite. Also, the formation of $Li_{4.4}Sn$ ("Sn + 4.4Li$^+$ + 4.4e $\leftrightarrow$ $Li_{4.4}Sn$") "Sn + xLi$^+$ + xe$^-$ $\leftrightarrow$ Li_xSn (0 < x $\leq$ 4.4)", shows an appeal for high-capability anode for LIB batteries [20–27]. When considering safety concerns, the Sn has a greater potential for discharge voltage around 0.38–0.66 V than the graphite anode 0.02–0.2 V as compared to metallic lithium [26,27]. Even when considering the electrical resistivity of Sn, it shows a value of 1.1×10^{-7} mΩ, which is a magnitude lower than the commercial graphite value of 0.00001 mΩ. However, there are some drawbacks like that of other Lithium alloy-based elements, particularly because of pure (bulk) Sn, which have large amounts of volume changes around 300% during lithiation and delithiation that are inherently prone to heavy capacity deterioration and decreased cycle life, ensuing pulverization, evolution of solid

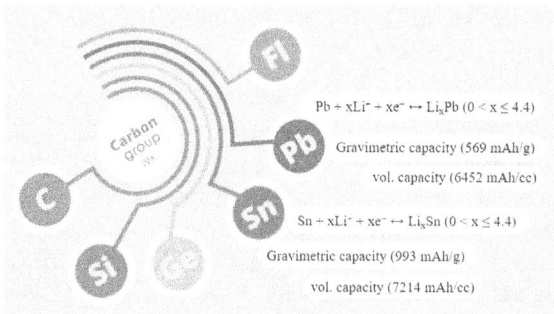

FIGURE 5.5   Schematic of carbon group IVA elements: Pb and Sn.

electrolyte interphase (SEI) that restricts electrons transport from current collectors [28–31]. In order to combat these issues linked with the lithiation and delithiation of pure Sn, huge amounts of Sn compounds have been investigated to enhance the performance of the electrode. Likewise, while preparing the tin-based materials with the help of structural merits and morphological characteristics, the 2D, nanosized, macro-sized, and micro-sized materials that reduce the volume expansion and volumetric contraction while lithiation and delithiation of electrons; the emergence of SEI will also be limited; therefore anode cycling ability will be improved [28,32]. Lithium and tin form intermetallic compounds of $Li_{22}Sn_5$, Li_7Sn_2, Li_3Sn, Li_5Sn, LiSn, and Li_2Sn_5 [33]. Electrolyte $LiClO_4$ in lithium can produce such intermetallics. Depending on the produced intermetallics, for example, $Li_{22}Sn_5$, can give an approximate value of up to 993 mAh/g; hence, the resultant Sn-based 2D material will be a neoteric anode for LIB [34,35].

Lead (Pb), another element in group IVA, shares many of the same physical properties as silicon (Si) and tin (Sn). In pursuit of a new anode material, lead (Pb) was not a significant competitor. Pb and its oxides do, however, provide certain benefits, including cheap cost and high energy density. In addition, the lead-acid battery business has built a reliable supply chain for Pb, which is earth-abundant, affordable, and readily available. In addition, it has a high rate of recycling (e.g. 99% in the United States) and is thus one of the most recycled materials worldwide. The problem of toxicity for Pb-based compounds is possibly solved with the implementation of a closed-loop recycling system, since LIB recycling technologies are advancing quickly. Therefore, recycling the lead-based materials with proper processing method advances the anodes, which also needs further research [19,36–42]. Pb demonstrates the ideal theoretical capacity of 569 mAh/g (Figure 5.4) higher than graphite, in the working mechanism "$Pb + xLi^+ + xe^- \leftrightarrow Li_xPb\ (0 < x \leq 4.4)$" that gives elevated electrical conductivity, safety, and reduction of the anode material cost. In terms of reversibility, cycle stability, and electrical conductivity, Pb (metal) anodes are superior to other lead compounds, thereby preventing unwanted irreversible reactions. Despite the utility of lead as an anode, the volume expansion that occurs during cycling remains a formidable obstacle. Designing a nanostructure and employing carbon composites are two possible methods for addressing this issue. Reducing the diffusion length of Li+ ions, increasing the electrode/electrolyte contact area, enhancing the activation of surface reactions, and increasing the number of active sites are all advantages of the nanostructure for lithium storage [43–45].

Recently, 2D halogen perovskites have received much attention due to their potential applications in the development of high-performance optoelectronic devices. The main feature of this 2D perovskite material is the atomically ordered layers of metal sheets in a crystalline stack with long organic bonds between the layers. A typical form of this type of 2D halogen perovskite is $(RNH_3)_2(CH_3NH_3BX_3)_{n-1}BX_4$, where R is a long alkyl chain or an aromatic group and n is the number of metal layers between the organic chain layers. Therefore, the composition and manufacturing process of the halogen perovskite has a great influence on the crystal structure, morphology, and optical properties. The photoactive perovskite layer (absorbing or emitting light) should cooperate with

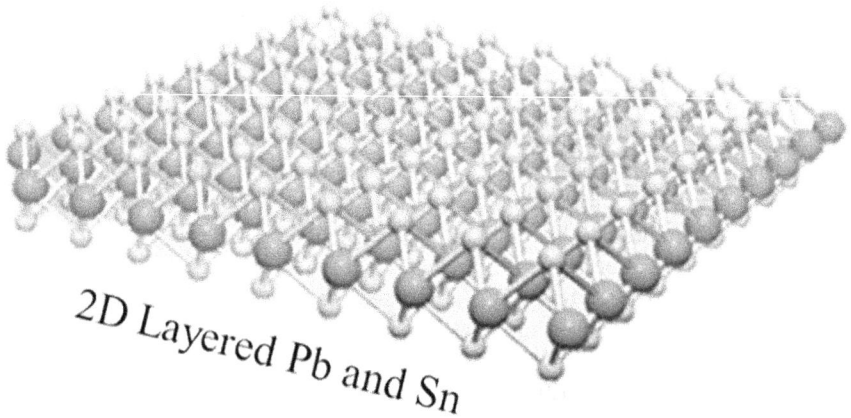

FIGURE 5.6   2D layered Pb and Sn.

other functional layers, such as electrodes, interfacial layers, and encapsulating films, to create inexpensive, efficient, and stable optoelectronic devices [46–49].

However, a wide variety of traditional 2D materials exhibit outstanding optical, electrical, thermal, mechanical, and/or catalytic capabilities. Because of their one-of-a-kind 2D structures, it is simple to process or combine these materials into ultrathin films that are homogeneous, flexible, and have highly aligned microstructures. As a consequence, 2D materials, mainly Pb and Sn (Figure 5.6), have received a lot of attention in recent years as potential candidates to be utilized as energy storage.

5.3   PROGRESS OF 2D ANODE MATERIALS PB AND SN ALLOYS FOR LI-ION

5.3.1   Pb-Based 2D Anode Materials

Pb is the chemical symbol for lead, which comes from the Latin word plumbum, which means lead. Since the initial reporting of the electrochemical lithiation of metallic lead (Pb) foil was published in 1971 [41], various research attempts have investigated the possibility of using lead and lead oxides as an anode for LIBs. LiPb compounds have also been identified, such as $LiPb$, $Li_{2.6}Pb$, Li_3Pb, $Li_{3.5}Pb$, and $Li_{4.4}Pb$, which all employ lead as the anode 2D material [50]. This shows that lead-based anode materials are energy storage materials from long many years; thus, so much research has been done in 3D lead material, but in 2D it's a new exploring element for storage purposes.

The most recent 2D substance in this class is Group 14 element Plumbene, a monolayer of lead with a honeycomb structure. Because of this, a monolayer of a lead film may be referred to as plumbene, comparable to graphene and silicone [51–55]. Researchers are showing a growing interest in investigating 2D halide perovskites lead based as potential materials for use in LIB electrodes [56–58]. As the latest photovoltaic research field has seen a boom in interest in HOIPs (hybrid organic-inorganic perovskites), particularly 3D AMX_3 (A = organic cation, M = divalent group 14 element, X = halogen) elements, owing to their readily available high power conversion efficiency (22.1%), that are of interest. [59,60]. However, environmental instability, including that caused by moisture, oxygen, and ultraviolet light, continues to be a significant impediment for 3D HOIPs. In contrast, 2D HOIPs display better stability and more promising performance for solar devices [61–65]. Due to their exceptional optoelectronic performance, hybrid lead halide perovskites have already attracted considerable interest in the field of solar cells. 3D metal halide-based materials may achieve a maximum efficiency of 25.2% in power conversion [58,66]. Moreover, the battery performance is improved by the effective adsorption and diffusion of lithium ions on the surface of the 2D nanomaterials [67–70]. The crystal and surface structures of halide perovskites are one-of-a-kind, which allows them to absorb a significant number of ions and lowers the energy required for the diffusion process. It's interesting to note that thin film photovoltaic cells' efficacy is often assumed to be substantially hindered by systematic ion diffusion such as hysteresis. [71–73]. Despite this, halide perovskite materials are suitable in the realm of batteries owing to their ion diffusion and storage capabilities. Battery capacities of up to 100 mAh/g and overall efficiencies equivalent to those offered by hybrid electrodes that utilize a blend of battery and solar cell materials are achieved by using a strongly photoactive 2D lead halide perovskite as the principal photo-rechargeable electrode material [74–76]. As summarized, Table 5.1 shows the Pb anode-based 2D material in the case of LIB batteries.

Duyen H. Cao et al. set out in 2015 to create 2D perovskite materials by reforming 3D perovskite materials with the help of iodide salts. Ammonium functional groups have a positive charge which can encapsulate the halide atoms using hydrogen bonding and vice versa in negatively charged ions; utilizing this concept, lead halides (PbX_2) react with iodide salts to form 2D layered perovskite [81]. The 2D hybrid halide perovskite concept anticipates to minimize the volumetric changes during the charge-discharge cycles, making this electrode more efficient. This chemical composition of the hybrid halide perovskite material based on Pb and Br is predicted to have a major effect on the interfacial interactions with the solid electrolyte, and the modification of the metal halide perovskite's structure is predicted to improve the battery's electrochemical performance [78]. Later, lead-bromide-based electrode material electrochemical properties investigated by Yuta Fujii et al. and Daniel Ramirez et al. show the improved intercalation ions capacities with improving lithium extraction. An intercalation mechanism in the hybrid material may accompany the alloying dealloying process of the Li_xPb intermetallic compounds in the mechanism of lithium insertion extraction [77,78]. Electrode stability is needed to further investigate for improved high efficiency for this PbBr.

The Pb-based 2D halide perovskite material was modelled using first-principles calculations by Mu He et al., which investigates the Li-ions transportation and interlayer diffusion. These 2D lead halide layers contribute the theoretical capacity of 83.1 mAh/g without structural distortion and degradation. The transportation between the lithium ions on the halide surface can easily migrate to the organic functional group. The minimum energy barrier is 150.1 meV, which is viable for the adsorption and release of lithium ions at an optimal band gap [79].

TABLE 5.1

Two-Dimensional Pb-Based Anode Materials

2D material	Alloy	Cycle Number	C (mAh/g)	Applications	Ref
Pb	$(CH_3NH_3)_2(CH_3(CH_2)_2NH_3)_2Pb_3Br_{10}$	10	375	**LIBs**	Daniel Ramirez et al., [77]
Pb	$(CH_3(CH_2)_2NH_3)_2(CH_3NH_3)_2Pb_3Br_{10}$	30	242	**Solid-State Lithium Secondary Batteries**	Yuta Fujii et al., [78]
Pb	$(C_6H_9C_2H_4NH_3)2PbI_4$		83.1	**LIBs**	Mu He et al., [79]
Pb	$(C_8H_{17}NH_3)_2PbI_4$			**LIBs-Solar Cells**	Teck Ming Koh et al., [80]
Pb	$(CH_3(CH_2)_3NH_3)_2(CH_3NH_3)_{n-1}Pb_nI_{3n+1}$			**Solar Cells**	Duyen H. Cao et al., [81]
Pb	$(C_4H_9NH_3)2PbX_4$ (X=Cl, Br, I or mixed halide)			**LIBs-Solar Cells**	Junxiong Wang et al., [82]
Pb	$PEA_xMA_{0.4-x}FA_{0.6}Pb_{0.9}Ba_{0.1}yCl_{3-y}$			**LIBs-Solar Cells**	Shun-Hsiang Chan et al., [83]

Combination of 2D and 3D approaches of perovskite film gives the benefit of both, like enhanced stability, reduced trap state densities, and increased photo-luminescence lifetimes. This is also an alternative solution that is used to overcome modulating defects, as propounded by Teck Ming Koh et al. [80].

Commercialization of perovskite solar cells may be slowed by the presence of hazardous lead [83]. Lead pessimistic can be overcome by recovering the spent lead acid battery paste; this can be used in 2D hybrid perovskite materials (Figure 5.7).

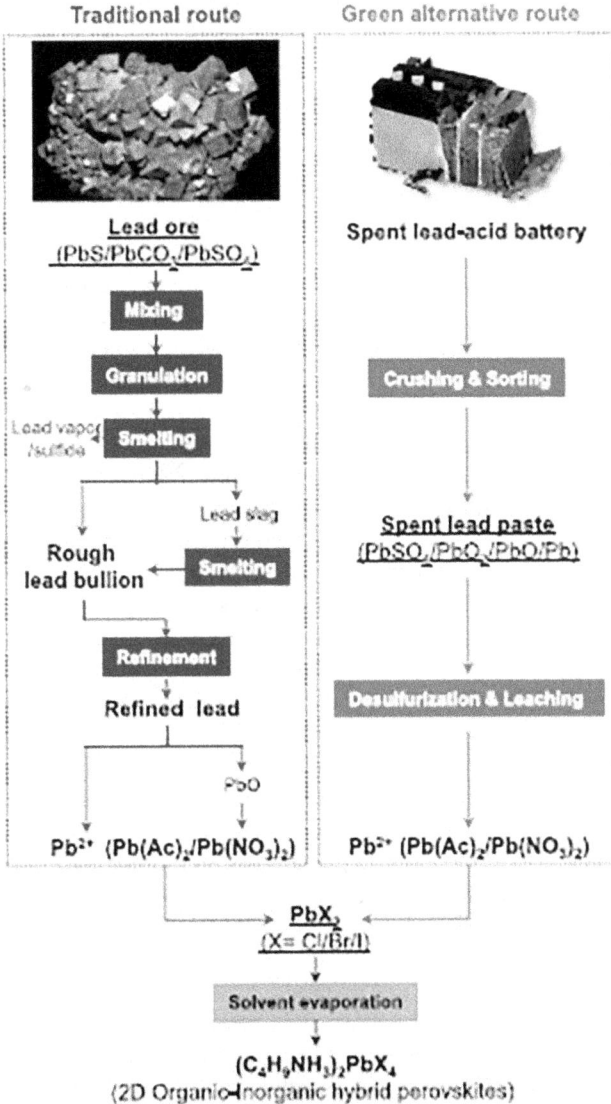

FIGURE 5.7 Schematic of two-dimensional hybrid pervskites produced from lead ores vs lead acid batteries [82] (Copyright: ©Elsevier).

More than 95% of the total lead in the lead paste can be recovered and converted into a low-contamination ion solution using a simple regeneration process at room temperature [82].

Lead as 2D batteries have many endless benefits as the material is dominating the battery field in the 20th century, and as a key factor, the recovery property of lead makes it endless. Researchers need to overcome the problem of reducing the toxicity of lead material. As a consequence, the invention of a technology that can stabilize this while maintaining its electrochemical efficiency is very significant and critical. Also, lead can be utilized as the anode for photovoltalic pervoskite-based lithium-ion batteries.

5.3.2 Sn-Based 2D Anode Materials

Tin atoms in a buckled honeycomb lattice (called stanene) have recently attracted a lot of interest as a potential large-gap 2D topological insulator for realizing the room-temperature quantum-spin-Hall effect. This is because the electronic structures of Sn allotropes are sensitive to lattice strain [84–86]. Tin-based bulk or micron-sized materials have been widely researched as potential replacements for currently available carbon electrodes. The fundamental disadvantage of this system derives from the substantial volume changes and accompanying rapid losses in the capacity that occur throughout electrochemical cycles. 2D-layered nanostructures are gaining popularity because of their exclusive nanoscale phenomena and prospective usage for energy storage applications and this also eliminates volumetric loss [87]. 2D materials are adopted to several advanced techniques. Table 5.2 shows a summarization of 2D Sn-based anode material of LIB batteries.

2D materials and nanostructures are growing in popularity due to their unique nanoscale phenomena and promising applications ranging from electronics and energy to catalysis. In 2008, Jung-wook Seo et al. showed laterally confined layered nanocrystals (LCLN) to have significantly improved electrochemical performance compared to bulk nanocrystals. Tin sulfides are of particular interest due to their unusual structural properties, such as the CdI2-type structure in which tin atoms are stacked between two hexagonal layers of closely gathered sulfur atoms [88]. Tin sulfide (SnS) is a binary metal chalcogenide semiconductor with an orthorhombic layered crystal structure. SnS layers are held together by mild van der Waals forces, resulting in a chemically inert surface free of suspended bonds and Fermi-level pinning at the semiconductor surface. This affords SnS a high level of chemical and environmental stability, even in acidic solutions. [88,105]. According to Aifantis et al., active spots, spherical, low-electron active sites may improve the advantages of anode materials in terms of electrochemical performance and standpoint of mechanical failure [106–108]. Lin Mei et al., developed an efficient and abundant hybrid structure of ultrafine SnS_2 nanocrystals decorated with low-modulus reduced graphene oxide (RGO), which proved to be an excellent anode material, delivering up to 1034 mAh/g showing a very high specific capacity even after 200 cycles with temperature range up to 0.1°C that give excellent cycle speed and stability. The charge transfer process for inserting and extracting lithium ions was quickest in SnS_2 nanocrystals @RGO, and the material had the greatest electrical conductivity overall.

TABLE 5.2
Two-Dimensional Sn-Based Anode Materials

2D material	Alloy	Cycle number	C (mAh/g)	Batteries	Ref
Sn	SnS2	30	323	LIB	Jung-wook Seo et al., [88]
Sn	Graphene-confined Sn nanosheets (G/Sn/G)	60	590	LIB	Bin Luo et al., [89]
Sn	SnS2/RGO nanocomposites	200	1034	LIB	Lin Mei et al., [90]
Sn	SnS2/GNS's nanocomposites	100	704	LIB	Shuangyu Liu et al., [91]
Sn	Co2SnO4 hollow cubes@RGO nanocomposites	100	1000	LIB	Jingjing Zhang et al., [92]
Sn	SnO2/graphene composites	1000	1813	LIB	Yu Chen et al., [93]
Sn	SnS NS/RGO	100	560		Shuankui Li et al., [94]
Sn	Hollow Zn2SnO4@graphene composites	45	752	LIB	Yang Zhao et al., [18]
Sn	F-G/Sn@C		506	LIB & SIB	Bin Luo et al., [95]
Sn	Sn@PNGC		500	LIB	Dan Zhou et al., [96]
Sn	SnS	30	826	LIB & SIB	Che-Ya Wu et al., [97]
Sn	SnSe2/graphene		490	LIB	Hongwen Chen et al., [98]
Sn	C@SnO2-rGO-SnO2		525	LIB	Weiqi Yao et al., [99]
Sn	SnO2/C	100	707.8	LIB	Qinghua Tian et al., [100]
Sn	SnSe Nanonetworks	50	1000	LIB	Fionán Davitt et al., [101]
Sn	Tin nano-platelets	100	820	LIB	S. Khabazian et al., [102]
Sn	SnO/p-G		633.9	LIB	Pinxian Jiang et al., [103]
Sn	SnS2 / SnS		510	LIB	Junjun Zhang et al., [104]
Sn	SnSe2/CNTs	100	210.3		Hongwen Chen et al., [98]

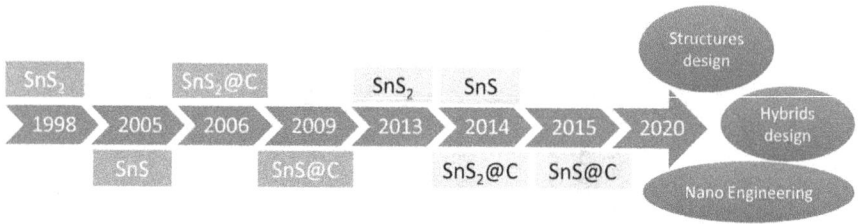

FIGURE 5.8 Chronology of important techniques and outcomes for SnSx materials as LIB (green) and SIB (red) anodes (yellow) [109].

During lithiation and delithiation, RGO nanosheets may flex to meet the volume change that occurs. The conductivity of SnS_2 nanocrystals @RGO is enhanced because the RGO nanosheets serve as electron transporting paths, making the system more conductive than pure SnS_2 [90]. Later on, Shuangyu Liu et al., Shuankui Li et al., Che-Ya et al., and Junjun Zhang et al. created SnS_2 but were able to get a capacity in the range of 560–826 mAh/g using various buffering nanostructures SnS_x that have significant promise as anode materials. Figure 5.8 depicts the history of major techniques and outcomes for SnS_x anodes for LIBs (green) and SIBs (yellow). The uses of SnSx materials in various novel rechargeable batteries (e.g., lithium-sulfur batteries, aluminum-ion batteries) are also being researched as part of the development of new battery systems [90,91,94,97,104,109].

SnSe is a metal chalcogenide with strong chemical interactions along the b and c crystal axes, and weak van der Waals bonds along the longer crystal axis [110]. This results in its tendency to develop in 2D plate structures due to slower crystal growth rate and easier cleavage [111]. Fionán Davitt et al. have reported the growth of complex 2D nanonetworks of crystalline SnSe as an anode material for Li-ion batteries, which provides valuable insights into this 2D material [101]. Recent studies by Hongwen Chen et al. reveal the interface coupling effect between $SnSe_2$ and graphene/CNTs in the hybrid and play a triple role in electrodes, preventing agglomeration, forming pore-rich nanostructures for electrolytes penetration and diffusion, and improving cycle stability [98,112].

Graphene is a monolayer of carbon atoms that has a special 2D nanostructure due to its honeycomb-like arrangement of carbon atoms. It has a large theoretical surface area of 2600 m^2g^{-1}, great electron conductivity, and good chemical stability. The addition of Sn shows surface-to-surface "binding" between graphene nanosheets and both sides of a 2D second phase. The volume shift of tin during Li-Sn alloying-dealloying processes is well accommodated by graphene nanosheets on both sides of the tin, thanks to the 2D structure of the tin phase and the elasticity of graphene nanosheets. Graphene-coated 2D SnO_2/SnO particles show exceptional Li-ion storage capability. The structural benefits of SnO nanosheets may include reduced ionic diffusion lengths and an abundance of easily accessible reactive sites. The volume shift of SnO_2 may be mitigated by the buffer space provided by the mesoporous microstructure. Graphene acts as a miniature current collector to facilitate quick Li+ transfer and diffusion. In order to produce a stable SEI layer during cycling, the carbon-coated protective layer efficiently suppresses SnO_2

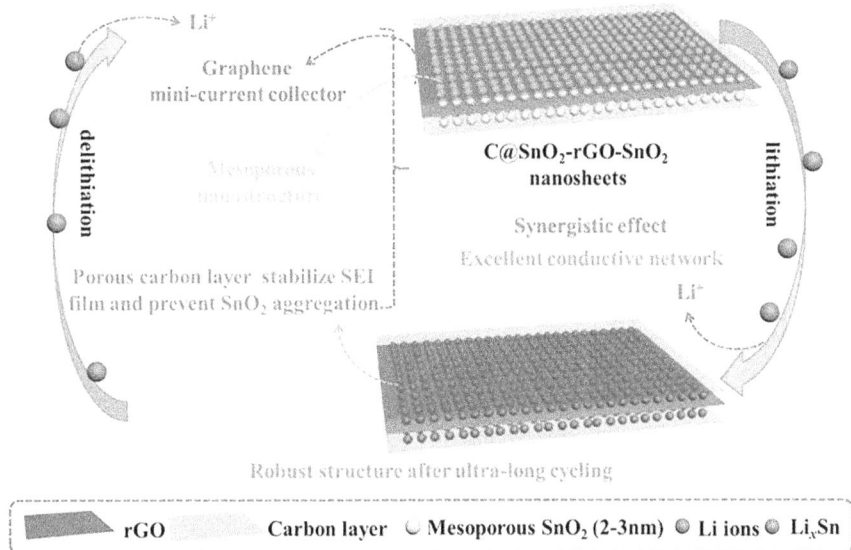

FIGURE 5.9   Two-dimensional SnO2/graphene/mesoporous [99] [copyright: © 2018 Elsevier].

aggregation. Together, graphene and the carbon-coated porous layer form a very effective conductive network that makes up for SnO_2 otherwise insufficient electrical conductivity. Weiqi Yao et al., Pinxian Jiang et al., Jingjing Zhang et al., Yu Chen et al., and Bin Luo et al. were all contributors to the development of 2D batteries based on tin and graphene and provided a result of their exceptional structural and surface properties, for great potential in lithium storages [89,92,93,95,99,103], Figure 5.9 shows the combination of graphene and Sn 2D batteries. Qinghua Tianthin et al. developed a 2D hollow nanostructures of SnO_2/carbon (SnO_2/C) LIB composite anodes that exhibit improved electrochemical kinetics and strong structural stability through two key variables; On the one hand, good conductivity and a flexible carbon coating alone cannot improve conductivity of the whole electrode but also buffer the large volume change; On the other hand, thin 2D hollow nanosheet structures alone can no longer deliver lithium [103]. Nanoparticles of tin (Sn) with uniform distribution are studied by Dan Zhou et al. in the context of a porous N-doped graphene-like carbon network (Sn@PNGC) indicating enhanced Li^+ accessibility and electrical conductivity, as well as room to accommodate volume change during lithiation/delithiation. High specific capacity, outstanding cycle stability, and improved rate capability were all features of the as-assembled electrode's promising Li storage performance [96]. Recently, Khabazian et al. have advanced the preparation of 2D tin platelets through template-free electrodeposition, resulting in high-capacity retention. The electrodes made from Sn platelets exhibit a specific capacity of 820 mAh/g for over 100 cycles. Moreover, nano-platelets display excellent rate capability and can operate up to 20°C without damage [102]. In 2022, Runsheng Gao et al. develop binary Sn-Co alloy that was embedded in a carbon matrix derived from a nitrogen-doped 2D leaf-like metal-organic framework (MOF). This composite electrode has outstanding conductivity, quick electrochemical response, and good Li storage due to its unique structural characteristics and homogeneous active material distribution. The carbon-shell prevents electrolyte side reactions. Sn-Co@C had an improved ICE, high specific capacity, and excellent rate performance as an LIB electrode, even with a 50-fold rise in current density, demonstrating that the 2D carbon framework accommodated volume variations and decreased Li-ion transmission distance [113].

Tin materials as 2D entities were employed for LIB systems. As a first material, tin sulfate 2D elements show higher electrochemical performance. Later, the combination of tin and graphene (carbon) composites approach was followed to improve the capacity. Researchers have shown more focus on this material, so in future, their capacity will be enhanced compared to the present with different chemical compositions and combinations.

5.3.3 PB-SN-BASED 2D ANODE MATERIALS

Tin (Stanene) and lead (Plumbene) have been members of group IVA and are potential contenders for future energy storage devices, according to theoretical and experimental results [13]. They have only been reported in solar cells based on 2D perovskites (Figure 5.10); printing, coating, or vacuum deposition are all methods used to create thin-film materials [114]. These materials have a variety of useful operations, involving solar cell tandem manufacturing, building-integrated photovoltaic cells, aerospace engineering, rechargeable batteries, capacitors, and PV catalysts [115].

Lingling Mao et al. (2016) report on the shifting band gap trends found in Pb and Sn, 2D system solid solutions, which is a first. Surprisingly, the solid solutions have no room-temperature phase transition, suggesting that the main cause is SOC. Furthermore, modifying organic spaced functional groups significantly impacts the general characteristics of 2D perovskite iodide materials made from lead and tin. This is because such modifications influence the geometric distortion levels of inorganic layers, namely $(PbI_4)_2$ and $(SnI_4)_2$ resulting in changes in their physical and electrical characteristics via stimulation [116]. Mixed Sn/Pb perovskite has the narrowest band gap and is made up of stoichiometrically identical Sn^{2+} and Pb^{2+} [117,118]. Sn-based perovskites have been utilized to wholly or partly replace Pb^{2+}, resulting in reduced optical band gaps (Egs). Sn minimizes the consumption of Pb to some level, which is better for the environment. Compared to analogous Pb-based 2D perovskites, it also significantly reduces Eg, improves light absorption,

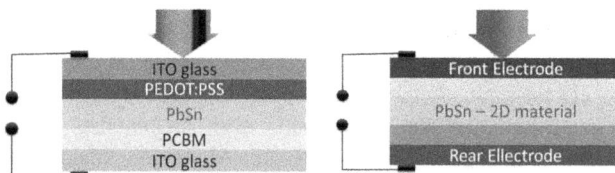

FIGURE 5.10 Device architecture of a PbSn perovskite solar cell [115].

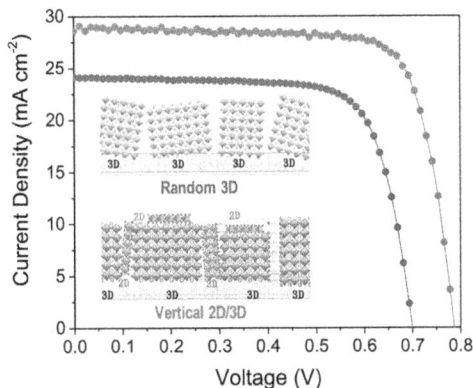

FIGURE 5.11 Vertically aligned 2D/3D Pb–Sn perovskites [119] (Copyright © 2020, American Chemical Society).

and improves charge transport [119]. For the production of efficient single-junction and tandem solar systems, mixed metal PbSn perovskites with variable band gaps (1.2–1.6 eV) are appealing. However, due to the oxidation of Sn^{2+} to Sn^{4+}, Pb Sn perovskites are prone to introducing high trap densities. Although compositional engineering and the incorporation of antioxidant agents have actively proposed strategies for preparing high-quality PbSn perovskite films, the extension of the mixed 2D/3D concept to low band gap PbSn perovskites would offer an alternative approach to address both the efficiency and stability of PbSn, which has, however, rarely been reported [120–122]. Figure 5.11 illustrates the concept of hybrid 2D and 3D heterostructures, which have emerged as an effective approach to improve the stability of lead halide perovskite solar cells. However, this concept has only been infrequently reported in lead-tin (Pb-Sn) composite perovskite devices.

In accordance with the investigations that were discussed before, the 2D materials tin and lead have shown good performance when utilized as materials for the negative electrode. Nevertheless, their electrochemical capacities and efficiencies mostly depend on the procedures employed to synthesize them, as well as their chemical structures and elemental makeup. Research on 2D materials of Group IV A mainly Pb and Sn has been severely limited up to this point. Further research is needed to establish a link between chemical composition, geometrical structure, and electrochemical performance; this would provide a solid theoretical basis for producing exceptional and high-performance anode materials for futuristic batteries.

5.4 SYNTHESIS OF PB-SN-BASED TWO-DIMENSIONAL ANODE MATERIALS

Various techniques are used to synthesize 2D halide perovskites Pb-Sn materials or 2D Pb-Sn materials with varied dimensions and chemical constituents. There are reports available that discuss the synthesis, characteristics, and uses of 2D materials composed of lead and tin. In the next section of this chapter, we will talk about the various synthesis techniques.

5.4.1 Mechanical Exfoliation

Mechanical exfoliation is also known as micro-mechanical exfoliation due to the use of tiny mechanical forces throughout the procedure. When the tape adhesive force is larger than the newly created surface energy, 2D mechanical exfoliation may be performed by breaking Van der Waals forces between neighboring layers using mechanical forces. The first mechanical exfoliation example is monolayer graphene exfoliated from graphite [123]. Figure 5.9 depicts the usual steps of the mechanical exfoliation approach for single-layer 2D transition metal dichalcogenides crystals. The bulk crystal is pushed onto 3 M scotch tape and folded multiple times. Appropriate pressure is applied to ensure that the tape containing samples adheres to the substrate successfully. The scotch tape is then removed, leaving numerous TMD flakes on the SiO_2/Si substrate's surface [124]. Junze Li et al. used mechanical exfoliation (Figure 5.12) to create single n-number phase 2D perovskites $(C_4H_9NH_3)_2(CH_3NH_3)_{n1}$ Pb_nI_{3n+1} microplates with n = 1–5. Using the mechanical exfoliation method, 2D perovskite microplates with thicknesses of less than 20 nm that have been shown to be single n-number perovskite phases can be used to improve the performance of optoelectronic devices and design new device architectures [125].

For Sn materials, a mechanical exfoliation process (Figure 5.12) was used to cleave the original bulk single crystal into 2D SnS_2 nanosheets. The bulk SnS_2 was first put on an adhesive scotch tape, and then a thin SnS_2 crystal was pulled off of it. The freshly cleaved thin crystals were then rubbed with a plastic spatula on a custom-designed substrate Si/ SiO_2 (300 nm)/Cr (3 nm)/Au (5 nm) to cleave the sliced SnS_2 further. After the Scotch tape was taken off, the few layered SnS_2 nanosheets were still on the substrate [126]; hence, the 2D synthesis was done through this method.

5.4.2 Liquid-Phase Exfoliation (LPE)

The isolation of 2D layers using the well-known "scotch tape" mechanical exfoliation process yields high-quality materials for basic physics research [122]. It is, however, time-consuming and yield-limited; hence, a variety of different techniques for producing 2D layers for large-scale applications, such as inks, coatings, membranes,

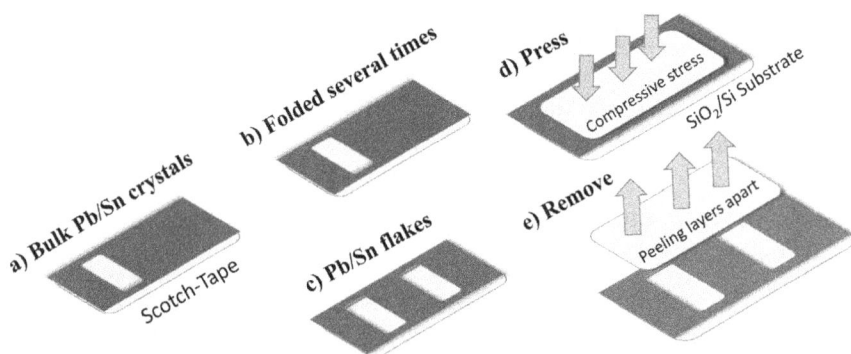

FIGURE 5.12 Illustration of the mechanical mechanism for exfoliation via original scotch tape [126].

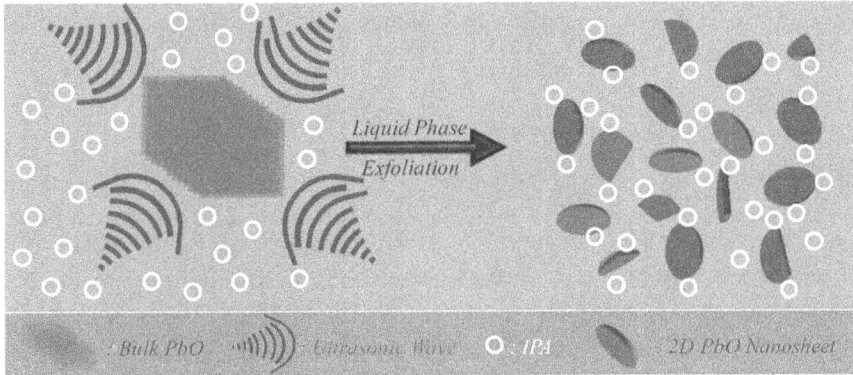

FIGURE 5.13 LPE to fabricate 2D PbO nanosheets from their layered bulk counterpart [130] Copyright © 2018, American Chemical Society.

and nanocomposites, have been developed [127]. Among these technologies, LPE, which involves exfoliating 2D layered materials into thinner and smaller flakes with the use of sonication in solvents, has been shown to be effective as it provides upscaling production and excellent quality at a reasonable cost [128,129]. Figure 5.13 shows the LPE method to fabricate 2D lead-based materials.

5.4.3 SOLUTION-PHASE GROWTH

The mechanical exfoliation method synthesizes materials for 2D perovskite structures; however, this method can only be applied to halide perovskites with large organic cations. In addition, the current approach for mechanical exfoliation has other constraints such as uncontrollable particle sizes, layer quantities, and material composition of the layers. Research is now being redirected toward creating monolayers and a few layers of 2D halide perovskites using a straightforward solution-phase synthesis to circumvent the constraints imposed by mechanical exfoliation [131]. The synthesis of 2D halide perovskite-like nanosheets uses a re-evaporation method by casting mixed halide perovskite solutions, where the organic solvents are acetonitrile, chlorobenzene, and dimethylformamide. This induces uniform liquid evaporation and crystallization, leading to the formation of ultrathin 2D halide perovskite nanosheets. Photoluminescence (PL) is measured to verify layer numbers and chemical composition [132]. Yuta Fujii et al. and Ziming Chen et al. have used this synthesis method. Figure 5.14 [133] shows the idea of solution-

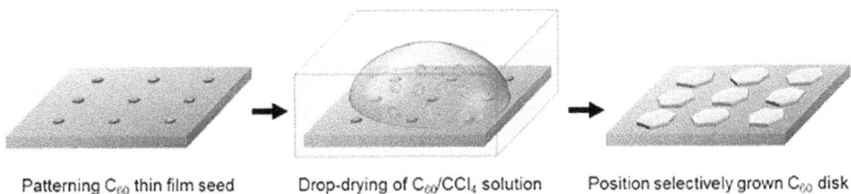

Patterning C_{60} thin film seed   Drop-drying of C_{60}/CCl$_4$ solution   Position selectively grown C_{60} disk

FIGURE 5.14 Schematic of solution-phase growth [133] © 2019 Elsevier Ltd.

phase growth of thin film materials. The solution phase growth method is a popular method in recent times.

5.4.4 Chemical Vapor Deposition

Chemical vapor deposition, often known as CVD, is a process that involves the condensation of components that are in the vapor phase in order to produce material that is in the solid phase [130]. In halide perovskites, non-Vander Waal CVD procedures might be used to cultivate different kinds of 2D halide perovskites. The development of halide perovskite nanoplatelets with a variety of compositions was reported by Xiong et al. using a developed CVD process [134]. This preparation was a two-step process, and the synthesis of 2D perovskite nanoparticles [135] is shown in Figure 5.15. CVD is a popular method for the synthesis of 2D films as reported by Daniel Ramirez et al., Duyen H. Cao et al., Junjun Zhang et al., and so on. The main benefits of this CVD are producing 2D films with good reproducibility, minimal cost, and rapid development [136].

5.5 CONCLUSION

Diversification of the market and the rapid development of electronic devices are driving the large and growing demand for energy-efficient and cost-effective products. Due to their advantages such as lightweight, efficiency and energy field, and extended life, rechargeable batteries can be utilized in a vast array of renewable energy storage devices. Existing 2D materials, however, are naturally rigid and do not satisfy the mechanical strength requirements of conventional batteries. Therefore, it is a significant challenge to develop advanced 2D anodes with new materials (Group IVA- Pb and Sn) that combine good structural and stability properties with good electrochemical performance.

This chapter discusses recent developments in 2D materials based on Pb and Sn. Moreover, battery recycling can decrease the toxicity of Pb (lead). A combination of Sn and Pb is very useful as a battery electrode material. However, the process of identifying and utilizing these resources is currently one of the most pressing requirements. Moreover, numerous anode standards are restricted to particular regions, substances, and materials. For the analysis of 2D anode materials, it is essential to implement specialized research methods such as machine learning (ML) and deep learning (HDS) with application models.

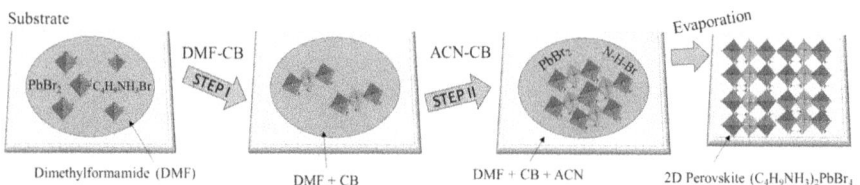

FIGURE 5.15 Schematic of 2D chemical vapor deposition synthesis [134].

ACKNOWLEDGEMENTS

Authors would like to thank Indian Institute of Technology Jodhpur for resources and financial support for this work.

CONFLICTS OF INTEREST

The authors declare no competing interest.

REFERENCES

1. Tarascon, J.-M., and Armand, M. Issues and challenges facing rechargeable lithium batteries. *Nature* 414 (2001): 359–367.
2. Armand, M., and Tarascon, J.-M. Building better batteries. *Nature* 451 (2008): 652–657.
3. Larcher, D., and Tarascon, J.-M. Towards greener and more sustainable batteries for electrical energy storage. *Nature Chemistry* 7 (2015): 19–29.
4. Peng, Lele, Zhu, Yue, Chen, Dahong, Ruoff, Rodney S., and Yu, Guihua. Two-dimensional materials for beyond-lithium-ion batteries. *Advanced Energy Materials* 6 (2016): 1600025.
5. Cheng, F, Liang, J, Tao, Z, and Chen, J. Functional materials for rechargeable batteries. *Advanced Materials* 23, no. 15 (2011 Apr 19): 1695–1715. doi: 10.1002/adma.201003587. Epub 2011 Mar 11. PMID: 21394791.
6. Arico, Antonino Salvatore, Bruce, Peter, Scrosati, Bruno, Tarascon, Jean-Marie, and Van Schalkwijk, Walter. Nanostructured materials for advanced energy conversion and storage devices. *Nature Materials* 4, no. 5 (2005): 366–377.
7. Wang, Yonggang, Li, Huiqiao, He, Ping, Hosono, Eiji, and Zhou, Haoshen. Nano active materials for lithium-ion batteries. *Nanoscale* 2, no. 8 (2010): 1294–1305.
8. Liu, Jiehua, and Liu, Xue-Wei. Two-dimensional nanoarchitectures for lithium storage. *Advanced Materials* 24, no. 30 (2012): 4097–4111.
9. Guo, Shaojun, and Dong, Shaojun. Graphene nanosheet: synthesis, molecular engineering, thin film, hybrids, and energy and analytical applications. *Chemical Society Reviews* 40, no. 5 (2011): 2644–2672.
10. Ming, Fangwang, Hanfeng Liang, Gang Huang, Zahra Bayhan, and Husam N. Alshareef MXenes for rechargeable batteries beyond the lithium-ion. *Advanced Materials* 33no.1 (2021): 2004039.
11. Guo, Qiang, Chen, Nan, and Qu, Liangti. Two-dimensional materials of group-IVA boosting the development of energy storage and conversion. *Carbon Energy* 2 (2020): 54–71.
12. Rojaee, Ramin, and Reza Shahbazian-Yassar. Two-dimensional materials to address the lithium battery challenges. *ACS Nano* 14, no. 3 (2020): 2628–2658.
13. Chronis, A. G., Karantagli, E., Michos, F. I., Garoufalis, Christos S., and Sigalas, M. M. Exotic nanoparticles of group IV monochalcogenides as anode materials for Li-Ion batteries. *Solid State Communications* 332 (2021): 114326.
14. Pomerantseva, E., and Gogotsi, Y. Two-dimensional heterostructures for energy storage. *Nature Energy* 2 (2017): 17089.
15. Tian, Huajun, Xin, Fengxia, Wang, Xiaoliang, He, Wei, and Han, Weiqiang. High capacity group-IV elements (Si, Ge, Sn) based anodes for lithium-ion batteries. *Journal of Materiomics* 1 (2015): 153–169.
16. Goriparti, Subrahmanyam, et al. Review on recent progress of nanostructured anode materials for Li-ion batteries. *Journal of Power Sources* 257 (2014): 421–443.

17. Sengupta, Srijan, Patra, Arghya, Akhtar, Mainul, Das, Karabi, Majumder, Subhasish Basu, and Das, Siddhartha. 3D microporous Sn-Sb-Ni alloy impregnated Ni foam as high-performance negative electrode for lithium-ion batteries. *Journal of Alloys and Compounds* 705 (2017): 290–300.

18. Zhao, Yang, Li, Xifei, Yan, Bo, Li, Dejun, Lawes, Stephen, and Sun, Xueliang. Significant impact of 2D graphene nanosheets on large volume change tin-based anodes in lithium-ion batteries: a review. *Journal of Power Sources* 274 (2015): 869–884.

19. Martos, M., Morales, J., and Sanchez, L. Lead-based systems as suitable anode materials for Li-ion batteries. *Electrochimica Acta* 48, no. 6 (2003): 615–621.

20. Liu, L., Lyu, J., Zhao, T., and Li, T. Preparations and properties of porous copper materials for lithium-ion battery applications, *ChemEngCommun* 203 (2016): 707e713.

21. Zhong, Y., Yang, M., Zhou, X., and Zhou, Z. Structural design for anodes of lithiumion batteries: emerging horizons from materials to electrodes. *Materials Horizons* 2 (2015): 553e566.

22. Wang, J., Raistrick, I. D., and Huggins, R. A. Behavior of some binary lithium alloys as negative electrodes in organic solvent-based electrolytes. *Journal of The Electrochemical Society* 133 (1985): 457e460.

23. Winter, M., and Besenhard, J. O. Electrochemical lithiation of tin and tin-based intermetallics and composites. *Electrochimica Acta* 45 (1999): 31e50.

24. Winter, M., Besenhard, J. O., Spahr, M. E., and Novak, P. Insertion electrode materials for rechargeable lithium batteries. *Advanced Materials* 10 (1998): 725e763.

25. Guo, J., Yang, Z., and Archer, L. A. Aerosol assisted synthesis of hierarchical tin–carbon composites and their application as lithium battery anode materials. *Journal of Materials Chemistry A* 1 (2013): 8710–8715.

26. Kepler, K. D., Vaughey, J. T., and Thackeray, M. M. Copperetin anodes for rechargeable lithium batteries: an example of the matrix effect in an intermetallic system. *Journal of Power Sources* 81e82 (1999): 383e387.

27. Sengupta, Srijan, Patra, Arghya, Deo, Yash, Das, Karabi, Majumder, Subhasish Basu, and Das, Siddhartha. A novel multiphase Sn-Sb-Cu alloy electrodeposited on 3D interconnected microporous Cu current collector as negative electrode for lithium ion battery. *Metallurgical and Materials Transactions E* 4 (2017): 51–59.

28. Liu, Lehao, Xie, Fan, Lyu, Jing, Zhao, Tingkai, Li, Tiehu, and Choi, Bong Gill. Tin-based anode materials with well-designed architectures for next-generation lithium-ion batteries. *Journal of Power Sources* 321 (2016): 11–35.

29. Dang, H. X., Klavetter, K. C., Meyerson, M. L., Heller, A., and Mullins, C. B. Tin microparticles for a lithium ion battery anode with enhanced cycling stability and efficiency derived from Se-doping. *Journal of Materials Chemistry A* 3 (2015): 13500e13506.

30. Beaulieu, L. Y., Eberman, K. W., Turner, R. L., Krause, L. J., and Dahn, J. R. Colossal reversible volume changes in lithium alloys. *Electrochemical and Solid-State Letters* 4 (2001): A137eA140.

31. Yang, Z., Meng, Q., Yan, W., Lv, J., Guo, Z., Yu, X., Chen, Z., Guo, T., and Zeng, R. Novel three-dimensional tin/carbon hybrid core/shell architecture with large amount of solid cross-linked micro/nanochannels for lithium ion battery application. *Energy* 82 (2015): 960e967.

32. Zhu, X., Zhu, Y., Murali, S., Stoller, M. D., and Ruoff, R. S. Reduced graphene oxide/tin oxide composite as an enhanced anode material for lithium ion batteries prepared by homogenous coprecipitation. *Journal of Power Sources* 196 (2011): 6473e6477.

33. Lee, W. W., and Lee, J.-M. Novel synthesis of high performance anode materials for lithium-ion batteries (LIBs). *Journal of Materials Chemistry A* 2 (2014): 1589e1626.

34. Tin-Based Materials As Advanced Anode Materials For Lithium Ion Batteries: A Review Ali Reza Kamali and Derek J. Fray Department of Materials Science and Metallurgy, University of Cambridge, Pembroke Street Cambridge CB2 3QZ, U.K. 14 A.R. Kamali and D.J. Fray. *Reviews on Advanced Materials Science* 27 (2011): 14–24.
35. Im, Hyung Soon, et al. Phase evolution of tin nanocrystals in lithium ion batteries. *ACS Nano* 7, no. 12 (2013): 11103–11111.
36. Han, Jinhyup, Park, Jehee, Bak, Seong-Min, Son, Seoung-Bum, Gim, Jihyeon, Villa, Cesar, Hu, Xiaobing, Dravid, Vinayak P., Su, Chi Cheung, Kim, Youngsik, Johnson, Christopher, and Lee, Eungje. New High-Performance Pb-Based Nanocomposite Anode Enabled by Wide-Range Pb Redox and Zintl Phase Transition. *Advanced Functional Materials* 31 (2021): 2005362.
37. Park, Cheol-Min, Kim, Jae-Hun, Kim, Hansu, and Sohn, Hun-Joon. Li-alloy based anode materials for Li secondary batteries. *Chemical Society Reviews* 39, no. 8 (2010): 3115–3141.
38. Wei, Tian-Ran, Wu, Chao-Feng, Li, Fu, and Li, Jing-Feng. Low-cost and environmentally benign selenides as promising thermoelectric materials. *Journal of Materiomics* 4, no. 4 (2018): 304–320.
39. Daniel, Stavros E., Pappis, Costas P., and Voutsinas, Theodore G. Applying life cycle inventory to reverse supply chains: a case study of lead recovery from batteries. *Resources, Conservation and Recycling* 37, no. 4 (2003): 251–281.
40. Gaines, Linda. The future of automotive lithium-ion battery recycling: Charting a sustainable course. *Sustainable Materials and Technologies* 1 (2014): 2–7.
41. Dey, A. N. Electrochemical alloying of lithium in organic electrolytes. *Journal of The Electrochemical Society* 118, no. 10 (1971): 1547.
42. Lipparoni, F. R., Bonino, F., Panero, S., and Scrosati, B. Electrochemical properties of metal oxides as anode materials for lithium ion batteries. *Ionics* 8 (2002): 177–182.
43. Wood, Sean M., Pham, Codey H., Heller, Adam, and Mullins, C. Buddie. Formation of an electroactive polymer gel film upon lithiation and delithiation of PbSe. *Journal of The Electrochemical Society* 163, no. 8 (2016): A1666.
44. Besenhard, J. O., Hess, M., and Komenda, P. Dimensionally stable Li-alloy electrodes for secondary batteries. *Solid State Ionics* 40 (1990): 525–529.
45. Li, Qing, Xu, Chunyang, Yang, Liting, Pei, Ke, Zhao, Yunhao, Liu, Xianhu, and Che, Renchao. Pb/C composite with spherical Pb nanoparticles encapsulated in carbon microspheres as a high-performance anode for lithium-ion batteries. *ACS Applied Energy Materials* 3 (2020): 7416–7426.
46. Zhang, W., Eperon, G., and Snaith, H. Metal halide perovskites for energy applications. *Nature Energy* 1 (2016): 16048.
48. Chen, Junnian, Zhou, Shasha, Jin, Shengye, Li, Huiqiao, and Zhai, Tianyou. Crystal organometal halide perovskites with promising optoelectronic applications. *Journal of Materials Chemistry C* 4, no. 1 (2016): 11–27.
47. Stoumpos, Constantinos C., and Kanatzidis, Mercouri G. Halide perovskites: poor man's high-performance semiconductors. *Advanced Materials* 28, no. 28 (2016): 5778–5793.
49. Chen, Shan, and Shi, Gaoquan. Two-dimensional materials for halide perovskite-based optoelectronic devices. *Advanced Materials* 29, no. 24 (2017): 1605448.
50. Martos, M., et al. Lead-based systems as suitable anode materials for Li-ion batteries. *Electrochimica Acta* 48 (2003): 615–621.
51. Yu, Xiang-Long, Huang, Li, and Wu, Jiansheng. From a normal insulator to a topological insulator in plumbene. *Physical Review B* 95, no. 12 (2017): 125113.
52. Kim, Kinam, Choi, Jae-Young, Kim, Taek, Cho, Seong-Ho, and Chung, Hyun-Jong. A role for graphene in silicon-based semiconductor devices. *Nature* 479, no. 7373 (2011): 338–344.

53. Ding, Xili, Liu, Haifeng, and Fan, Yubo. Graphene-based materials in regenerative medicine." *Advanced Healthcare Materials* 4, no. 10 (2015): 1451–1468.
54. Tonelli, Fernanda M. P., Goulart, Vânia A. M., Gomes, Katia N., Ladeira, Marina S., Santos, Anderson K., Lorençon, Eudes, Ladeira, Luiz O., and Resende, Rodrigo R. Graphene-based nanomaterials: biological and medical applications and toxicity. *Nanomedicine* 10, no. 15 (2015): 2423–2450.
55. Kumar, Sachin, and Chatterjee, Kaushik. Comprehensive review on the use of graphene-based substrates for regenerative medicine and biomedical devices. *ACS Applied Materials & Interfaces* 8, no. 40 (2016): 26431–26457.
56. Xu, Jiantie, Chen, Yonghua, and Dai, Liming. Efficiently photo-charging lithium-ion battery by perovskite solar cell. *Nature Communications* 6, no. 1 (2015): 8103.
57. Vicente, Nuria, and Garcia-Belmonte, Germà. Methylammonium lead bromide perovskite battery anodes reversibly host high Li-ion concentrations. *The Journal of Physical Chemistry Letters* 8, no. 7 (2017): 1371–1374.
58. Dawson, James A., Naylor, Andrew J., Eames, Christopher, Roberts, Matthew, Zhang, Wei, Snaith, Henry J., Bruce, Peter G., and Islam, M. Saiful. Mechanisms of lithium intercalation and conversion processes in organic–inorganic halide perovskites. *ACS Energy Letters* 2, no. 8 (2017): 1818–1824.
59. Kojima, Akihiro, Teshima, Kenjiro, Shirai, Yasuo, and Miyasaka, Tsutomu. Organometal halide perovskites as visible-light sensitizers for photovoltaic cells. *Journal of the American Chemical Society* 131, no. 17 (2009): 6050–6051.
60. Green, Martin A., Emery, Keith, Hishikawa, Yoshihiro, and Warta, Wilhelm. "Dunlop." Solar cell efficiency tables (version 47). *Progress in Photovoltaics: Research and Applications* 24 (2016): 3–11.
61. Leijtens, Tomas, Eperon, Giles E., Noel, Nakita K., Habisreutinger, Severin N., Petrozza, Annamaria, and Snaith, Henry J. Stability of metal halide perovskite solar cells. *Advanced Energy Materials* 5, no. 20 (2015): 1500963.
62. Xiao, Zewen, Du, Ke-Zhao, Meng, Weiwei, Wang, Jianbo, Mitzi, David B., and Yan, Yanfa. Intrinsic instability of Cs2In (I) M (III) X6 (M = Bi, Sb; X = halogen) double perovskites: a combined density functional theory and experimental study. *Journal of the American Chemical Society* 139, no. 17 (2017): 6054–6057.
63. Xiao, Zewen, Meng, Weiwei, Wang, Jianbo, Mitzi, David B., and Yan, Yanfa. Searching for promising new perovskite-based photovoltaic absorbers: the importance of electronic dimensionality. *Materials Horizons* 4, no. 2 (2017): 206–216.
64. Stoumpos, Constantinos C., Cao, Duyen H., Clark, Daniel J., Young, Joshua, Rondinelli, James M., Jang, Joon I., Hupp, Joseph T., and Kanatzidis, Mercouri G.. Ruddlesden–Popper hybrid lead iodide perovskite 2D homologous semiconductors. *Chemistry of Materials* 28, no. 8 (2016): 2852–2867.
65. Tsai, Hsinhan, Nie, Wanyi, Blancon, Jean-Christophe, Stoumpos, Constantinos C., Asadpour, Reza, Harutyunyan, Boris, Neukirch, Amanda J. et al. High-efficiency two-dimensional Ruddlesden–Popper perovskite solar cells. *Nature* 536, no. 7616 (2016): 312–316.
66. Sahli, Florent, et al. Fully textured monolithic perovskite/silicon tandem solar cells with 25.2% power conversion efficiency. *Nature Materials* 17, (2018): 820–826.
67. Ponnada S, Maryam Sadat Kiai, Demudu Babu Gorle, Rapaka, SC Bose. "Recent status and challenges in multifunctional electrocatalysis based on 2D MXenes." *Catalysis Science & Technology* 12, no. 14 (2022): 4413–4441.
68. Zhang, Qianfan, Wang, Yapeng, Seh, Zhi Wei, Fu, Zhongheng, Zhang, Ruifeng, and Cui, Yi. Understanding the anchoring effect of two-dimensional layered materials for lithium–sulfur batteries. *Nano Letters* 15, no. 6 (2015): 3780–3786.
69. Pomerantseva, Ekaterina, and Gogotsi, Yury. Two-dimensional heterostructures for energy storage. *Nature Energy* 2, no. 7 (2017): 1–6.

70. Zhang, Lei, Wu, Bo, Li, Qingfang, and Li, Jingfa. Molecular engineering lithium sulfur battery cathode based on small organic molecules: An ab-initio investigation. *Applied Surface Science* 484 (2019): 1184–1190.

71. Eftekhari, Ali, and Kim, Dong-Won. Cathode materials for lithium–sulfur batteries: a practical perspective. *Journal of Materials Chemistry A* 5, no. 34 (2017): 17734–17776.

72. Son, Dae-Yong, Kim, Seul-Gi, Seo, Ja-Young, Lee, Seon-Hee, Shin, Hyunjung, Lee, Donghwa, and Park, Nam-Gyu. Universal approach toward hysteresis-free perovskite solar cell via defect engineering. *Journal of the American Chemical Society* 140, no. 4 (2018): 1358–1364.

73. Lee, Jin-Wook, Kim, Seul-Gi, Bae, Sang-Hoon, Lee, Do-Kyoung, Lin, Oliver, Yang, Yang, and Park, Nam-Gyu. The interplay between trap density and hysteresis in planar heterojunction perovskite solar cells. *Nano Letters* 17, no. 7 (2017): 4270–4276.

74. Ke, Weijun, Xiao, Chuanxiao, Wang, Changlei, Saparov, Bayrammurad, Duan, Hsin-Sheng, Zhao, Dewei, Xiao, Zewen et al. Employing lead thiocyanate additive to reduce the hysteresis and boost the fill factor of planar perovskite solar cells. *Advanced Materials* 28, no. 26 (2016): 5214–5221.

75. Ahmad, Shahab, George, Chandramohan, Beesley, David J., Baumberg, Jeremy J., and De Volder, Michael. Photo-rechargeable organo-halide perovskite batteries. *Nano Letters* 18, no. 3 (2018): 1856–1862.

76. Dawson, J. A., and Robertson, J. Improved calculation of Li and Na intercalation properties in anatase, rutile, and TiO2 (B). *The Journal of Physical Chemistry C* 120, no. 40 (2016): 22910–22917.

77. Ramirez, Daniel, Suto, Yusaku, Rosero-Navarro, Nataly Carolina, Miura, Akira, Tadanaga, Kiyoharu, and Jaramillo, Franklin. Structural and electrochemical evaluation of three- and two-dimensional organohalide perovskites and their influence on the reversibility of lithium intercalation. *Inorganic Chemistry* 57 (2018): 4181–4188.

78. Fujii, Yuta, Ramirez, Daniel, Rosero-Navarro, Nataly Carolina, Jullian, Domingo, Miura, Akira, Jaramillo, Franklin, and Tadanaga, Kiyoharu. Two-dimensional hybrid halide perovskite as electrode materials for all-solid-state lithium secondary batteries based on sulfide solid electrolytes. *ACS Applied Energy Materials* 2 (2019): 6569–6576.

79. He, Mu, Zhang, Lei, and Li, Jingfa. Theoretical investigation on interactions between lithium ions and two-dimensional halide perovskite for solar-rechargeable batteries. *Applied Surface Science* 541 (2021): 148509.

80. Koha, Teck Ming, Shanmugam, Vignesh, Guoc, Xintong, Limc, Swee Sien, Filonike, Oliver, Herzige, Eva M., Müller-Buschbaumg, Peter, Swamy, Varghese, Chiend, Sum Tze, Mhaisalkara, Subodh G., and Mathewsa, Nripan. Enhanced moisture tolerance in efficient hybrid 3D/2D perovskite photovoltaics. *Journal of Materials Chemistry A* (2017). no. 5 (2018): 2122–2128.

81. Cao, Duyen H., Stoumpos, Constantinos C., Farha, Omar K., Hupp, Joseph T., and Kanatzidis, Mercouri G. 2D homologous perovskites as light-absorbing materials for solar cell applications. *Journal of the American Chemical Society* 137 (2015): 7843–7850.

82. Wang, Junxiong, et al. A green strategy to synthesize two-dimensional lead halide perovskite via direct recovery of spent lead-acid battery. *Resources, Conservation and Recycling* 169 (2021): 105463.

83. Chan, Shun-Hsiang, Chang, Yin-Hsuan, Jao, Meng-Huan, Hsiao, Kai-Chi, Lee, Kun-Mu, Lai, Chao-Sung, and Wu, Ming-Chung. High efficiency quasi-2D/3D Pb–Ba perovskite solar cells via phenethylammonium chloride addition. *Solar RRL* 6 (2022): 2101098.

84. Uchiyama, Hiroaki, Hosono, Eiji, Honma, Itaru, Zhou, Haoshen, and Imai, Hiroaki. A nanoscale meshed electrode of single-crystalline SnO for lithium-ion rechargeable batteries. *Electrochemistry Communications* 10, no. 1 (2008): 52–55.

85. Park, Min-Gu, Lee, Dong-Hun, Jung, Heechul, Choi, Jeong-Hee, and Park, Cheol-Min. Sn-based nanocomposite for Li-ion battery anode with high energy density, rate capability, and reversibility. *ACS Nano* 12, no. 3 (2018): 2955–2967.

86. Park, Jae-Wan, and Park, Cheol-Min. A fundamental understanding of Li insertion/extraction behaviors in SnO and SnO₂. *Journal of The Electrochemical Society* 162, no. 14 (2015): A2811.

87. Lee, Kyu T., Jung, Yoon S., and Oh, Seung M. Synthesis of tin-encapsulated spherical hollow carbon for anode material in lithium secondary batteries. *Journal of the American Chemical Society* 125, no. 19 (2003): 5652–5653.

88. Seo, Jung-wook, Jang, Jung-tak, Park, Seung-won, Kim, Chunjoong, Park, Byungwoo, and Cheon, Jinwoo. Two-dimensional SnS2 nanoplates with extraordinary high discharge capacity for lithium ion batteries. *Advanced Materials* 20, no. 22 (2008): 4269–4273.

89. Luo, Bin, Wang, Bin, Li, Xianglong, Jia, Yuying, Liang, Minghui, and Zhi, Linjie. Graphene-confined Sn nanosheets with enhanced lithium storage capability. *Advanced Materials* 24, no. 26 (2012): 3538–3543.

90. Mei, Lin, Xu, Cheng, Yang, Ting, Ma, Jianmin, Chen, Libao, Li, Qiuhong, and Wang, Taihong. Superior electrochemical performance of ultrasmall SnS2 nanocrystals decorated on flexible RGO in lithium-ion batteries. *Journal of Materials Chemistry A* 1, no. 30 (2013): 8658–8664.

91. Liu, Shuangyu, Lu, Xiang, Xie, Jian, Cao, Gaoshao, Zhu, Tiejun, and Zhao, Xinbing. Preferential c-axis orientation of ultrathin SnS2 nanoplates on graphene as high-performance anode for Li-ion batteries. *ACS Applied Materials & Interfaces* 5, no. 5 (2013): 1588–1595.

92. Zhang, Jingjing, Liang, Jianwen, Zhu, Yongchun, Wei, Denghu, Fan, Long, and Qian, Yitai. Synthesis of Co2SnO4 hollow cubes encapsulated in graphene as high capacity anode materials for lithium-ion batteries. *Journal of Materials Chemistry A* 2, no. 8 (2014): 2728–2734.

93. Chen, Yu, Song, Bohang, Chen, Rebecca Meiting, Lu, Li, and Xue, Junmin. A study of the superior electrochemical performance of 3 nm SnO2 nanoparticles supported by graphene. *Journal of Materials Chemistry A* 2, no. 16 (2014): 5688–5695.

94. Li, Shuankui, Zheng, Jiaxin, Zuo, Shiyong, Wu, Zhiguo, Yan, Pengxun, and Pan, Feng. 2D hybrid anode based on SnS nanosheet bonded with graphene to enhance electrochemical performance for lithium-ion batteries. *RSC Advances* 5, no. 58 (2015): 46941–46946.

95. Luo, Bin, Qiu, Tengfei, Ye, Delai, Wang, Lianzhou, and Zhi, Linjie. Tin nanoparticles encapsulated in graphene backboned carbonaceous foams as high-performance anodes for lithium-ion and sodium-ion storage. *Nano Energy* 22 (2016): 232–240.

96. Zhou Dan, Wei-Li Song, Li, Xiaogang, Fan, Li-Zhen, and Deng, Yonghong. Tin nanoparticles embedded in porous N-doped graphene-like carbon network as high-performance anode material for lithium-ion batteries. *Journal of Alloys and Compounds* 699 (2017): 730–737.

97. Wu, Che-Ya, Yang, Hao, Wu, Cheng-Yu, and Duh, Jenq-Gong. Flower-like structure of SnS with N-doped carbon via polymer additive for lithium-ion battery and sodium-ion battery. *Journal of Alloys and Compounds* 750 (2018): 23e32.

98. Chen, Hongwen, Jia, Bei-Er, Lu, Xinsheng, Guo, Yichuan, Hu, Rui, Khatoon, Rabia, Jiao, Lei, Leng, Jianxing, Zhang, Liqiang, and Lu, Jianguo. Two-dimensional SnSe2/

CNTs hybrid nanostructures as anode materials for high-performance lithium-ion batteries. *Chemistry – A European Journal* 25, no. 42 (2019): 9973–9983.

99. Yao, Weiqi, Wu, Shengbo, Zhan, Liang, and Wang, Yanli. Two-dimensional porous carbon-coated sandwich-like mesoporous SnO2/graphene/mesoporous SnO2 nanosheets towards high-rate and long cycle life lithium-ion batteries. *Chemical Engineering Journal* 361 (2019): 329–341.

100. Tian, Qinghua, Zhang, Feng, and Yang, Li. Fabricating thin two-dimensional hollow tin dioxide/carbon nanocomposite for high-performance lithium-ion battery anode. *Applied Surface Science* 481 (2019): 1377–1384.

101. Davitt, Fionán, Stokes, Killian, Collins, Timothy W., Roldan-Gutierrez, Manuel, Robinson, Fred, Geaney, Hugh, Biswas, Subhajit et al. Two-dimensional SnSe nanonetworks: growth and evaluation for Li-Ion battery applications. *ACS Applied Energy Materials* 3, no. 7 (2020): 6602–6610.

102. Khabazian, Siavash, Sanjabi, S., and Tonti, Dino. Electrochemical growth of two-dimensional tin nano-platelet as high-performance anode material in lithium-ion batteries. *Journal of Industrial and Engineering Chemistry* 84 (2020): 120–130.

103. Jiang, Pinxian, Jing, Jialun, Wang, Yizhe, Li, Hongju, He, Xiaoying, Chen, Yungui, and Liu, Wei. Facilely transforming bulk materials to SnO/pristine graphene 2D-2D heterostructures for stable and fast lithium storage. *Journal of Alloys and Compounds* 812 (2020): 152114.

104. Zhanga, Junjun, Caob, Dongwei, Wua, Yang, Chenga, Xialan, Kangb, Wenpei, and Xua, Jun. Phase transformation and sulfur vacancy modulation of 2D layered tin sulfide nanoplates as highly durable anodes for pseudocapacitive lithium storage. *Chemical Engineering Journal* 392 (2020): 123722.

105. Norton, Kane J., Alam, Firoz, and Lewis, David J. A review of the synthesis, properties, and applications of bulk and two-dimensional tin (II) sulfide (SnS). *Applied Sciences* 11, no. 5 (2021): 2062.

106. Aifantis, Katerina E., Hackney, S. A., and Dempsey, J. P. Design criteria for nanostructured Li-ion batteries. *Journal of Power Sources* 165, no. 2 (2007): 874–879.

107. Aifantis, K. E., and Dempsey, J. P.. Stable crack growth in nanostructured Li-batteries. *Journal of Power Sources* 143, no. 1–2 (2005): 203–211.

108. Aifantis, K. E., Dempsey, J. P., and Hackney, S. A.. Cracking in Si-based anodes for Li-ion batteries. *Reviews on Advanced Materials Science* 10 (2005): 403–408.

109. Shan, Yuying, Li, Yan, and Pang, Huan. Applications of tin sulfide-based materials in lithium-ion batteries and sodium-ion batteries. *Advanced Functional Materials* 30, no. 23 (2020): 2001298.

110. Huang, Yajie, Li, Liangliang, Lin, Yuan-Hua, and Nan, Ce-Wen. Liquid exfoliation few-layer SnSe nanosheets with tunable band gap. *The Journal of Physical Chemistry C* 121, no. 32 (2017): 17530–17537.

111. Nicolosi, Valeria, Chhowalla, Manish, Kanatzidis, Mercouri G., Strano, Michael S., and Coleman, Jonathan N. Liquid exfoliation of layered materials. *Science* 340, no. 6139 (2013): 1226419.

112. Chen, Hongwen, Liu, Rumin, Wu, Yang, Cao, Junhui, Chen, Jian, Hou, Yang, Guo, Yichuan et al. Interface coupling 2D/2D SnSe2/graphene heterostructure as long-cycle anode for all-climate lithium-ion battery. *Chemical Engineering Journal* 407 (2021): 126973.

113. Gao, Runsheng, Tang, Jie, Tang, Shuai, Zhang, Kun, Ozawa, Kiyoshi, and Qin, Lu-Chang. Tin-cobalt bimetals in 2D leaf-like MOF-derived carbon for advanced lithium storage applications. *Electrochimica Acta* 410 (2022): 140036.

114. *Perovskite Solar Cells.* (n.d.). Energy.gov. https://www.energy.gov/eere/solar/perovskite-solar-cells.

115. Bati, Abdulaziz S. R., Zhong, Yu Lin, Burn, Paul L., Nazeeruddin, Mohammad Khaja., Shaw, Paul E., and Batmunkh, Munkhbayar. Next-generation applications for integrated perovskite solar cells. *Communications Materials* 4, no. 1 (2023): 2.

116. Mao, Lingling, Tsai, Hsinhan, Nie, Wanyi, Ma, Lin, Im, Jino, Stoumpos, Constantinos C., Malliakas, Christos D. et al. Role of organic counterion in lead- and tin-based two-dimensional semiconducting iodide perovskites and application in planar solar cells. *Chemistry of Materials* 28, no. 21 (2016): 7781–7792.

117. Chen, Ziming, Liu, Meiyue, Li, Zhenchao, Shi, Tingting, Yang, Yongchao, Yip, Hin-Lap, and Cao, Yong. Stable Sn/Pb-based perovskite solar cells with a coherent 2D/3D interface. *iScience* 9 (2018): 337–346.

118. Chen, Yani, Sun, Yong, Peng, Jiajun, Chábera, Pavel, Honarfar, Alireza, Zheng, Kaibo, and Liang, Ziqi. Composition engineering in two-dimensional Pb–Sn-alloyed perovskites for efficient and stable solar cells. *ACS Applied Materials & Interfaces* 10, no. 25 (2018): 21343–21348.

119. Li, Chaohui, Pan, Yamin, Hu, Jinlong, Qiu, Shudi, Zhang, Cuiling, Yang, Yuzhao, Chen, Shi et al. Vertically aligned 2D/3D Pb–Sn perovskites with enhanced charge extraction and suppressed phase segregation for efficient printable solar cells. *ACS Energy Letters* 5, no. 5 (2020): 1386–1395.

120. Tong, Jinhui, Song, Zhaoning, Kim, Dong Hoe, Chen, Xihan, Chen, Cong, Palmstrom, Axel F., Ndione, Paul F. et al. Carrier lifetimes of >1 μs in Sn-Pb perovskites enable efficient all-perovskite tandem solar cells. *Science* 364, no. 6439 (2019): 475–479.

121. Lu, Zhili, Li, Chaohui, Lai, Hongwei, Zhou, Xinming, Wang, Chunfeng, Liu, Xianhu, Guo, Fei, and Pan, Caofeng. Mixed 2D-Dion—Jacobson/3D Sn-Pb alloyed perovskites for efficient photovoltaic solar devices. *Nano Research* 16, no. 2 (2023): 3142–3148.

122. Novoselov, K. S., et al. Electric field effect in atomically thin carbon films. *Science* 306, no. 5696 (2004): 666–669. doi: 10.1126/science.1102896.

123. Radisavljevic, B., Radenovic, A., Brivio, J. Single-layer MoS$_2$ transistors. *Nature Nanotechnology* 6 (2011): 147–150. 10.1038/nnano.2010.279

124. Li, Junze, et al. Fabrication of single phase 2D homologous perovskite microplates by mechanical exfoliation. *2D Materials* 5, no. 2 (2018): 021001.

125. Li, Yangang, et al. Recent progress on the mechanical exfoliation of 2D transition metal dichalcogenides. *Materials Research Express* (2022).

126. Wang, Yichao, et al. Piezoelectric responses of mechanically exfoliated two-dimensional SnS2 nanosheets. *ACS Applied Materials & Interfaces* 12, no. 46 (2020): 51662–51668.

127. Novoselov, Konstantin S., et al. A roadmap for graphene. *Nature* 490, no. 7419 (2012): 192–200.

128. Coleman, Jonathan N., et al. Two-dimensional nanosheets produced by liquid exfoliation of layered materials. *Science* 331, no. 6017 (2011): 568–571.

129. Tahir, Muhammad Bilal, et al. Photocatalytic nanomaterials for degradation of organic pollutants and heavy metals. *Nanotechnology and Photocatalysis for Environmental Applications*. Elsevier (2020): 119–138.

130. Xing, Chenyang, et al. Two-dimensional lead monoxide: facile liquid phase exfoliation, excellent photoresponse performance, and theoretical investigation. *ACS Photonics* 5, no. 12 (2018): 5055–5067.

131. Cai, Xingke et al. Preparation of 2D material dispersions and their applications. *Chemical Society Reviews* 47, no. 16 (2018): 6224–6266.

132. Kim, Eun-Bi, et al. A review on two-dimensional (2D) and 2D-3D multi-dimensional perovskite solar cells: Perovskites structures, stability, and photovoltaic performances. *Journal of Photochemistry and Photobiology C: Photochemistry Reviews* 48 (2021): 100405.

133. Park, Chibeom, et al. Position-selective solution phase growth of fullerene crystals. *Carbon* 145 (2019): 31–37.
134. Ha, Son-Tung, et al. Metal halide perovskite nanomaterials: synthesis and applications. *Chemical Science* 8, no. 4 (2017): 2522–2536.
135. Chen, Junnian, et al. A ternary solvent method for large-sized two-dimensional perovskites. *Angewandte Chemie* 129, no. 9 (2017): 2430–2434.
136. Nam, D.-H., Kim, J. W., Lee, J.-H., Lee, S.-Y., Shin, H.-A. S., Lee, S.-H., and Joo, Y.-C. Tunable Sn structures in porosity-controlled carbon nanofibers for all-solid-state lithium-ion battery anodes. *Journal of Materials Chemistry A* 3 (2015): 11021–11030.

6 Two-Dimensional Materials in Metal-Air Batteries

Sankar Devi Vaithiyanathan, Tirupathi Rao Penki, and Krishna Harika Villa

6.1 INTRODUCTION

6.1.1 METAL-AIR BATTERIES

The contemporary global energy demand represents a growing concern, in conjunction with the depletion of conventional energy sources. In order to end this, the utilization of renewable and sustainable energy resources, such as solar, wind, and tidal energies, can provide viable solutions. Nevertheless, renewable energy production poses an intermittency challenge, requiring the development of advanced energy storage and conversion technologies to enable consistent utilization.[1,2] In this regard, efforts have been made to discover energy storage systems that are effective, dependable, economical, and durable. Electrochemical energy storage devices encompass a variety of systems, including metal-ion batteries, metal-air batteries, supercapacitors, and fuel cells.[1,2] Li-ion batteries are widely regarded as a significant and commercially accessible technology among all secondary batteries, owing to their exceptional energy density ($\sim$300 Wh kg^{-1}).[1–4] Considering the pressing demand for extended-range electric vehicles and portable electronic devices, metal-air batteries are projected to emerge as a groundbreaking substitute for the current lithium-ion batteries. This is mainly due to its high theoretical energy densities of 1300–1100 Wh kg^{-1} higher than the current Li-ion theoretical energy density of 450 Wh kg^{-1} and their ability to operate even in an open atmosphere has made them more attractive to the manufacturers. Various types of metal-air batteries are available, including Li-air, Zn-air, Al-air, and Fe-air.[5–7] Among them, Li-air and Zn-air batteries have gained more attention because of their better capacity and energy density.[8–10] As shown in Figure 6.1(a), a metal-air battery is composed of pure metal (Ca, Mg, Fe, Al, Li, and Zn) as an anode, (quasi-) solid-state electrolyte, separator, and oxygen as a cathode to provide an inexhaustible air supply. When assembled into a full battery, suitable current collectors (some electrodes also function as current collectors) and flexible packaging materials are also required. Specifically, a typical air cathode includes a current collector or gas diffusion layer (GDL), a hydrophobic protecting layer, a catalyst layer (2D materials shown in Figure 6.1 (a)), and binders. As tabulated in

DOI: 10.1201/9781003404729-6

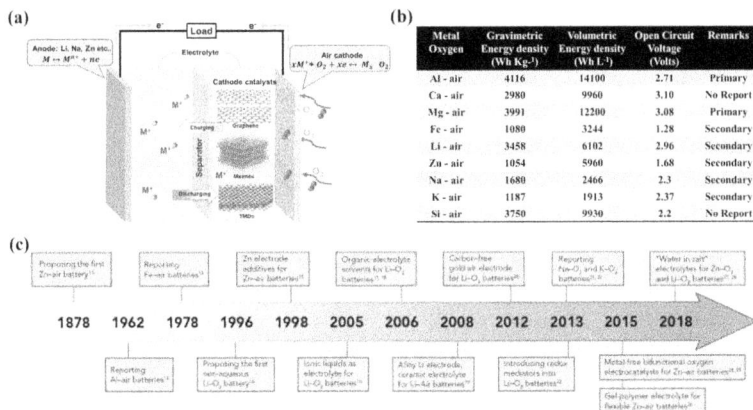

FIGURE 6.1 (a) Schematic representation 2D materials for metal-air batteries, (b) energy densities of different metal-air batteries,[8] and (c) a brief timeline of research progress in metal-air batteries.[9]

Figure 6.1(b), a calcium-air battery (Ca-air) operates at a high theoretical voltage of 3.10 V and has a volumetric and gravimetric energy density of 2980 Wh kg^{-1} and 9960 Wh L^{-1}, respectively. Despite their benefits, Ca-O$_2$ batteries are extremely expensive and their capacity gradually degrades. As a result, Ca is used in electrodes as a metal alloy to increase efficiency. Similarly, magnesium-air (Mg-O$_2$) batteries have an impressive theoretical voltage (3.08 V) and energy density of 3991 Wh kg^{-1} and 12200 Wh L^{-1}, as well as the fact that magnesium metal is inexpensive, lightweight, harmless, and readily available on both land and in the sea. While Mg-O$_2$ batteries are primarily used in underwater vehicles, they have the disadvantage of having a high self-discharge and heat generation, which limits their use in commercial applications. Iron-air (Fe-air) batteries have a very low specific energy of approximately 763 Wh kg^{-1} and a theoretical potential of approximately 1.35 V; they are cost-effective and have a long electrical service life cycle. Additionally, it can be used with fuel cells where hydrogen generated during iron oxidation is consumed by the fuel cell, while in reverse operation, hydrogen generated during oxidation is used for reducing iron oxide to metallic iron. Lithium-air batteries (Li-air) have a high theoretical specific energy of 11140 Wh kg^{-1} (excluding air), which is nearly 100 times superior to other battery systems while petrol has 10150 Wh kg^{-1} specific energy. As can be seen from Table in Figure 6.1 (b), the practical values are about half of that, approximately 3500 Wh kg^{-1}. However, due to a safety concern regarding the risk of flames resulting from Li's high reactivity with air and humidity, its applications were limited.[8] Although the research on metal-air batteries was launched quite earlier than Li-ion batteries, it is still in the nascent stages of research and development. Metal-air batteries have the potential to hold several times the energy (ten times in the case of Li) and are much lighter in weight compared with metal-ion batteries. The primary zinc-air battery was introduced by Maiche dating back to 1878 and its products were commercially available in 1932. Later on, aqueous Fe-air, Al-air, and Mg-air batteries were developed in the 1960s followed by the first emergence of non-aqueous metal-air batteries like Li-air, Na-air, Mg-air and K-air about two decades ago.[11–17] Research work on Si-air batteries started after 2009.[18] Each metal (Ca, Al, Zn, Mg, Li, and Si) has its own set of benefits and drawbacks when it comes to being used as anode electrodes. Al-, Zn-, Si-, and Fe–air batteries are also the research hotspots because of economic and safety considerations.

6.1.2 Challenges of Metal-Air Batteries

In essence, a standard metal-air battery comprises metal (Li, Zn, Al, Fe, Mg, Ca, and Si) employed as the anode, while oxygen (O_2) derived from the surrounding atmosphere serves as the cathode and an organic solvent consisting of metal salt as shown in Figure 6.1(a). The operational mechanism of the rechargeable metal-air batteries involves the oxidation and reduction of metal at the anode, leading to stripping and plating, while the cathode undergoes the oxygen evolution reaction (OER) and oxygen reduction reactions (ORR) during the discharge and charge cycles, respectively.[5–7,19] Throughout the discharge process of metal-air batteries, the chemical reaction involves the oxidation of the metal resulting in the formation of both metal ion and e^-. Following this reaction, the metal ions proceed to migrate from the anode to the cathode region, where they ultimately react with O_2. This reaction leads to the formation of either metal oxides or metal peroxides, which serves as the resulting discharge product. However, in the charging process, an inverse occurrence takes place wherein metal ion migrates toward the anode and is then electroplated as lithium with the release of O_2. During the process, metal-air batteries face several challenges, such as the presence of O_2 and electrolyte impurities yield side reactions (e.g. redox reactions that produce reduced oxygen species such as superoxides) on the metal electrode, which usually result in the corrosion of the metal anode by the electrolyte, electrolyte degradation, or blockage of cathode reaction sites because of electrolyte decomposition. Consequently, metal electrodes suffer from metal dendrite growth, shape change, corrosion, and surface passivation. Despite possessing a high theoretical specific energy for metal-air batteries, achieving its practical capacity is extremely difficult due to unstable electrolytes, poor cyclability, high overpotential for ORR/OER, and sluggish kinetics of the ORR/OER. In order to enhance the kinetics of surface oxygen reactions, developing a proficient bifunctional electrocatalyst is crucial. According to this perspective, numerous types of electrocatalysts have been explored, including precious metals and metal oxides. As of present, the most commonly employed catalysts for benchmarking OER and ORR consist of noble metal oxides. Nonetheless, the widespread use of these catalysts is limited due to their exorbitant cost and limited availability. Therefore, designing and researching ORR/OER catalysts that are cost-effective, environmentally friendly, and incredibly efficient are of high interest and necessity.

Synthetic catalysts based on two-dimensional (2D) nanostructures and carbon-based materials, like doped graphene, transition metal dichalcogenides (TMDCs), and transition metal carbides, nitrides, and carbonitrides, have been extensively examined for the ORR and OER. In fact, the use of innovative catalysts created from 2D materials has already shown promising results by the enhancement of the overall efficiency and charging capacity in the case of metal-air batteries and could be further optimized by employing more advanced catalysts derived from 2D materials and for other metal-air battery systems. Catalysts play a vital role in enhancing the speed of chemical reactions within batteries while the catalyst's composition greatly increases the capacity of the battery to store and release energy. This chapter provides detailed mechanisms, classification, and fundamental prerequisites for ORR/OER electrocatalysts, along with their role with respect to 2D materials are extensively discussed.

6.1.3 MECHANISM OF ORR/OER

The reduction of O_2 in an alkaline solution can be divided into two electrons and a four-electron process as shown in Figure 6.2. Figure 6.2(a) explains ORR process, where two-electron processes are characterized by the breakage of O-O bonds only after the formation of hydrogen peroxide molecules ($M + O_2 + H_2O + 2e^- \rightarrow MO_2H^- + OH^-$). Unlike the two-electron reaction, the four-electron reaction is characterized by the first adsorption of O_2 molecules during the reaction and the subsequent dissociation into adsorbed oxygen atoms M–O after adsorption ($2 M + O_2 \rightarrow 2 MO$). There is usually a restriction on electrocatalytic ORR imposed by O_2 reduction and OH* reduction. The activation barrier of OOH* and OH* intermediates directly affects the rate of the ORR reaction. Figure 6.2(b) shows two approaches that are generally taken as part of the OER process: the approach involves forming the O_2 molecules as a result of combining two M–Os ($2MO \rightarrow 2M + O_2$). In contrast, another approach generates MOOH intermediates that break down to produce oxygen ($MOOH + OH^- \rightarrow M + O_2 + H_2O + e^-$). Surface metal cations are considered the active site for OER, and the number of d-electrons in

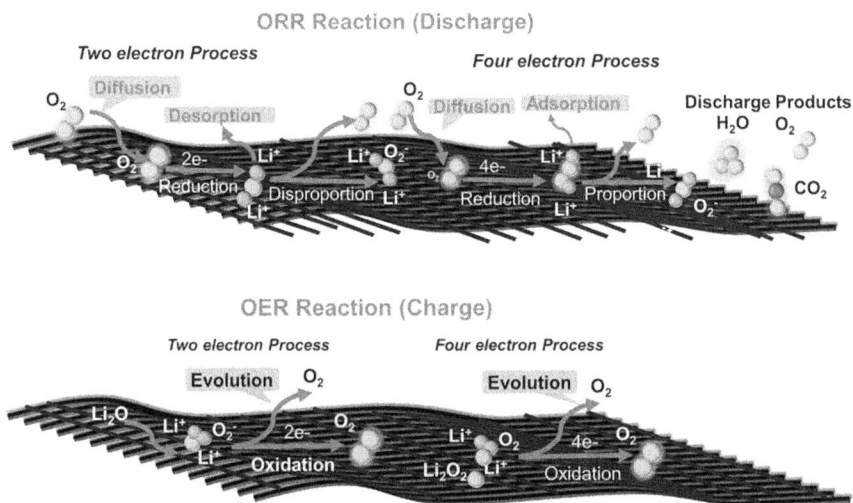

FIGURE 6.2 Schematic of oxygen diffusion, reduction, and formation of discharge products along with other side reactions.

their orbits strongly affects OER activity. Furthermore, the strength of the M–O bond and the activation barrier between intermediates determine the overall rate of OER reaction. Generally, the OER and ORR processes produce intermediates such as per-oxygen or oxygen bound to the catalyst surface. In each step, the reaction rate is affected by the relative stability of these intermediates and the activation barrier between them. This determines the total OER/ORR rate. It is important to note that the properties of these intermediates are affected by the catalyst material and reaction conditions (such as pH).

6.1.4 REQUIREMENTS FOR ORR/OER ELECTROCATALYST IN METAL-AIR BATTERIES

Several factors need to be considered when designing an ORR/OER electrocatalyst. The three most critical features of a superior catalyst are its catalytic activity, electronic conductivity, and porosity. Due to the slow reaction rate of ORR, fuel cell/metal-air batteries also have limited performance. The reaction occurs primarily at the triple-phase boundary (TPB), where the solid electrode is linked to the liquid electrolyte and gaseous oxygen from the atmosphere. An air electrode consists of three layers in real devices: a GDL, a conductive current collector, and an oxygen electrocatalyst layer. The GDL faces an outer atmosphere, and the catalyst layer meets an internal liquid electrolyte. Since all of the components of the air electrode have distinct and equal roles, it is rational to modify its overall structure for a robust structural design that can provide effective active sites for ORR and OER. Furthermore, some major modifications of the air electrode are described to overcome the difficulty of deciphering the electrode architecture responsible for device performance. Heteroatom doping via improved oxygen chemisorption, electron transfer, and defect structures aids in charge transferability by increasing the number of low-coordinated sites that can act as active sites. ORR electrocatalysts should meet the following requirements as shown in the pictorial representation in Figure 6.3.

- High catalytic activity,
- Good electrical conductivity,
- Excellent wettability with electrolytes,
- Acceleration of ORR and OER,

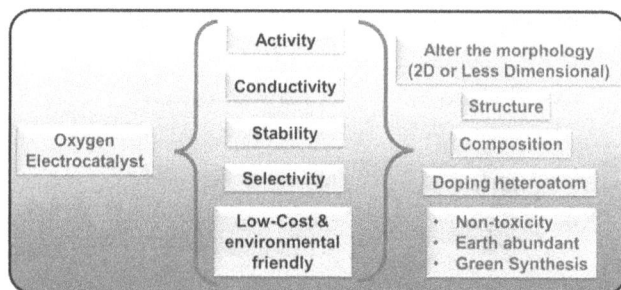

FIGURE 6.3   Pictorial representation of requirements for ORR and OER electrocatalysts.

- Outstanding electrochemical stability (does not oxidize at high electrode potentials),
- Excellent chemical stability (insoluble in acidic and basic aqueous solutions; not oxidized with protons or oxygen),
- A structure with optimal composition, constructive morphology, large specific surface areas (SSA), high porosity, small particle sizes, and uniform catalytic particle distribution on the support,
- Cohesive surface-catalyst bonding.

It is impossible for a single electrocatalyst to meet all of these requirements simultaneously. Some tradeoffs and tolerances will be based on application type and conditions. Designing and applying the most appropriate catalyst are vital to increasing the efficiency of OER and ORR reactions. Below are several bifunctional catalyst materials.

6.1.5 CLASSIFICATION OF ELECTROCATALYSTS FOR ORR/OER

The electrode performance of metal-air batteries is determined by the electrocatalyst in the air electrode, as the cathode reaction is generally catalytic electrochemical in nature. Over the past few decades, ORR catalysts have been extensively studied in fuel cell technologies, which require oxygen reduction at the cathode. In fact, the cathode catalyst is largely responsible for fuel cell cost and determines energy conversion efficiency. Most catalytic materials found in fuel cells can also serve as cathodes in metal-air batteries. Similarly, the techniques and strategies used in fuel cells can also be applied to improve cathode efficiency in metal-air batteries. Materials ranging from noble metals and metal alloys to carbons, transition-metal oxides, and metal macrocyclic compounds have been utilized as cathode catalysts.

  i. Noble metals and alloys
 ii. Carbonaceous materials
iii. Transition-metal oxides
 iv. Inorganic–organic composites

6.2 2D MATERIALS AS ELECTROCATALYSTS FOR METAL-AIR BATTERIES

Many unique properties, such as their controllable structure, ultrathin thickness, electrical/thermal/ionic conductivity, hydrophilicity/hydrophobicity, and defect sites, make 2D materials ideal for use in metal-air batteries. In particular, the high surface area of 2D materials enhances the functionality to promote electrocatalytic reactions and lower overpotentials during the charge and discharge processe. This promotes metal-air batteries' round-trip efficiency. A new era of materials research has opened up with the demonstration of graphene, and extensive effort has been made to study novel 2D materials. A variety of 2D materials have been developed following graphene's footprints, including mono elemental analogs of graphene (MEAs), TMDCs, and transition metal carbides, nitrides, and carbonitrides (MXenes) as

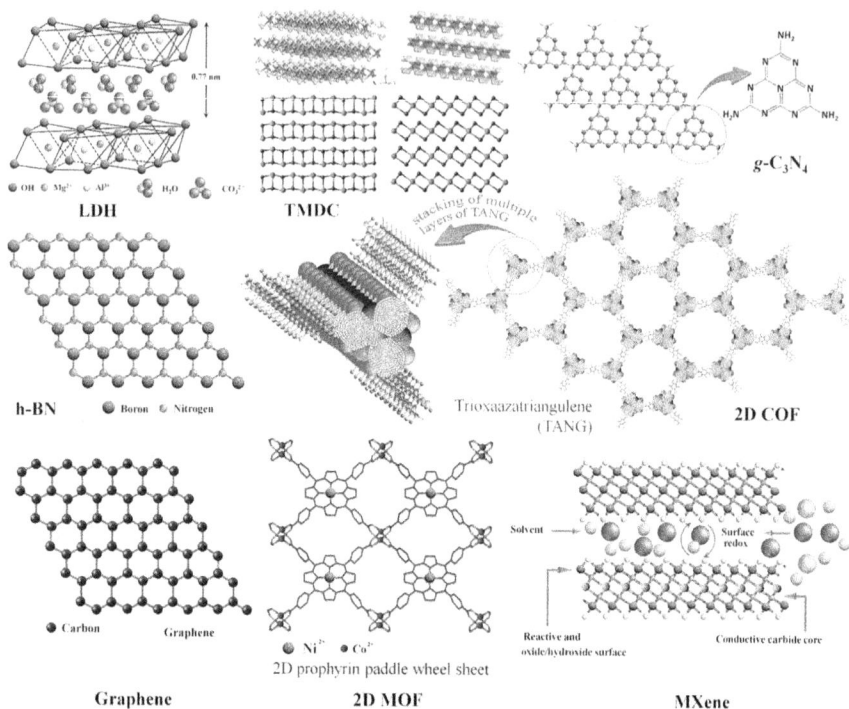

FIGURE 6.4 Schematic of typical 2D materials applied in energy storage devices.[20]

shown in Figure 6.4.[20] Several MEAs have been identified to date, including silicene, germanene, stanene, phosphorene, arsenene, and antimonene, all of which have 2D structures similar to graphene. A TMDC consists of MX_2, where M is a transition metal atom (such as Mo, V, Nb, or W) and X is a chalcogen atom (such as S, Se, or Te).[20,21] Unlike graphene, 2D TMDCs have direct band gaps. Featuring strong spin-orbit couplings and favorable electronic and mechanical properties, 2D TMDCs are potential candidates for high-end electronics, spintronics, and opto-electronics.[20–23] The literature reports that Salehi-Khojin and his colleagues synthesized 15 different types of 2D TMDCs, capable of serving as catalysts. Several of their 2D materials were found to possess up to ten times the energy capacity of lithium-air batteries when added as catalysts, compared with batteries containing traditional catalysts.[24] As reported by Wang et al., 2D bimetallic Co_3FeN nanosheets have been found to enhance the electrocatalytic activity of Zn-air batteries.[25] Thus, the Co_3FeN-based 2D material showed a maximum power density of 108 mW cm^{-2} and a better life cycle of 900 cycles. Furthermore, Xia et al. constructed Zn-air batteries with a higher power density (223 mW cm^{-2}) and a longer life cycle using metal-organic framework-derived 2D nitrogen-doped carbon nanotubes/graphene hybrid electrocatalysts.[26] Normally, MXenes are prepared by selectively removing the "A" elements from the MAX phase, which is stoichiometrically $M_{n+1}AX_n$, where M is an early transition metal, A is an element from groups 12–16, and X is carbon or nitrogen.[27] The relatively strong bonds that bind M and A in the MAX phase are replaced by weaker hydrogen bonds with OH, O, and F in MXenes. This is done by etching and exfoliation. Several outstanding properties of MXenes have been discovered since their discovery in 2011, including excellent electrical conductivity, hydrophilic properties, good thermal stability, and large interlayer spacing. These features make MXenes promising for electrochemical energy storage.[28]

6.3 EXPERIMENTAL AND CASE STUDIES OF 2D MATERIALS FOR METAL-AIR BATTERIES

6.3.1 PREPARATION METHODS OF 2D MATERIALS

As a result of the unique properties and structures of 2D materials, they have attracted considerable attention, which has led to the rapid development of complementary preparation techniques. 2D materials have been prepared using a variety of approaches, which can be categorized as top-down and bottom-up approaches. Top-down methods typically involve exfoliation, including mechanical exfoliation, intercalation exfoliation, and etching exfoliation.

In the case of bottom-up methods, chemical vapor deposition (CVD) and epitaxial growth are commonly employed. A detailed description of the reported preparation methods for graphene, MEAs, TMDs, and MXenes is presented in this section.

6.3.1.1 Exfoliation-Related Methods

Exfoliation methods weaken the interaction between different layers. Exfoliation methods can be classified into mechanical exfoliation, intercalation exfoliation, and etching exfoliation based on their exfoliation mechanisms as shown in Figure 6.5.[29]

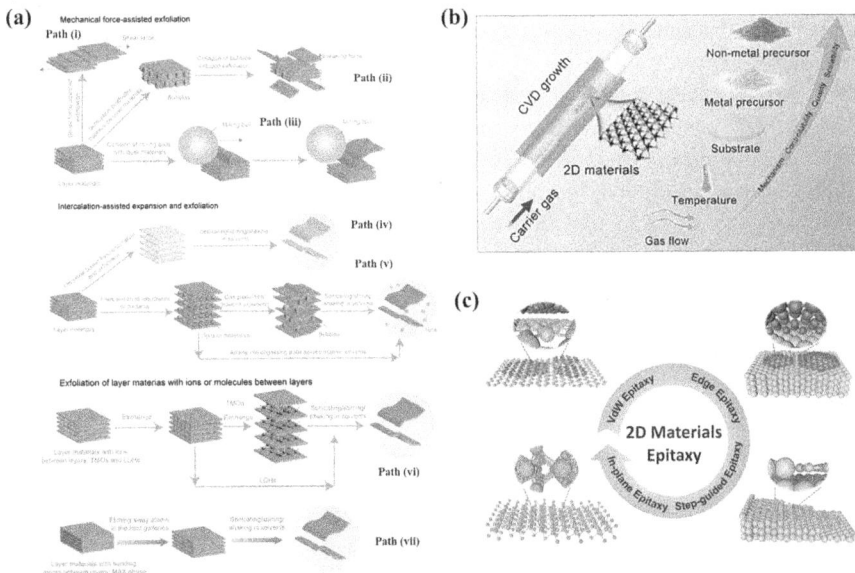

FIGURE 6.5 Schematic illustrations of (a) exfoliation methods,[29] (b) CVD growth,[30] and (c) epitaxial growth mode for 2D materials.[31]

Mechanical exfoliation is accomplished by applying an extra shear force to break the interlayer interaction (Figure 6.5 path (i)). One example of a 2D material prepared through this method is graphene, which was produced first by mechanically exfoliating highly oriented pyrolytic graphite with Scotch tape. Furthermore, single-layer MoS_2 was prepared using scotch tape-based micromechanical exfoliation. A transistor made from MoS_2 single layers demonstrated an on/off ratio of 1 × 10(8) at room temperature and an ultralow standby power consumption. High-quality 2D materials can be produced with this method, which can be used to study their inherent properties. The low yield, however, makes practical applications unlikely. As a result, other approaches have been developed to prepare 2D materials over the years such as sonication-assisted exfoliation and shear force-assisted liquid phase exfoliation and ball milling-assisted exfoliation (Figure 6.5 path (ii) and (iii)). Intercalation-assisted expansion and exfoliation are carried out by oxidation and reduction. In intercalation and exfoliation, a strong oxidative agent has been used to intercalate the host gallery of layer materials, thus expanding the interlayer spacing of the material. The material exfoliated into thin flakes or even monolayers via the gas production reaction occurs or an external shear force is applied (Figure 6.5 path (iv)). The best examples are the classical method for the production of oxidized graphene, i.e., GO, which was developed by Hummers and coworkers in 1958 and is still widely used nowadays because this method allows the production of monolayer GO with a high yield at low cost. In case of Reduction-based intercalation and exfoliation, alkali metal atoms or ions are intercalated into the layered materials (Figure 6.5 path (v)). This is due to small size of alkali metal atoms or ions, which makes it easier for them to enter the gallery of the host material which has a small spacing between the layers. By adding alkali metal-intercalated layer materials to water, hydrogen bubbles produced by reactions between the alkali metals and water will expand the interlayer spacing and exfoliate them into 2D materials in solvents. The most-used alkali metal is lithium (Li), which has the smallest size and the highest reactivity among all the alkali metals. A series of layer materials have been exfoliated to 2D materials, including a variety of metal chalcogenides (e.g., MoS_2, WS_2, $NbSe_2$, VSe_2, Bi_2Te_3, WSe_2, Sb_2Se_3, Ta_2NiS_5, and Ta_2NiSe_5), graphite, and h-BN. Exfoliation of layer materials containing ions or molecules between the layers is carried out via ion exchange-assisted intercalation and exfoliation and etching and intercalation-assisted exfoliation (Figure 6.5 path (vi)). Ion exchange-induced intercalation can be used to exfoliate specific types of layer materials such as oxides and hydroxides, which have ions or molecules between the layers. These materials are difficult to exfoliate by the oxidation or reduction-based exfoliation methods, due to their sensitivity to oxidative acids and reductive alkali metals. In addition, the ions between the layers have a strong electrostatic attraction with atoms in the layers, and this is comparable to the in-plane bonding force, making them difficult to exfoliate by mechanical force-assisted methods. Exfoliation of these materials is based on the principle of expanding their interlayer gallery to weaken the interactions between layers. So far, a large number of 2D oxide and LDH have been prepared by this method, such as titanium oxide, manganese oxide, calcium niobate, tungsten oxide, ruthenium oxide, and $M_{1-x}^{2+}M_x^{3+}(OH)_2A_{x/n}^{n-}\cdot yH_2O$ (here $M^{2+} = Mg^{2+}$, Fe^{2+}, Co^{2+}, Ni^{2+},

Zn^{2+}, etc.; $M^{3+} = Al^{3+}$, Fe^{3+}, Co^{3+}, etc.). whereas etching and intercalation-assisted exfoliation has been used to prepare Mxene from the ceramic MAX phase materials (Figure 6.5 path (vii)). These MAX phase materials cannot be exfoliated by the ion exchange process due to very strong bonding between ions and the host layers, some of them being strong covalent bonds. To break the bonds connecting the layers, a strong force or chemical reaction is needed.

6.3.1.2 Chemical Vapor Deposition (CVD) Methods

CVD is a typical bottom-up technique for synthesizing high-quality 2D materials from atom assembly, which is suitable for producing graphene and TMDs. In general, CVD is the process of sulfurizing/selenizing thin films of metals/metal oxides. The CVD process of 2D materials is one of the promising methods because it can grow a series of 2D compound materials with high quality as well as reasonable cost. So far, many efforts have been made in the CVD growth of 2D compound materials (with large domain size, controllable number of layers, fast growth rate, and high-quality features). Therefore, the CVD method has shown much potential for the commercialization of 2D compound materials. However, due to the complicated growth mechanisms like sublimation and diffusion processes of multiple precursors, maintaining the controllability, repeatability, and high quality of CVD-grown 2D compound materials is still a big challenge, which prevents their widespread use.[30] Figure 6.3(b) shows a schematic of the key parameters for the CVD growth of 2D materials. Gao et al. reported the controlled synthesis of 2D TiS_2 via the NH_4Cl-assisted ambient pressure CVD method. The NH_4Cl promoter can react with Ti powders and convert the solid-phase sulfuration reaction into a CVD process, thereby improving control. Using molten glass as a growth substrate, Chen et al. successfully achieved millimeter-sized monolayer $MoSe_2$ and MoS_2 crystals. It was found that the isotropic surface of molten glass suppressed nucleation events and significantly enhanced the growth of large crystalline domains. The CVD process can be defined as a metal-organic CVD method when the precursor is a metal-organic compound. Kang et al. reported the preparation of monolayer MoS_2 and WS_2 films with high mobility using Mo $(CO)_6$, $W(CO)_6$, and $(C_2H_5)_2S$ and H_2 as gas-phase precursors. A high-mobility monolayer TMD film may be fabricated immediately for batch processing of integrated circuits based on TMDs.

6.3.1.3 Epitaxial Growth Methods

To promote epitaxial growth, the substrate and overlayer must be of the same lattice, or a commensurate condition must be met. This technique prepares graphene, MEAs, and TMDs. Group IV MEAs contain silicene, germanene, and stanene. Because of their preferential sp3 bonding, they exhibit sp2-sp3 hybridization, unlike graphene with sp2 hybridization. Due to their structure, they cannot be produced through exfoliation methods like graphene. This technique utilizes unique substrates for epitaxial growth. Figure 6.5(d) shows the schematic illustration of four types of epitaxial growth mode for 2D materials, such as van der Waals (VdW) epitaxy, edge epitaxy, in-plane epitaxy, and step-guided epitaxy, respectively. In 2012, silicene was epitaxially grown on Ag(111) surface to produce silicene for the first time. Since then, several substrates such as ZrB_2, Ir, and ZrC have been used to grow silicene. In a similar way to silicene, germanene was also produced on Au and Al surfaces. For example, Derivaz et al. demonstrated the formation of germanene on Al(111), providing a uniform coating of the substrate with a (3×3) superstructure. Stanene preparation has also followed in the footsteps of silicene and germanene in recent years. Zhu et al., in 2015 achieved ultrathin Sn films with 2D stanene structures on a Bi_2Te_3 substrate. According to Yuhara et al.'s recent study, epitaxial 2D stanene sheets lay over 2D Ag_2Sn surface alloys, but not directly on bulk-terminated Ag111 surfaces. As reported by Zhang et al., low-temperature molecular beam epitaxy (MBE) was used to fabricate monolayer stanene on PbTe (111) films. By doping PbTe with Sr, a decorated stanene sample with true insulating bulk was produced. It is also possible to synthesize phosphophorene by epitaxial growth, which is the typical representative of MEAs from group V. Zhang et al. developed blue phosphorene on Au (111) using MBE growth. The phosphorus layer formed a (5×5) superstructure with a band gap of at least 0.8 eV. Except for MEAs, MBE also achieves TMDs. Using MBE, Wong et al., recently developed monolayer VSe_2 films on graphite.

6.3.2 Performance Evaluation of 2D Material-Based Metal-Air Batteries

6.3.2.1 Graphene-Based Materials

2D carbon allotrope graphene has gained significant attention for its unique properties. Incorporating graphene into metal-air batteries has potential benefits. An important function of graphene in metal-air batteries is to support ORR and OER. The surface of graphene supports metal catalysts such as platinum or other transition metals, which enhances catalytic activity leading to improved battery performance. Due to its high electrical conductivity and large surface area, graphene is also used as an electrode material in metal-air batteries. Electrodes based on graphene improve battery performance by increasing oxygen transport, improving charge transfer kinetics, and improving electrochemical activity. Composite structures with graphene materials, such as graphene oxide or reduced graphene oxide, have been developed. In metal-air batteries, these composite materials are combined with other functional materials that exhibit synergistic effects, which enhance electrochemical performance, stability, and durability.[32,33] Graphene electrodes are highly electrically conducting and can also serve as a barrier layer between anodes and cathodes to prevent crossover reactions. This can be accomplished by selectively blocking gas diffusion but allowing ion transport, which reduces self-discharge while improving battery efficiency. Some of the recent literature on graphene-based electrocatalysts examined for metal-air battery is tabulated in Table 6.1, providing a detailed explanation below (Figure 6.6).

In the work reported by D. Y. Kim et al., macropores are introduced into graphene paper by using polystyrene colloidal particles as a sacrificial template to create free-standing graphene paper.[34] The as-prepared macro-porous graphene paper has a high porosity, a large surface area, and a large pore volume. It exhibited

TABLE 6.1

List of Graphene-Based Electrocatalysts Examined for Metal-Air Batteries

S. No	Catalyst	Method	Device	Current Density	Specific Capacity mAh g^{-1}	Cycles	Ref
1	Graphene paper	Modified dispersion polymerization	Li-O$_2$	200 mAg^{-1}	12,200	78	34
2	RuO$_2$/graphene/N-doped porous carbon	Stirring followed by calcination	Li-O$_2$	500 mAg^{-1}	19358.6	75	35
3	Holey graphene	Annealing	Li-O$_2$	100 mAg^{-1}	16546	91	36
4	B, N-hG	Hummers' method followed by freeze-drying	Li-O$_2$	500 mAg^{-1}	15340	117	37
5	CoN/UNG	Hydrothermal	Zn-O$_2$	10 mA cm^{-2}	917.2	350 h	38
6	Co/Co$_3$O$_4$/CoN/NG)	Adsorption-complexation-calcination method	Zn-O$_2$	10 mA cm^{-2}	843.0	20 h	39
7	Graphene nanosheets	Hydrothermal	Na-O$_2$	0.1 mA cm^{-2}	7574	10	40

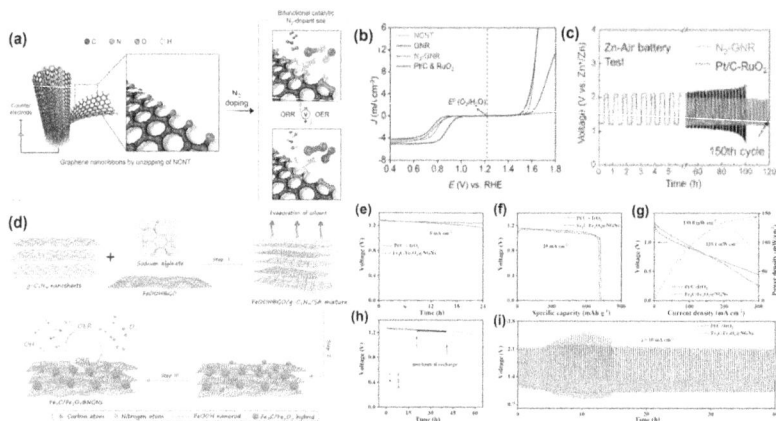

FIGURE 6.6 (a) Schematic depiction of synthetic procedure of N_2 - doped graphene nanoribbons via intact crystalline unzipping and hydrazine treatment (b) linear scan voltammograms of N_2-GNR, RuO_2 and (c) commercial Pt/C catalyst on an RDE (1600 rpm) in 1 M KOH with a scan rate of 5 mV s^{-1}, showing bifunctional electrocatalytic activity towards both ORR and OER.[35] (d-i) schematic illustration of the preparation procedures for Fe_2C/Fe_2O_3@NGNs and Electrochemical characterization of the Zinc-air battery using Fe_3C/Fe_2O_3@NGN air electrode at a discharging current density of 5mA cm^{-2} for 60 h (i) Long-term cycling performances at a current density of 10 mA cm^{-2}.[60]

a high specific capacity of ca. 12,200 mAh g^{-1} at a current density of 200 mA g^{-1} in non-aqueous Li-O$_2$ batteries when used as an air cathode. At a current density of 500 and 2000 mA g^{-1}, the macro-porous graphene paper has good stability up to 100 and 78 cycles, respectively, with a limiting capacity of 1000 mAh g^{-1}.[34] Using N-doped porous carbon, Wang et al. developed a composite hydrophobic catalyst comprising RuO_2 and graphene as shown in Figure 6.6.[35]

The catalyst was capable of better catalytic performance and had a low affinity for water vapor. This novel composite catalyst was successfully used to optimize the cycling of Li-air batteries in pure oxygen for 470 hours, humid oxygen for 310 hours, and ambient air for 330 hours at a current density of 500 mA g^{-1}, leading to a significant increase in discharge specific capacity from 13122.1 to 19358.6 mAh g^{-1}.[35]

Further, Yuhui Tian et al. have developed a novel N-doped graphene wrapped Fe_3C/Fe_2O_3 heterostructure, serving as a versatile oxygen catalyst for zinc–air batteries. During the catalyst synthesis process, graphene oxide (GO) interacts with graphitic carbon nitride (g-C_3N_4) and FeOOH nanorods through π–π stacking and hydrogen bonds, respectively. This unique assembly forms a nanostructure of carbon-coated iron species, resulting in exceptional durability in both alkaline and acidic environments. The nitrogen doping and Fe_3C/Fe_2O^+ heterostructure synergistically enhance the catalyst's oxygen reduction and evolution reaction activities. When integrated into a liquid zinc–air battery, the catalyst demonstrates remarkable peak power density (139.8 mW cm^{-2}), large specific capacity (722 mAh g^{-1}), and outstanding cycling performance. Moreover, in quasi-solid-state zinc–air batteries, the catalyst exhibits impressive open-circuit voltage and peak power density. These findings suggest that the

Fe_3C/Fe_2O_3@NGNs catalyst holds great promise for practical applications in energy conversion devices, as illustrated in Figure 6.6 (d-g).[60] A series of holey graphene (hG-x) with different pore sizes were synthesized by Y. Meng et al. where phosphorus was further introduced into hG-2.[36] As a result of the synergistic effect of holey structures and P doping (P-hG-2) as a cathode for Li-oxygen batteries exhibits excellent lithium storage properties with a max discharge of 16546 mAhg^{-1} at 100 mA g^{-1} and good cycle life of 91 cycles at 1000 mA g^{-1}.[36] Due to its nanoholes through the graphene nanosheet basal plane, holey graphene is becoming increasingly popular as a cathode for Li-O_2 batteries. Aside from facilitating rapid diffusion channels for mass (such as O_2 and Li$^+$), nanoholes offer rich edges for mass adsorption and storage as well as heteroatom doping. Following this through the direct oxidation of reduced graphene oxides (rG) by a controlled flow of hydrogen, Zhang et al. report an easy and environment-friendly approach to producing hG with abundant nanoholes.[37] The obtained hG from the above approach is co-doped with 3.0 at% B and 2.1 at% N atoms to form B, N-hG. Due to its holey structure and synergistic effect of B and N, B, N-hG displays promising properties for Li-O_2 batteries with a maximum discharge capacity of 15340 mAh g^{-1} and long cycling stability of 117 cycles.[37] Transition metal nitrides anchored on carbon are commonly used as oxygen reduction catalysts in zinc-air batteries, but their size and thickness can be difficult to control. Using the combination strategy of 1, 10-phenanthroline coordination with polyethylene imine interaction, Suqin Wu et al. synthesized CoN nanoparticles anchored to ultra-thin nitrogen-doped graphene electrocatalysts (CoN/UNG).[38] With a half-wave potential of 0.87 V versus RHE, the CoN/UNG exhibits excellent electrochemical ORR capabilities. The CoN/UNG-based zinc-air battery has 149.3 mW cm^{-2} power density, 917.2 mAh g^{-1} specific capacity, and 350 h cycle life.[38] Zinc-air batteries can be improved using the combination strategy for small-sized transition metal nitrides anchored on nitrogen-doped graphene catalysts. Further, Shanjing Liu et al. proposed the adsorption–complexation–calcination method to improve the efficiency of zinc-air batteries (ZABs) cathodic ORRs.[39] Here, graphene nanosheets were used to synthesize multicomponent cobalt nanoparticles containing Co, Co_3O_4, and CoN, along with numerous N heteroatoms (Co/Co_3O_4/CoN/NG). These Co/Co_3O_4/CoN nanoparticles have sizes less than 50 nanometers, which are homogeneously dispersed on N-doped graphene (NG) substrate, which enhances ORR catalysis. In this study, the half-wave potential was 0.80 V vs. RHE, and the limiting current density was 4.60 mA cm^{-2}, which is in line with commercially available platinum/carbon (Pt/C) catalysts.[39] Utilized as a cathodic catalyst for ZABs, the battery exhibits excellent performance with large specific capacity and open circuit voltage, with 843.0 mAh g^{-1} and 1.41 V, respectively. This is due to the highly scattered Co/Co_3O_4/CoN particles and the doped nitrogen on the carbon matrix in combination.[39] Furthermore, graphene nanosheets containing platinum nanoparticles (Pt@GNS) with a particle size of 5 nm were successfully synthesized using a simple hydrothermal method.[40] Using Pt@GNS as a cathode resulted in a high discharge capacity of 7574 mAh g^{-1} at a current density of 0.1 mA cm^{-2} and good cycling performance with a limited discharge capacity of 1000 mAh g^{-1}. Thus, Pt@GNS demonstrated strong electrocatalytic activity in sodium-oxygen rechargeable batteries for cathode reactions.[40] As a fascinating and inspiring result, platinum exhibits an extremely complex catalytic character in Na-O_2 cells.

6.3.2.2 Transition Metal Chalcogenides

Transition metal chalcogenides (TMCs) have attracted considerable attention in recent years due to their high electrical conductivity, catalytic activity, and large surface area and making them potential energy storage materials, including metal-air batteries. Cathodes containing TMCs can reduce oxygen and enhance the overall efficiency of metal-air batteries. The improved ORR kinetics of TMCs result in a reduction in the overpotential required for the cathodic reaction, resulting in higher discharge capacities and better cycling stability. A number of TMCs have been explored for metal-air batteries, including molybdenum sulfide (MoS_2), tungsten sulfide (WS_2), and cobalt diselenide ($CoSe_2$). Despite their potential, TMCs still pose some challenges in metal-air batteries. There are several challenges associated with TMC catalyst development, including the synthesis of high-quality catalysts of controlled morphology and composition, their integration into battery electrode architectures, and their long-term stability under harsh battery operating conditions. There are a few examples discussed below as well as tabulated and also discussed in Figure 6.7 about the hierarchical Co and Nb dual-doped MoS2 nanosheets shelled micro-TiO2 hollow spheres as effective multifunctional electrocatalysts for HER, OER, and ORR.

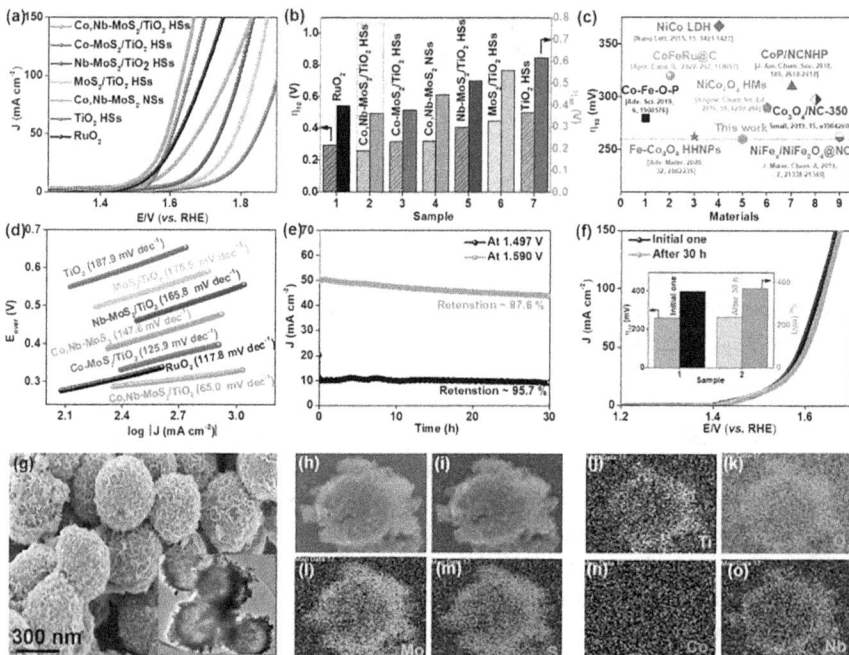

FIGURE 6.7 The Co and Nb dual-doped MoS2 nanosheets shelled micro-TiO2 hollow spheres and corresponding electrochemical activities (a)–(f) and (g)–(o) SEM image and elemental mapping.[61]

An efficient oxygen electrocatalyst for rechargeable Zn-air batteries (ZABs) was demonstrated by W.-J. Niu et al. using an ordinary solid-phase pyrolysis procedure to synthesize hybrid transition metal nanocrystal-embedded graphitic carbon nitride nanosheets, namely M-CNNs. Metal acetylacetonates and the g-C3N4 precursor can be controlled by optimizing Fe-CNNs-0.7, Ni-CNNs-0.7, and Co-NNs-0.7 composites to exhibit superior ORR/OER bifunctional electrocatalytic activities. Specifically, Co-CNNs–0.7 has comparable half-wave potential to the commercial Pt/C catalyst (0.832 V vs RHE) as well as a lower overpotential (440 mV) toward the OER compared to the commercial IrO_2 catalyst (460 mV), demonstrating impressive applications in rechargeable ZABs. Consequently, Co-CNNs-0.7 cathodes outperformed the commercially available Pt/C + IrO_2 catalysts in terms of peak power density (85.3 mW cm^{-2}) and specific capacity (675.7 mAh g^{-1}) as well as cycling stability with remarkable 1000 cycles. A promising approach for advanced metal-air cathode materials is proposed by this study, which highlights the synergy of heterointerfaces in oxygen electrocatalysis. A bifunctional oxygen catalyst that can be fabricated with high efficiency and use non-precious materials offers a promising alternative to precious metals for the development of real-time MABs. As a bifunctional oxygen catalyst for MABs, Co_3O_4 embedded in nitrogen defect-rich g-C_3N_4 is prepared by graphitizing zeolitic imidazolate framework (ZIF)-67@ND-CN by Wenhao Tang et al. Co_3O_4@ND-CN has superb bifunctional catalytic performance, facilitating metal-air battery development. In summary, the zinc-air battery based on Co_3O_4 @ND-CN achieves a peak round-trip efficiency of 60% with long-term durability (over 340 cycles), surpassing the battery utilizing noble metals at the forefront of technological advancements. Similarly, the corresponding lithium-oxygen battery utilizing Co_3O_4@ND-CN exhibits surprisingly high maximum discharge/charge capacities (9838.8/9657.6 mAh g^{-1}), impressive overpotentials (1.14/0.18 V), and good cycling performance. Co_3O_4@ ND-CN exhibits compelling electrocatalytic processes and device performances due to concurrent compositional and structural synthesis (wrinkled morphology with abundant porosity) as well as synergistic interaction, which produces advantageous surface electronic environments. Furthermore, it is also possible to design and craft a rich diversity of metal oxides@ND-CN that have adjustable defects, architectures, and enhanced activities for research and development in catalysis and renewable energy. Several transition metal-based catalysts for ORR and OER have been developed; however, most are based on oxides, which hinder catalytic performance. Using 2D Ni-Fe nitride nanoplates strongly coupled with graphene support, Yuchi Fan et al developed a metallic Ni-Fe nitride/nitrogen-doped graphene hybrid. Ni-Fe nitride's electronic structure is changed by hybridizing with nitrogen-doped graphene. This hybrid catalyst has the lowest onset overpotential (150 mV) reported for OER activity and exhibits ORR activity similar to commercial Pt/C. This bifunctional catalyst exhibits high activity and durability in rechargeable zinc-air batteries that are stable for 180 cycles with an overall overpotential of 0.77 V at 10 mA cm^{-2}. Further, it has been demonstrated that two-dimensional metalorganic frameworks (2D MOFs) with high oxygen accessibility, large surface areas, and open catalytic sites may enhance Li-O_2 batteries' performance. In order to obtain highly efficient cathode catalysts for aprotic Li–O_2 batteries, Mengwei Yuan et al.

described a facile ultrasonication method to synthesize three classes of 2D MOFs (2D Co-MOF, Ni-MOF, and Mn-MOF). The inherent open active sites of the Mn–O framework of the 2D Mn-MOF cathode contribute to a discharge-specific capacity of 9464 mAh g^{-1} which is higher than those of the 2D Co-MOF and Ni-MOF cathodes. In the cycling test, the 2D Mn-MOF cathode operated over 200 cycles at 100 mA g^{-1} with a curtailed discharge capacity of 1000 mAh g^{-1}, quite longer than others. The 2D Mn-MOF outperforms 2D Ni-MOF and Co-MOF due to superior ORR and OER, especially the efficient oxidation of both LiOH and Li$_2$O$_2$. The promising results of the cycling test demonstrate the superior cycling stability of the 2D Mn-MOF cathode, which makes it a great candidate for a wide range of practical applications in high energy density and long cycling cycle cells. As low-cost and high-performance substitutes for noble metal-based oxygen electrocatalysts, Ni$_x$Co$_{3-x}$N/NCNTs hybrids have been developed. Ni and Co solid solution nitrides were analyzed, optimized, and controlled atomistically by coupling theoretical and experimental approaches. Catalytic processes identified multiple active sites and critical limiting steps. By optimizing Ni/Co ratios in nitrides via bulk-surface responses and separated surface charge distributions, specific ORR and OER activities were tailored. Moreover, the residual Ni$_x$Co$_{3-x}$N acts as a conduit for metallic electron conductivity, thus improving the effectiveness of the catalyst. It has been shown that surface oxidation increases charge-deficient states of transition metal atoms while weakening TM-N bonds simultaneously. In terms of rechargeability and durability, SS/NCNT catalysts designed for Zn air batteries outperformed both commercial noble metal catalysts and electrodes. Clearly, these findings should encourage future research on the use of multiple-component nitrides as electrocatalysts for energy storage and conversion devices, as well as studies of their catalytic mechanisms in situ. A facile strategy for scalable production of vertically aligned hybrid nanosheets on nickel foam was reported by Jixin Zhu et al. for lithium-oxygen batteries. This resulting architecture consists of very thin metal hydroxides and nitrogen-doped graphitic carbon nanosheets as combined functional components that were integrated into one system enabling activity for both OER and ORR thus forming a bifunctional catalyst. Strong mechanical connections between hybrids and nickel substrates prevent binders and facilitate good electrical contact. Through the use of a nanoporous card-house structure, Li$_2$O$_2$ growth can be confined throughout the battery charging process, allowing oxygen to be transported most effectively. In conclusion, high specific capacities for the metal hydroxide hybrid (5403 mAh g^{-1} at 4.15 V), improved oxygen activation properties, low charge potentials, and excellent reversibility are observed. In spite of reversibility, side reactions contribute at potentials below 4.15 V (i.e., before carbon oxidation occurs). Stable cycling was achieved at a voltage cutoff of 3.9 V, at a capacity of 2496 mAh g^{-1}. This is the first report on metal hydroxides and carbon hybrids for LOB. Using this method, ORR and OER bifunctional components can be developed into one single catalyst with a favorable pore structure. In light of current electrolyte developments, the electrode architecture is promising for LOB applications. Following this, researchers developed a template-free approach to fabricating a hybrid composite of graphitic-shell encapsulated FeNi alloy nanocrystals and biomass-doped porous carbon (Fe$_x$Ni$_y$N@C/NC). A porous architecture offers many active sites and improves mass transfer. In the hybrid material (FexNiyN@C/NC), the strong catalytic interactions between graphite-coated Fe_xNi_yN nanocrystals and porous carbon enable it to maximize the effects of the unique structures for both ORR and OER. XAS analysis showed that the bulk structure of the Fe_xNi_yN nanocrystals remained virtually unchanged after the harsh OER electrochemical corrosion process. Hybrid catalyst demonstrated excellent stability over 400 hours at 5.0 mA cm^{-2} when integrated into a Zn air electrode. For metal-air batteries and beyond, these findings pave a new way to design highly efficient bifunctional electrocatalysts. A porous composite of g-C_3N_4 and α-MnO_2 was synthesized by Yi Yang Hang et al. to serve as a catalyst for Li-air batteries. α-MnO_2 nanorods and porous g-C3N4 sheets combine synergistically to produce a hybrid catalyst with superior ORR and OER activities. When comparing the pure α-MnO_2 to the conventional XC-72 carbon catalysts, the as-synthesized g-C_3N_4/α-MnO_2 catalyst exhibits a better discharge capacity of 9180 mAh g^{-1} at 100 mA g^{-1}, lower voltage gap $\sim$1.33 V at 100 mA g^{-1} and $\sim$40 cycles of longer lifespan with a limited discharge depth of 1000 mAh g^{-1}. It should also be noted that a facile hydrothermal method can be used to deposit other nanoscale transition metal oxides on porous g-C_3N_4 sheets as cathodes for LOBs. In a study by Feng Zhang et al., urchin-like CoP microspheres were synthesized using cetrimonium bromide as a softer template. When compared with a commercial Pt/C catalyst, the as-prepared CoP exhibits an ORR onset potential of 0.11 V (vs. Ag/AgCl) with a potential gap of 62 mV. In addition, the CoP catalyst exhibits higher OER activities than the commercial RuO_2 catalyst. An interesting observation is that, when used as a cathodic catalyst for Li-air batteries, the battery is capable of discharging 2994 mAh g^{-1} at 100 mA g^{-1}, has a high round-trip efficiency of over 90%, and is capable of sustaining 80 cycles without capacity degradation at a discharge current density of 500 mA g^{-1} without any capacity degradation. A 2D sheetlike core-shell nanostructure was developed by Palanichamy Sennu et al. to ensure superior cycling performance and rate capability in OER and ORR systems. This pristine Co_3S_4 NS material was evaluated as a conversion type anode and demonstrated to maintain 600 mAh g^{-1} after 200 extremely stable cycles at a current density of 0.5 A g^{-1}. Ex-situ studies have proven that Co_3S_4 transitions to Co_9S_8 upon electrochemical cycling. It was more encouraging to find that bifunctional electrocatalytic activity in aqueous media was comparable to that of established catalysts such as Pt/C and RuO_2. Upon integrating the bifunctional catalyst into the Li–O_2 system, the cell delivered an impressive capacity of 6990 mAh g^{-1} at a current density of 0.2 mA cm^{-2}. Co_3S_4 NS also exhibited excellent capacity-limited cycling stability and a low potential gap, adding to its advantages. 2D mesoporous cobalt sulfides have been introduced as bifunctional electrocatalysts in LOBs. These materials are promising for the development of advanced MABs with high energy efficiency and long cycle life. Furthermore, the high surface area of these materials makes them an ideal choice for both oxygen evolution and reduction reactions (Table 6.2).

6.3.2.3   Transition Metal Carbides, Nitrides, and Carbon Nitirdes (Mxenes)

MXenes are a family of 2D transition metal carbides, nitrides, and carbonitrides. MXenes exhibit both high conductivity and high surface area, making them desirable

TABLE 6.2
List of 2D Transition Metal Chalcogenides Electrocatalysts Examined for Metal-Air Batteries

S. No	Catalyst	Method	Device	Current Density mAg^{-1}	Specific Capacity $mAh\ g^{-1}$	Cycles	Ref
1	Graphitic carbon nitride	Solid-phase pyrolysis	$Zn-O_2$	10 mA cm^{-2}	675.7	1000	41
2	Co_3O_4@ND-CN	Freeze-dried	$Li-O_2$	500 mA g^{-1}	9838.8	340	42
3	$Ni_3FeN/NRGO$	Facile ultrasonicated method	$Zn-O_2$	10 mA cm^{-2}	-	180	43
4	Co-MOF, Ni-MOF, and Mn-MOF	Facile ultrasonicated method	$Li-O_2$	100 mA g^{-1}	9464	120	44
5	Nitride/carbon nanotube	Hydrothermal	$Zn-O_2$	10 mA cm^{-2}	-	300 (5 h)	45
6	Graphene-Metal Hydroxide	Heat treatment	$Li-O_2$	75 mA g^{-1}	12123	40 h	46
7	Fe_xNi_yN@C/NC	Template-free approach	$Zn-O_2$	5.0 mA cm^{-2}	-	400 h	47
8	$g-C_3N_4/\alpha-MnO$	Hydro-thermal	$Li-O_2$	100 mA g^{-1}	9180	40	48
9	Co_2P	Hydrothermal	$Li-O_2$	100 mA g^{-1}	2994	80	49
10	Co_3S_4	Hydrothermal	$Li-O_2$	0.2 mA cm^{-2}	5917	60	50

materials for metal-air batteries. These materials have excellent catalytic activity for ORR and OER, which are vital processes in metal-air batteries. Metal-air batteries also benefit from MXenes as anodes where MXenes' conductive framework can enhance battery cycling stability and overall performance by maintaining anode structural integrity. Additionally, MXenes can be used as conductive additives in electrode formulations. This improves the electronic conductivity and ion diffusion within the electrode, enhancing the batteries' overall performance and power density. For this reason, these materials are considered promising for metal-air batteries, but they will probably require protective coatings or alterations to ensure their stability in the battery environment. Moreover, further development is necessary to develop cost-effective and scalable synthesis methods for MXenes. Some of the literature is as follows (Figure 6.8).

To create a high-rate cathode for LOBs, Gaoyang Li et al. created Nb_2C MXene nanosheets with fewer layers and an O-terminated surface using an etching-exfoliation-annealing route. Through theoretical calculations, it was shown that Nb_2C MXene's O-terminated surface was the most stable structure and tended to form first in various surface group structures. Designed as an oxygen electrode, it has a 19 785.5 mAh g^{-1} capacity at 200 mA g^{-1} and demonstrates high-rate stability of 130 cycles at 3 Ag^{-1}. This outstanding electrochemical performance is attributable to MXene's special lamellar structure and the efficient utilization of the O group active site. As a result of DFT calculations, the O-terminated Nb_2C MXene monolayer has a good affinity with LiO_2 and Li_2O_2, as well as promoting the accumulation of spatial orientation and stable decomposition of Li_2O_2 porous nanosheets. However, F, OH terminated, and bare surfaces produce high overpotentials and low cyclic stability because of insufficient/excessive binding to Li_2O_2. Similarly, through DFT calculations, J. Li et al., discovered that the Ti_2CO_2 surface facilies a highly efficient single electron reaction pathway during the adsorption-nucleation-decomposition process of Li_2O_2, serving as the primary catalytic sites compared to the Ti_2CF_2 surface. In contrast, the unmodified Ti_2C surface dimished the catalytic efficacncy of Ti_2C MXene, due to its strong chemical binding with Li_2O_2 (Figure 6.8(a)). For this study, Ti_2C MXene adorned with multifunctional groups such as -O and -F was synthesizes as shown in Figure 6.8(b). These heterogeneous surface condition was proposed to enhance polarized nucleation and growth of discharge products, resulting in the formation of a porous structure comprised of spatially accumulated nanoflackes, thereby facilitating efficient mass transfer pathways. Consequently, the Ti_2C MXene catode demonstrated outstanding performance in terms of high capacity and prolonged cycle stability. Furthermore, this study offers intrinsic insights into the catalytic mechanism and reaction kinetics of Ti_2C MXene adorned with multi-functional groups, thus aiding in the rational design of high performance MXene - based materials for Li-O_2 batteries. The cycle stablity of the Ti_2C MXene cathode was evaluted at various current densities while maintaining a fixed capacity, as depicted in Figure 6.8(c), which includes the terminal discharge/charge volatges for each cycle along with corresponding discharge/charge profiles. The Ti_2C cathode exhibited excellent galvanostatic cycling perfomance over 250 cycles at a current density of 200 mA g^{-1} with a fixed capacity of 600 mAh g^{-1}. Additionally, as shown in Figure 6.8(d), S. Tam et al., developed a hybrid material comprising $NiCo_2O_4$ nanocrystals and

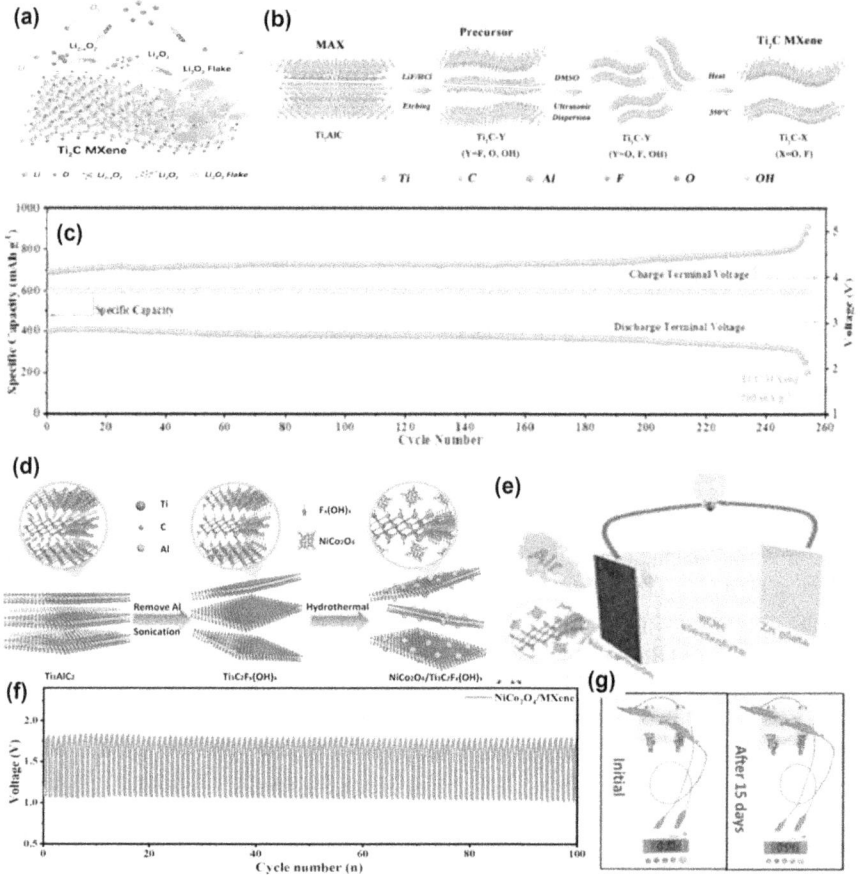

FIGURE 6.8 Schematic illustration for synthesis process of the Ti_2CMXene and the efficient single electron reaction pathway for theadsorption-nucleation-decomposition process of Li_2O_2 (c) Terminaldischarge/charge voltages of the Ti_2C MXene at a current density of 200 mA g^{-1} with a limited specific capacity of 600 mAh g^{-1}.51 (d) Schematic ofthe Preparation Route of $NiCo_2O_4$/ MXene (e) Schematicphotography of the liquid ZAB (f) Galvanostatic discharge-charge cycling curvesat 5.0 mA cm^{-2} (g) Images of an electronic timer powered by one ZABat initial and after 15 days.52

MXene featuring robust Ni/Co-F bonds. These MXene based composites exhibit impressive electrocatalytic activity for oxygen reduction reaction (ORR) and oxygen evolution reaction (OER) for Zn-air batteries (Figure 6.8 (c)). Consequently, the solid-state Zn-air battey constructed with these materials attains notable performance metrics, including an open circuit volatge of 1.40 V, peak power density of 55.1 mW cm^{-2} and energy efficincy of 66.1% at 1.0 mA cm^{-2}. Furthermore, the flexibility of the Zn-air batteries devices allows for easy tailoring without compromissing performance, even after cutting and reassembling after 15 days (Figure 6.8 (g)).

Following this, Yu Wang et al. constructed TiCN, MXene hybrid BCN nanotubes with trace level Co catalyst (TiCNeBCNeCo). This is achieved by using spray-lyophilization of 2D $Ti_3C_2T_x$ sheets and in situ growth of BCN nanotubes between Mxene sheets which increases the specific area of the catalyst with more active sites. This hybrid catalyst showed superior rate capability and charge-discharge stability for ZABs over commercial Pt/C. Later, Na Li et al. reported an impressive high-rate and long-cycling cathode catalyst for LOBs. These results were achieved using 2D $Ti_3C_2T_x$ MXene with high conductivity, larger specific areas, and more active sites. Ex-situ SEM, XRD, and DEMS analysis indicated Li_2O_2 as the discharged product. The discharge process first forms LiO_2 in the electrolyte, followed by converting it into a composite of $Li_{2-x}O_2$ and Li_2O_2 through a disproportionate reaction. It is then adsorbed to the MXene phase. A heterogeneous surface condition with O and F anchored to the surface of MXene contributes to the porous structure of discharge products. $Li_{2-x}O_2$ is first decomposed at a lower charging voltage (3.7 V) in the charging process. The release of O_2 becomes smooth when the voltage exceeds 4.0 Volts, resulting in uniform decomposition of Li_2O_2 via delithiation. According to Monireh Faraji et al., a graphene aerogel/MXene heterostructure containing N-doped graphene has a higher onset potential than commercial 10% Pt/C catalysts. The hybrid Mxene/NGA structure demonstrated remarkable stability after 40,000 seconds with only 10% decay. The excellent catalytic performance of MXene/NGA is attributed primarily to its micro-mesoporous graphitized structure as well as its highly conductive MXene nanosheets. The MXene provides the 2D platform and tremendous electronic channels in this heterostructure. NGAs have an abundance of N-rich catalytic sites that are capable of adsorbing oxygenated species and facilitating ORRs. In addition, the Zn-air battery fabricated on Mxene/NGA catalyst has greater power density, a higher charge-discharge potential, and a longer battery life than the 20 wt% Pt/C IrO_2 battery. An MXene-based hierarchical pore structure containing CoNi/CoNiP heterostructure called H-CNP@M was successfully synthesized by Jingyuan Qiao et al. Heterostructures exhibit excellent intrinsic bifunctionality because of their fast electron transfer and rearrangement. This explains H-CNP@M's high half-wave potential (0.833 V) and low overpotential and Tafel slope (294 mV at 10 mA cm^{-2}, 63.7 mV dec^{-1}). When assembled into Zn-air batteries, H-CNP@M exhibits a high peak power density of 166.5 mW cm^{-2} and a good cycle stability. This makes H-CNP@M a promising candidate for cost-effective, bifunctional electrocatalysts for sustainable energy conversion and storage.

Haoyang Xu et al. report the successful development of V_2O_5@V_2C MXene as an oxygen electrode catalyst in the Li-O_2 battery. Applied to Li-O_2, the experimental results suggest that MXene compounds can produce lithium-ion batteries with low overpotential and high efficiency of 83.4%. MXene has abundant vanadium vacancies, which contributes to its superior catalytic activity. The DFT calculations show that vanadium vacancies lead to enhanced adsorption energies for Li_2O_2 and intermediate LiO_2 on MXene, which has the most significant effect on electrochemical performance. Additionally, its 2D layer structure maintains mechanical integrity during ion transport and provides abundant ion transport channels. Thus, vanadium vacancies enhance the catalytic performance of MXene, providing information on how to design advanced catalysts. Using solvothermal treatment, an epitaxial Co@Mn@TCPP bimetal organic framework was grown on

$Ti_3C_2T_x$ MXene sheets with grafted surface terminators for optimum antioxidation. According to Sanghee Nam et al., through Fischer esterification and fluorine substitution, the carboxyl group of the TCPP molecule forms a chemical bond with the surface terminators (-OH and -F) of MXene. The coordination bonds formed by Co and Mn with the porphyrin's N and carboxyl's O contribute to enhanced electrocatalytic performance. $Li-O_2$ batteries are fueled by this CMT@MXene bifunctional electrocatalyst. Compared to conventional electrocatalysts, CMT@ MXene exhibits improved overpotential and cycling stability in $Li-O_2$ batteries with a specific capacity of ≈6850 mAh g^{-1} CMTM+SuperP and cyclic retention of 94.1% for 312 charging cycles. In total, 247 cycles of long-term cycling stabilities are observed with a limited specific capacity (1001 mAh g^{-1} CMTM+SuperP) and current density (500 mA g^{-1} CMTM+SuperP). In OER and ORR, the overpotentials were 0.9–1.01 V and 0.22–0.24 V, respectively, versus Li^+/Li. For ORR and OER, the unique metalloporphyrin structures and unpaired electrons between CMT and MXene contribute to enhanced electrochemical performance. Through facile ultrasonication, the researchers developed NiO/Ti_3C_2 nanomaterials for $Li-O_2$ batteries. Due to MXene's advantageous electronic conductivity and the interaction of NiO nanoparticles with Ti_3C_2, Ti_3C_2 MXene nanosheets layered with NiO nanoparticles have excellent $Li-O_2$ battery performance. Ti_3C_2 MXene provides quick electron transfer, while NiO nanoparticles provide enough active area for catalysis. When NiO nanocrystals are dispersed on Ti_3C_2, they significantly enhance the active site concentration. This improves the catalytic activity. Among NiO/Ti_3C_2 nanomaterials, NiO/Ti_3C_2-2 cathodes exhibit the greatest initial discharge capacity. They also exhibit the lowest overpotentials at 500 mAg^{-1} for OER and ORR. They also perform over 90 cycles at 500 mAg^{-1} with outstanding cycle performance. As a result of its excellent electrochemical properties, NiO/Ti_3C_2 is capable of serving as a promising cathode material for $Li-O_2$ batteries.

Yingxinjie Wang et al. successfully prepared a new in situ graphited carbon-coated $FeTiO_3$ nanosheet ($FeTiO_3$@C) electrocatalyst with an excellent OER/ORR performance for applying in RZABs. The excellent electrochemical performance originates from the following points. First, the unique 2D structure may facilitate the rapid transmission of ions and electrons and expose more catalytically active sites. Second, in situ $Ti_3C_2T_x$-derived graphited carbon interacting with $FeTiO_3$ nanosheets is beneficial for forming a conductive network to prevent the aggregation of $FeTiO_3$ nanosheets and improve its electron transfer. Third, $FeTiO_3$@C can be converted into highly active iron oxyhydroxide by in situ electrochemical reconstruction in the alkaline OER process. As expected, $FeTiO_3$@C achieves a low overpotential (323 mV at 10 mA cm^{-2}) for the OER and a positive half-wave potential of 0.502 V for the ORR. Moreover, the $FeTiO_3$@C-based RZAB has a high trip efficiency of 63.4% and long-time cycling stability. Ultimately, synthetic modification strategies and combined modulations of catalytic capability are key factors in developing MABs and high-specific energy storage and conversion systems using MXenes. Through these strategies, MXenes and other 2D materials can be tailored to meet specific needs, allowing them to be used in a wide range of energy storage and conversion systems that require a high-specific energy capacity. This can help to improve the efficiency and performance of these systems (Table 6.3).

TABLE 6.3

List of Mxene-Based Electrocatalysts Examined for Metal-Air Batteries

S. No	Catalyst	Method	Device	Current Density mAg^{-1}	Specific Capacity $mAh\ g^{-1}$	Cycles	Ref
1	Nb_2C MXene	Etching-exfoliating annealing route	$Li-O_2$	200	19 785.5	130	51
2	TiCNeBCNeCo	Spray-lyophilization	Zn-air	$10\ mA\ cm^{-2}$	791	200 h	52
3	$Ti_3C_2T_x$	Magnetic stirring	$Li-O_2$	200	15468	80	53
4	N-doped graphene/$Ti_3C_2T_x$ MXene	Chemical etching	$Zn-O_2$	$50\ mA\ cm^{-2}$	$53.6\ mW\ cm^{-2}$	400 h	54
5	CoNi/CoNiP	Pyrolysis-phosphorization	$Zn-O_2$	$10mA\ cm^{-2}$	$166.5\ mW\ cm^{-2}$	120 h	55
6	V_2C MXene	Chemical etching	$Li-O_2$	100	1000	500	56
7	CMT and Ti_3C_2Tx MXene	Solvothermal treatment	$Li-O_2$	500	1000	247	57
8	NiO/Ti_3C_2	Ultrasonication	$Li-O_2$	100	13350	-	58
9	Graphited carbon-coated $FeTiO_3$ nanosheet	In situ annealing	$Zn-O_2$	$10\ mA\ cm^{-2}$	-	100 h	59

6.4   OUTLOOK ON FUTURE RESEARCH DIRECTIONS

In the field of metal-air batteries, substantial progress has been made, but still there are extensive challenges related to the development of cathodes, catalysts, anodes, and compactable electrolytes. The development of new flexible and self-supporting cathode electrode materials, insight into electrocatalytic mechanisms, and recognition of suitable materials for the synthesis of flexible electrodes with outstanding catalytic performance are very crucial in the development of metal-air batteries. In the case of catalysts, extensive efforts are made in the development of bifunctional ORR/OER electrocatalysts; however, imminent efforts have to be made in the development of advanced electrocatalysts that have a good electroactive surface area, porosity electrical conductivity, less particle agglomeration can be tuned, which is closely related to the synthesis. Construction of a GDL is also very important; the particle agglomeration blocks the diffusion path and in turn limits the catalytic activity. So, the development of binder-free or direct growth of air electrodes is a promising solution as electroactive nanoparticles, directly grown or deposited on the framework of the highly porous and conductive substrate, prevent the particle's rapid agglomeration. In the case of anode materials, reversible metal electrodes (Li, Na, Zn, Al, Mg, etc.) possess a high proportion of functional active sites, high rechargeable efficiency, and sustainable electroactivity over long cycles. In addition, new technological developments have to be made in the design of new electrolytes that operate for the long term. Aqueous and non-aqueous electrolytes are well known until now; the main challenges of electrolyte drying and carbonate formation have to be addressed. Polymer, solid, and ionic liquids are better alternatives to these liquid electrolytes due to their limited electrolyte evaporation and resistance toward CO_2 poisoning and hydrogen evolution, so the electrochemical performance in terms of battery efficiency and life cycle is improved. However, the development of these electrolytes requires a constant and systematic study.

ACKNOWLDGEMENTS

The Authors would like to thank Prof. P. Elumalai, Dept. of Green Energy technology, Pondicherry University for his expertise and insightful feedback, which significantly contributed to the development of this chapter. We are also grateful to Dr. Srikanth Ponnada for his dedication and support throughout this process. Dr Villa Krishna Harika would like to thank DST-India for DST-INSPIRE faculty fellowship grant.

CONFLICTS OF INTEREST

The authors declare no competing interest.

REFERENCES

1. Thomas Reddy, D. L. Linden's Handbook of Batteries, 4/e (SET 2). 2011.
2. Winter, M.; Brodd, R. J. What Are Batteries, Fuel Cells, and Supercapacitors? *Chem. Rev.* 2004, *104* (10), 4245–4269. 10.1021/CR020730K/ASSET/IMAGES/MEDIUM/ CR020730KE00044.GIF.

3. Li, M.; Lu, J.; Chen, Z.; Amine, K. 30 Years of Lithium-Ion Batteries. *Adv. Mater.* 2018, *30* (33), 1800561. 10.1002/ADMA.201800561.

4. Etacheri, V.; Marom, R.; Elazari, R.; Salitra, G.; Aurbach, D. Challenges in the Development of Advanced Li-Ion Batteries: A Review. *Energy Environ. Sci.* 2011, *4* (9), 3243–3262. 10.1039/C1EE01598B.

5. Li, T.; Huang, M.; Bai, X.; Wang, Y. X. Metal–Air Batteries: A Review on Current Status and Future Applications. *Prog. Nat. Sci. Mater. Int.* 2023, *33* (2), 151–171. 10.1016/J.PNSC.2023.05.007.

6. Chawla, N. Recent Advances in Air-Battery Chemistries. *Mater. Today Chem.* 2019, *12*, 324–331. 10.1016/J.MTCHEM.2019.03.006.

7. Olabi, A. G.; Sayed, E. T.; Wilberforce, T.; Jamal, A.; Alami, A. H.; Elsaid, K.; Rahman, S. M. A.; Shah, S. K.; Abdelkareem, M. A. Metal-Air Batteries—A Review. *Energies* 2021, *14* (21), 7373. 10.3390/EN14217373.

8. Neburchilov, V.; Zhang, J. Metal-Air and Metal-Sulfur Batteries: Fundamentals and Applications. *Met. Met. Batter. Fundam. Appl.* 2016, 1–195. 10.1201/97813153722 80/METAL-AIR-METAL-SULFUR-BATTERIES-VLADIMIR-NEBURCHILOV-JIUJUN-ZHANG.

9. Wang, H. F.; Xu, Q. Materials Design for Rechargeable Metal-Air Batteries. *Matter* 2019, *1* (3), 565–595. 10.1016/J.MATT.2019.05.008.

10. Choi, J. W.; Aurbach, D. Promise and Reality of Post-Lithium-Ion Batteries with High Energy Densities. *Nat. Rev. Mater.* 2016, *1* (4), 1–16. 10.1038/natrevmats.2016.13.

11. Hartmann, P.; Bender, C. L.; Vračar, M.; Dürr, A. K.; Garsuch, A.; Janek, J.; Adelhelm, P. A Rechargeable Room-Temperature Sodium Superoxide (NaO2) Battery. *Nat. Mater.* 2012, *12* (3), 228–232. 10.1038/nmat3486.

12. Ren, X.; Wu, Y. A Low-Overpotential Potassium-Oxygen Battery Based on Potassium Superoxide. *J. Am. Chem. Soc.* 2013, *135* (8), 2923–2926. 10.1021/JA312059Q/ SUPPL_FILE/JA312059Q_SI_001.PDF.

13. Sathyanarayana, S.; Munichandraiah, N. A New Magnesium – Air Cell for Long-Life Applications. *J. Appl. Electrochem.* 1981, *11* (1), 33–39. 10.1007/BF00615319/METRICS.

14. Abraham, K. M.; Jiang, Z. A Polymer Electrolyte-Based Rechargeable Lithium/ Oxygen Battery. *J. Electrochem. Soc.* 1996, *143* (1), 1–5. 10.1149/1.1836378/XML.

15. Inozemtseva, A.; Rulev, A.; Zakharchenko, T.; Isaev, V.; Yashina, L.; Itkis, D. Chemistry of Li-Air Batteries. *Compr. Inorg. Chem. III* 2023, 324–362. 10.1016/ B978-0-12-823144-9.00055-8.

16. Zaromb, S. The Use and Behavior of Aluminum Anodes in Alkaline Primary Batteries. *J. Electrochem. Soc.* 1962, *109* (12), 1125. 10.1149/1.2425257.

17. Öjefors, L.; Carlsson, L. An Iron—Air Vehicle Battery. *J. Power Sources* 1978, *2* (3), 287–296. 10.1016/0378-7753(78)85019-8.

18. Cohn, G.; Starosvetsky, D.; Hagiwara, R.; Macdonald, D. D.; Ein-Eli, Y. Silicon–Air Batteries. *Electrochem. Commun.* 2009, *11* (10), 1916–1918. 10.1016/J.ELECOM.2 009.08.015.

19. Li, Y.; Lu, J. Metal-Air Batteries: Will They Be the Future Electrochemical Energy Storage Device of Choice? *ACS Energy Lett.* 2017, *2* (6), 1370–1377. 10.1021/ ACSENERGYLETT.7B00119/ASSET/IMAGES/LARGE/NZ-2017-00119S_0005.JPEG.

20. Raghavan, P.; Ahn, J. H.; Shelke, M. The Role of 2D Material Families in Energy Harvesting: An Editorial Overview. *J. Mater. Res.* 2022, *37* (22), 3857–3864. 10.155 7/S43578-022-00721-Z/FIGURES/2.

21. Manzeli, S.; Ovchinnikov, D.; Pasquier, D.; Yazyev, O. V.; Kis, A. 2D Transition Metal Dichalcogenides. *Nat. Rev. Mater.* 2017 28 2017, *2* (8), 1–15. 10.1038/ natrevmats.2017.33.

22. Han, S. A.; Bhatia, R.; Kim, S. W. Synthesis, Properties and Potential Applications of Two-Dimensional Transition Metal Dichalcogenides. *Nano Converg.* 2015, *2* (1), 1–14. 10.1186/S40580-015-0048-4/FIGURES/10.

23. Joseph, S.; Mohan, J.; Lakshmy, S.; Thomas, S.; Chakraborty, B.; Thomas, S.; Kalarikkal, N. A Review of the Synthesis, Properties, and Applications of 2D Transition Metal Dichalcogenides and Their Heterostructures. *Mater. Chem. Phys.* 2023, *297*, 127332. 10.1016/J.MATCHEMPHYS.2023.127332.

24. Majidi, L. et al. New Class of Electrocatalysts Based on 2D Transition Metal Dichalcogenides in Ionic Liquid. *Adv. Mater.* 2019, *31* (4), 1804453. 10.1002/ADMA.201804453.

25. Guo, H. P.; Gao, X. W.; Yu, N. F.; Zheng, Z.; Luo, W. Bin; Wu, C.; Liu, H. K.; Wang, J. Z. Metallic State Two-Dimensional Holey-Structured Co3FeN Nanosheets as Stable and Bifunctional Electrocatalysts for Zinc–Air Batteries. *J. Mater. Chem. A* 2019, *7* (46), 26549–26556. 10.1039/C9TA10079B.

26. Xu, Y.; Deng, P.; Chen, G.; Chen, J.; Yan, Y.; Qi, K.; Liu, H.; Yu Xia, B.; Xu, Y.; Deng, P.; Chen, G.; Chen, J.; Qi, K.; Liu, H.; Xia, B. Y.; Yan, Y. 2D Nitrogen-Doped Carbon Nanotubes/Graphene Hybrid as Bifunctional Oxygen Electrocatalyst for Long-Life Rechargeable Zn–Air Batteries. *Adv. Funct. Mater.* 2020, *30* (6), 1906081. 10.1002/ADFM.201906081.

27. Lim, K. R. G.; Shekhirev, M.; Wyatt, B. C.; Anasori, B.; Gogotsi, Y.; Seh, Z. W. Fundamentals of MXene Synthesis. *Nat. Synth.* 2022, *1* (8), 601–614. 10.1038/s441 60-022-00104-6.

28. Gogotsi, Y.; Anasori, B. The Rise of MXenes. *ACS Nano* 2019, *13* (8), 8491–8494. 10.1021/ACSNANO.9B06394/ASSET/IMAGES/MEDIUM/NN9B06394_0005.GIF.

29. Cai, X.; Luo, Y.; Liu, B.; Cheng, H. M. Preparation of 2D Material Dispersions and Their Applications. *Chem. Soc. Rev.* 2018, *47* (16), 6224–6266. 10.1039/C8CS00254A.

30. Tang, L.; Tan, J.; Nong, H.; Liu, B.; Cheng, H. M. Chemical Vapor Deposition Growth of Two-Dimensional Compound Materials: Controllability, Material Quality, and Growth Mechanism. *Accounts Mater. Res.* 2021, *2* (1), 36–47. 10.1021/ACCOUNTSMR.0C00063/ASSET/IMAGES/MEDIUM/MR0C00063_0011.GIF.

31. Zhang, Z.; Yang, X.; Liu, K.; Wang, R. Epitaxy of 2D Materials toward Single Crystals. *Adv. Sci.* 2022, *9* (8), 2105201. 10.1002/ADVS.202105201.

32. Lim, J.; Jung, J. W.; Kim, N. Y.; Lee, G. Y.; Lee, H. J.; Lee, Y.; Choi, D. S.; Yoon, K. R.; Kim, Y. H.; Kim, I. D.; Kim, S. O. N2-Dopant of Graphene with Electrochemically Switchable Bifunctional ORR/OER Catalysis for Zn-Air Battery. *Energy Storage Mater.* 2020, *32*, 517–524. 10.1016/J.ENSM.2020.06.034.

33. Wang, N. et al. Hydrophobic RuO2/Graphene/N-Doped Porous Carbon Hybrid Catalyst for Li-Air Batteries Operating in Ambient Air. Electrochim. *Electrochim. Acta.,* 2022, *428*, 140894. 10.1002/ADMA.201803588.

34. Kim, D. Y.; Kim, M.; Kim, D. W.; Suk, J.; Park, J. J.; Park, O. O.; Kang, Y. Graphene Paper with Controlled Pore Structure for High-Performance Cathodes in Li–O2 Batteries. *Carbon N. Y.* 2016, *100*, 265–272. 10.1016/J.CARBON.2016.01.013.

35. Wang, N.; Fu, J.; Cao, X.; Tang, L.; Meng, X.; Han, Z.; Sun, L.; Qi, S.; Xiong, D. Hydrophobic RuO2/Graphene/N-Doped Porous Carbon Hybrid Catalyst for Li-Air Batteries Operating in Ambient Air. *Electrochim. Acta* 2022, *428*, 140894. 10.1016/J.ELECTACTA.2022.140894.

36. Meng, Y.; Zhang, J. K.; Lu, H. Y.; Chen, X. H.; Xu, J. T. High Performance Lithium Oxygen Batteries Based on a Phosphorous-Doped Holey Graphene Cathode. *Rare Met.* 2022, *41* (12), 4027–4033. 10.1007/S12598-022-02089-9/FIGURES/4.

37. Zhang, J.; Chen, X.; Lei, Y.; Lu, H.; Xu, J.; Wang, S.; Yan, M.; Xiao, F.; Xu, J. Highly Rechargeable Lithium Oxygen Batteries Cathode Based on Boron and Nitrogen Co-Doped Holey Graphene. *Chem. Eng. J.* 2022, *428*, 131025. 10.1016/J.CEJ.2021.131025.

38. Wu, S.; Deng, D.; Zhang, E.; Li, H.; Xu, L. CoN Nanoparticles Anchored on Ultra-Thin N-Doped Graphene as the Oxygen Reduction Electrocatalyst for Highly Stable Zinc-Air Batteries. *Carbon N. Y.* 2022, *196*, 347–353. 10.1016/J.CARBON.2022.04.043.

39. Liu, S.; et al. Cobalt-Based Multicomponent Nanoparticles Supported on N-Doped Graphene as Advanced Cathodic Catalyst for Zinc–Air Batteries. *Int. J. Miner. Metall. Mater.* 2022, *29* (12), 2212–2220. 10.1007/S12613-022-2498-0.

40. Zhang, S.; Wen, Z.; Rui, K.; Shen, C.; Lu, Y.; Yang, J. Graphene Nanosheets Loaded with Pt Nanoparticles with Enhanced Electrochemical Performance for Sodium–Oxygen Batteries. *J. Mater. Chem. A* 2015, *3* (6), 2568–2571. 10.1039/C4TA05427J.

41. Niu, W. J.; He, J. Z.; Wang, Y. P.; Sun, Q. Q.; Liu, W. W.; Zhang, L. Y.; Liu, M. C.; Liu, M. J.; Chueh, Y. L. A Hybrid Transition Metal Nanocrystal-Embedded Graphitic Carbon Nitride Nanosheet System as a Superior Oxygen Electrocatalyst for Rechargeable Zn–Air Batteries. *Nanoscale* 2020, *12* (38), 19644–19654. 10.1039/D0NR03987J.

42. Tang, W.; Teng, K.; Guo, W.; Gu, F.; Li, B.; Qi, R.; Liu, R.; Lin, Y.; Wu, M.; Chen, Y. Defect-Engineered Co3O4@Nitrogen-Deficient Graphitic Carbon Nitride as an Efficient Bifunctional Electrocatalyst for High-Performance Metal-Air Batteries. *Small* 2022, *18* (27), 2202194. 10.1002/SMLL.202202194.

43. Fan, Y.; Ida, S.; Staykov, A.; Akbay, T.; Hagiwara, H.; Matsuda, J.; Kaneko, K.; Ishihara, T. Ni-Fe Nitride Nanoplates on Nitrogen-Doped Graphene as a Synergistic Catalyst for Reversible Oxygen Evolution Reaction and Rechargeable Zn-Air Battery. *Small* 2017, *13* (25), 1700099. 10.1002/SMLL.201700099.

44. Yuan, M.; Wang, R.; Fu, W.; Lin, L.; Sun, Z.; Long, X.; Zhang, S.; Nan, C.; Sun, G.; Li, H.; Ma, S. Ultrathin Two-Dimensional Metal-Organic Framework Nanosheets with the Inherent Open Active Sites as Electrocatalysts in Aprotic Li-O 2 Batteries. *ACS Appl. Mater. Interfaces* 2019, *11* (12), 11403–11413. 10.1021/ACSAMI.8B21 808/ASSET/IMAGES/LARGE/AM-2018-21808P_0008.JPEG.

45. He, G.; Han, X.; Moss, B.; Weng, Z.; Gadipelli, S.; Lai, F.; Kafizas, A. G.; Brett, D. J. L.; Guo, Z. X.; Wang, H.; Parkin, I. P. Solid Solution Nitride/Carbon Nanotube Hybrids Enhance Electrocatalysis of Oxygen in Zinc-Air Batteries. *Energy Storage Mater.* 2018, *15*, 380–387. 10.1016/J.ENSM.2018.08.020.

46. Zhu, J.; Metzger, M.; Antonietti, M.; Fellinger, T. P. Vertically Aligned Two-Dimensional Graphene-Metal Hydroxide Hybrid Arrays for Li-O2 Batteries. *ACS Appl. Mater. Interfaces* 2016, *8* (39), 26041–26050. 10.1021/ACSAMI.6B08222/ SUPPL_FILE/AM6B08222_SI_001.PDF.

47. Graphitic-shell encapsulated FeNi alloy/nitride nanocrystals on biomass-derived N-doped carbon as an efficient electrocatalyst for rechargeable Zn-air battery – Wu - 2021 – Carbon Energy – Wiley Online Library https://onlinelibrary.wiley.com/doi/10. 1002/cey2.52 (accessed Jul 21, 2023).

48. Hang, Y.; Zhang, C.; Luo, X.; Xie, Y.; Xin, S.; Li, Y.; Zhang, D.; Goodenough, J. B. α-MnO2 Nanorods Supported on Porous Graphitic Carbon Nitride as Efficient Electrocatalysts for Lithium-Air Batteries. *J. Power Sources* 2018, *392*, 15–22. 10.1016/J.JPOWSOUR.2018.04.078.

49. Zhang, F.; Wei, M.; Sui, J.; Jin, C.; Luo, Y.; Bie, S.; Yang, R. Cobalt Phosphide Microsphere as an Efficient Bifunctional Oxygen Catalyst for Li-Air Batteries. *J. Alloys Compd.* 2018, *750*, 655–658. 10.1016/J.JALLCOM.2018.04.070.

50. Ji, X.; Copenhaver, D.; Sichmeller, C.; Peng, X. Ligand Bonding and Dynamics on Colloidal Nanocrystals at Room Temperature: The Case of Alkylamines on CdSe Nanocrystals. *J. Am. Chem. Soc.* 2008, *130* (17), 5726–5735. 10.1021/JA710909F/ SUPPL_FILE/JA710909F-FILE003.PDF.

51. Li, G.; Li, N.; Peng, S.; He, B.; Wang, J.; Du, Y.; Zhang, W.; Han, K.; Dang, F. Highly Efficient Nb2C MXene Cathode Catalyst with Uniform O-Terminated Surface for Lithium–Oxygen Batteries. *Adv. Energy Mater.* 2021, *11* (1), 2002721. 10.1002/ AENM.202002721.

52. Wang, Y.; Gu, F.; Cao, L.; Fan, L.; Hou, T.; Zhu, Q.; Wu, Y.; Xiong, S. TiCN MXene Hybrid BCN Nanotubes with Trace Level Co as an Efficient ORR Electrocatalyst for Zn-Air Batteries. *Int. J. Hydrogen Energy* 2022, *47* (48), 20894–20904. 10.1016/J.IJHYDENE.2022.04.211.

53. Li, N.; Wang, Y.; Peng, S.; Yuan, Y.; Wang, J.; Du, Y.; Zhang, W.; Han, K.; Ji, Y.; Dang, F. Ti3C2Tx MXene Cathode Catalyst with Efficient Decomposition Li2O2 and High-Rate Cycle Stability for Li-O2 Batteries. *Electrochim. Acta* 2021, *388*, 138622. 10.1016/J.ELECTACTA.2021.138622.

54. Faraji, M.; Parsaee, F.; Kheirmand, M. Facile Fabrication of N-Doped Graphene/ Ti3C2Tx (Mxene) Aerogel with Excellent Electrocatalytic Activity toward Oxygen Reduction Reaction in Fuel Cells and Metal-Air Batteries. *J. Solid State Chem.* 2021, *303*, 122529. 10.1016/J.JSSC.2021.122529.

55. Qiao, J.; Bao, Z.; Kong, L.; Liu, X.; Lu, C.; Ni, M.; He, W.; Zhou, M.; Sun, Z. MOF-Derived Heterostructure CoNi/CoNiP Anchored on MXene Framework as a Superior Bifunctional Electrocatalyst for Zinc-Air Batteries. *Chin. Chem. Lett.* 2023, 108318. 10.1016/J.CCLET.2023.108318.

56. Xu, H.; Zheng, R.; Du, D.; Ren, L.; Li, R.; Wen, X.; Zhao, C.; Zeng, T.; Zhou, B.; Shu, C. Cationic Vanadium Vacancy-Enriched V2–xO5 on V2C MXene as Superior Bifunctional Electrocatalysts for Li-O2 Batteries. *Sci. China Mater.* 2022, *65* (7), 1761–1770. 10.1007/S40843-021-1959-1/METRICS.

57. Nam, S.; Mahato, M.; Matthews, K.; Lord, R. W.; Lee, Y.; Thangasamy, P.; Ahn, C. W.; Gogotsi, Y.; Oh, I. K. Bimetal Organic Framework–Ti3C2Tx MXene with Metalloporphyrin Electrocatalyst for Lithium–Oxygen Batteries. *Adv. Funct. Mater.* 2023, *33* (1), 2210702. 10.1002/ADFM.202210702.

58. Li, X.; Wen, C.; Yuan, M.; Sun, Z.; Wei, Y.; Ma, L.; Li, H.; Sun, G. Nickel Oxide Nanoparticles Decorated Highly Conductive Ti3C2 MXene as Cathode Catalyst for Rechargeable Li–O2 Battery. *J. Alloys Compd.* 2020, *824*, 153803. 10.1016/J.JALLCOM.2020.153803.

59. Wang, Y.; Zhu, J.; Jiang, Y.; An, T.; Huang, J.; Jiang, M.; Cao, M. Highly Graphited Carbon-Coated FeTiO3 Nanosheets in Situ Derived from MXene: An Efficient Bifunctional Catalyst for Zn–Air Batteries. *Dalt. Trans.* 2022, *51* (14), 5706–5713. 10.1039/D2DT00114D.

60. Tian, Y.; Xu, L.; Qian, J.; Bao, J.; Yan, C.; Li, H.; Li, H.; Zhang, S. Fe3C/Fe2O3 heterostructure embedded in N-doped graphene as a bifunctional catalyst for quasi-solid-state zinc–air batteries. *Carbon.* 2019, *146*, 763. 10.1016/j.carbon.2019.02.046.

61. Nguyen, D. C.; Doan, T. L. L.; Prabhakaran, S.; Tran, D. T.; Kim, D. H.; Lee, J. H.; Kim, N. H. Hierarchical Co, and Nb dual-doped MoS2 nanosheets shelled micro-TiO2 hollow spheres as effective multifunctional electrocatalysts for HER, OER, and ORR. *Nano Energy.* 2021, *82*, 105750. 10.1016/j.nanoen.2021.105750.

7 Two-Dimensional Materials in Na-Ion Batteries

Trapa Banik, Ankit Gupta, Indranil Bhattacharya, and Ismail Fidan

7.1 BACKGROUND AND RECENT PROGRESSES IN SODIUM-ION BATTERIES

Increasing dependence on fossil fuels to maintain economic and population development is a major global concern. Changing over from fossil fuels to renewable energies like wind and solar is necessary to meet increased energy demand while reducing environmental concerns. Energy storage devices on a massive scale are essential for tapping into these resources. Massive and inexpensive energy storage is necessary due to the distributed and intermittent nature of renewables. In addition, electrifying the transportation sector, which is a significant contributor to greenhouse gas emissions, highlights the need for energy storage once again. To achieve a cleaner, greener world, the continuous shift from fossil fuels to renewable energy plays a crucial role, and rechargeable electrochemical batteries, as one of the most adaptable energy storage technologies, are at the heart of this shift. In a wide variety of important fields, they play a crucial role as the primary instruments for reducing the CO_2 footprint of the transportation and electricity grid sectors. Lithium-ion batteries (LIBs) are currently the most popular portable electronic device power source due to their high voltage, long cycle life, and ability to function in ambient temperature; however, it will be difficult to meet the demands of large-scale applications due to the scarcity of lithium resources. The growing reliance on batteries in our society has raised concerns about the availability of materials and the need for sustainability. This has resulted in an explosion of battery research into novel chemistries such as Li-S, Li(Na/Al)-air, Mg (Ca)-ion, Na(K)-ion, or aqueous $ZnMnO_2$ systems. While none of the aforementioned "beyond-Li-ion" battery technologies have yet achieved maturity, Na-ion is the closest to doing so, having spawned a number of firms throughout the world, including Faradion (UK), Novasis (USA), HiNa (China), and Tiamat (France). By far the most comparable to LIBs in terms of typical electrode materials and electrolyte compositions are sodium-ion batteries (NIBs). Although the journey of Na-ion and LIBs started together in the same decade back in the 1970s, it was Li-ion's superior performance that propelled it to widespread adoption. NIBs were

DOI: 10.1201/9781003404729-7

formerly dismissed because of their limited storage capacity, but their natural abundance brought them back into consideration in light of the rising demand for rechargeable batteries and the worry for continuous supply of lithium. The Na-ion technology has advanced rapidly over the past eight years by essentially emulating the chemistry of Li-ion. As sodium is so readily available in the environment, NIBs are seen as a potential replacement for LIBs. However, NIBs have a lower energy density than LIBs due to the higher standard reduction potential of Na^+/Na than Li^+/Li, the heavier atomic mass, and the larger ionic radius of Na ions [1]. Because Na-ions are 1.34 times larger in diameter than Li-ions, NIBs exhibit andante kinetics and poor cyclic performance when compared to LIBs. To get beyond this inherent constraint and create usable NIBs, structural modification is a viable approach. It is generally recognized that electrode materials are crucial to the success of developing improved NIBs to satisfy the needs of the renewable energy sector. For this reason, low-cost electrode materials with high energy density, cycle stability, and rate capability must be developed and used. As a result, there has been a flurry of new research into NIBs in recent years.

In recent years, 2D materials for cathode and anode have gained popularity due to their ultrathin atomic/molecular layered structure with more exposed interior atoms and wider planar lengths, which facilitates the tuning of chemical and structural properties. The use of 2D materials in electrodes offers several advantages, including increased contact area between active materials and electrolytes, as well as reduced diffusion length for sodium ions. Consequently, electrodes incorporating 2D materials have a higher theoretical capacity compared to their bulk counterparts. Moreover, mechanical flexibility is more in the case of 2D materials as the weak van der Waals bond facilitates the materials to accommodate larger Na-ion during the cycling when volume change occurs. Despite the numerous advantages of 2D materials, there are several obstacles that need to be addressed before they can be effectively used in NIBs. These challenges include complex synthesis methods, unclear sodiation/desodiation mechanisms, and significant issues with layer stacking and aggregation.

7.2 SODIUM-ION BATTERIES – WORKING PRINCIPLE, ADVANTAGES, AND DISADVANTAGES

The NIB is seen as a viable contender for next-generation large-scale energy storage technology, not only because it has comparable chemical characteristics to lithium but also because of its massive mineral deposit, cheap cost, and environmental aspect. Sodium is the fourth most prevalent element in the earth's crust, with an apparently limitless distribution. With 23 billion tons of soda ash, the United States alone has a mountain of sodium-containing precursor sources. Furthermore, as alternatives to LIBs, NIB can be made using significantly cheaper Trona ($135–165/ ton), from which sodium carbonate is formed, and when compared to lithium carbonate ($5000/ton), it gives a sensible basis to consider the development of NIBs as a replacement for LIBs [1,2]. Furthermore, since sodium is positioned right below lithium in the same group of the periodic table (Group 1), it has many chemical and physical characteristics with Li [3]. For all of these reasons, researchers worldwide are drawn to the creation of NIB, despite the fact that studies for sodium batteries began in the 1970s and 1980s. However, because of significant advances in the area of LIB, the research for NIB is discontinued after that period. After 2010, scientists renewed their interest in NIB technology [4].

When compared to LIBs, NIBs have a very similar intercalation method. According to recent studies, several electrode materials for NIBs have preferable activation energy and migration barriers compared to LIBs [5]. This is because of the voltage, stability, and diffusion of sodium-ion intercalation materials. Electrode materials for both the cathode and anode that can accept Na ions, and perform quick Na-ion insertion and extraction, are more difficult to discover since the Na ion (r = 1.02 Å, CN = 6) is roughly 34% bigger than the Li ion (r = 0.76 Å, CN = 4). Because of its larger ionic radius, Na presents two major difficulties for anode materials in SIBs. The first is a decrease in cell voltage due to the high potentials at which the conversion events occurred, which was notably true for oxides and halides. Second, the intercalation process causes a volume increase of more than 10%, which is more than that for lithium-based counterparts due only to the higher ionic radius of Na. As a result of this mechanical deterioration of anode materials, low cycle life becomes readily apparent [4]. Slower reaction kinetics and modifications to phase behavior (coordination, lattice constants, crystal structure) make the bigger Na+ ion less polarizing, which has a significant impact on diffusion characteristics. Since this structural stability is fragile, poor rate behavior and inadequate cyclability follow. In comparison with lithium, for example, whose desolvation energy in diverse organic solvents is 218 kJ/mole, the desolvation energy for Na ions in organic solvents is 157.3 kJ/mole, according to certain research. As a result, the kinetics of the electrode are improved due to the decreased charge transfer resistance at the electrolyte contact [6]. Moreover, Na+ has a greater standard electrode potential (−2.71 V vs SHE as opposed to −3.02 V vs SHE for lithium), which results in lower capacity and energy density, and a larger atomic weight (23 g mol^{-1} vs 6.9 g mol^{-1}). When compared to lithium (gravimetric capacity: 3829 mA h g^{-1}), sodium's value of 1165 mA h g^{-1} is much lower. Nevertheless, the capacity is mostly determined by the features of the host structures, i.e., the electrodes, since only a tiny percentage of the mass of the components originates from the weight of cyclable Li or Na (for example, 7% mass of $LiCoO_2$, 20% mass of $NaCoO_2$). As a result, in this scenario, the change from LIBs to SIBs should not have any repercussions on energy density. Table 7.1 provides an overview of the differences and similarities between sodium and lithium.

The use of Al metal as a negative current collector rather than Cu is perhaps the most significant benefit of Na-ion chemistry, since Al is not only cheaper and lighter than Cu but also serves to make the battery safer. When a LIB is overcharged, Cu oxidizes and begins dissolving into the electrolyte when the potential of the Cu current collector climbs over around 3.5 V vs Li+/Li. Dendrites, formed when dissolved Cu is reduced onto the anode during subsequent charging, may create severe short circuits and, in extreme cases, flames or explosions [7]. However, Al does not dissolve at voltages larger than 4 V vs Li$^+$/Li because it is passivated by the electrolyte. Since the Al negative current collector is so much more reliable, the NIB may be safely depleted to 0 V ($E_{cathode} - E_{anode} \approx 2.8V$ vs Na$^+$/Na) while not in use. However, aluminum is used as a current collector for LIB

TABLE 7.1

Sodium Versus Lithium Characteristics [55]

Category	Lithium	Sodium
Cation radius (Å)	0.76	1.02
Atomic weight (g mol^{-1})	6.9	23
$E°$ vs. SHE (V)	0	0.3
Cost, carbonates ($ ton^{-1})	5000	150
Capacity (mA h g^{-1}), metal	3829	1165
Coordination preference	Octahedral and tetrahedral	Octahedral and trigonal prismatic

despite the fact that it has an alloy reaction with lithium below 0.1 V vs Li/Li$^+$ [8], but not with sodium. By using aluminum instead of copper as the current collector, NIB may save roughly 8% of the cost over LIB as shown in Figure 7.1 [54].

Besides, sodium-based electrolyte has higher ionic conductivity (7.16 S cm^2/mol) than the Li+ electrolyte (6.54 S cm^2/mol) which is beneficial for ion and electron movement and increases the battery performance. Another advantage with Na-based layered oxides is that a wide range of cheap and abundant transition metals varying from Ti to Cu can be deployed for battery synthesis, while in the case of their Li analog, synthesis is limited to Mn, Ni, and Co [6]. For instance, in the case of LiFeO$_2$ synthesis, a disordered rock salt phase appears that is electrochemically inactive as high-spin Fe3+ (0.645 Å) has a relatively large ionic radius, close to that of Li$^+$ (0.76 Å). In contrast, the larger ionic radius of Na$^+$ (1.06 Å) facilitates the formation of an ideal layered structure for NaFeO$_2$. A similar effect is seen for NaMO$_2$ compounds containing Cr and Ni [7]. Komaba et al. showed that LiCrO$_2$ and NaCrO$_2$ both possess quite similar crystal structures, but the former is inactive in Li cells while the latter is active in Na cells.

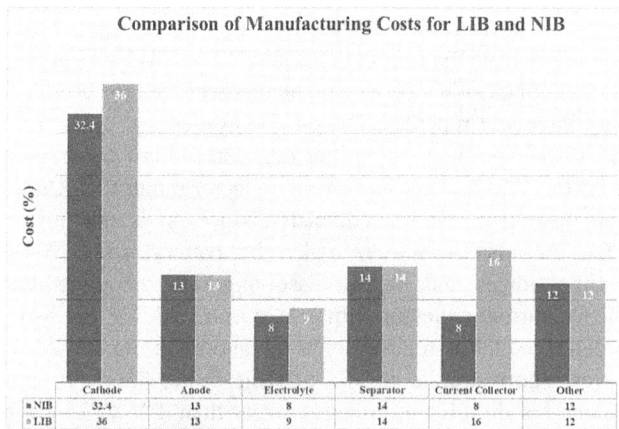

Comparison of Manufacturing Costs for LIB and NIB

	Cathode	Anode	Electrolyte	Separator	Current Collector	Other
NIB	32.4	13	8	14	8	12
LIB	36	13	9	14	16	12

FIGURE 7.1 Comparison of manufacturing cost between lithium-ion battery (LIB) and sodium-ion battery (NIB).

Schauffautl made a groundbreaking discovery in 1841 by demonstrating the process of intercalating sulfate ions into graphite, marking a significant milestone in history. In the 1960s, researchers developed a keen interest in intercalation materials. They were particularly intrigued by the potential of these materials to alter the electrochemical and optical properties of other substances through the process of guest ion intercalation. The initial concept of a rechargeable lithium battery using the intercalation theory was introduced by Whittingham at Exxon Corporation in the United States [9].

Armand suggested the concept of rocking chair or shuttle battery technology in the 1970s [10–15]. The NIB operates on a similar principle as the LIB. In both types of batteries, only the cations (specifically, Na+ ions in the case of NIB) are involved in reversible electrochemical reactions. These reactions occur at the interfaces between the positive electrode and the electrolyte, as well as between the negative electrode and the electrolyte. The cations shuttle back and forth between the intercalation electrodes during these reactions [14]. Typically, sodium ions are extracted from the positive (cathode) source; then they move through the electrolyte and become incorporated into the negative (anode). The discharging process is the reverse of the charging process. Figure 7.2 shows the detailed working principle of NIBs during cycling.

The equation for intercalation in the case of transition metal oxide (TMO) can be rewritten as:

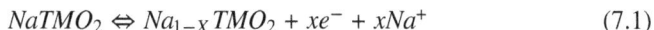

$$NaTMO_2 \Leftrightarrow Na_{1-X}TMO_2 + xe^- + xNa^+ \tag{7.1}$$

Electricity is transferred to loads through an external circuit as electrons escape from the anode. Like LIBs, NIBs have an anode (where Na+ ions are stored during charging and released during discharge) and a cathode (where Na+ ions are produced during charging and stored as reserves during discharge), which are separated by a polymer membrane and combined with an aprotic electrolyte. Na-ion and LIBs are similar to standard batteries, or galvanic cells, in that the redox processes and ion diffusion in the electrolyte are facilitated by the interfaces between the electrolyte and the electrodes.

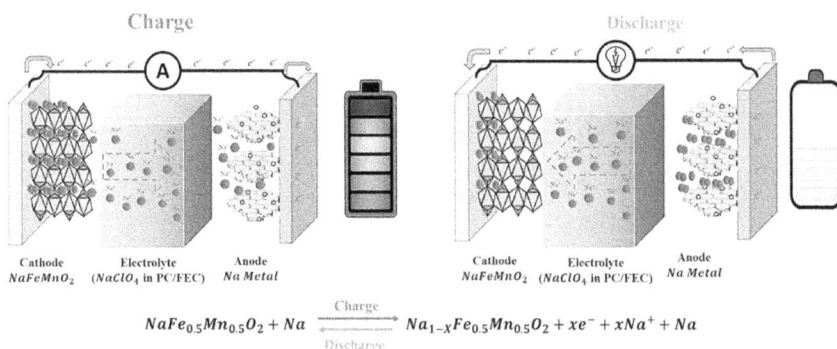

Cathode	Electrolyte	Anode
$NaFeMnO_2$	$(NaClO_4$ in PC/FEC)	$Na\ Metal$

Cathode	Electrolyte	Anode
$NaFeMnO_2$	$(NaClO_4$ in PC/FEC)	$Na\ Metal$

$$NaFe_{0.5}Mn_{0.5}O_2 + Na \xrightleftharpoons[\text{Discharge}]{\text{Charge}} Na_{1-x}Fe_{0.5}Mn_{0.5}O_2 + xe^- + xNa^+ + Na$$

FIGURE 7.2 Working principle of sodium-ion battery (NIB) during cycling.

7.3 TWO-DIMENSIONAL (2D) NANOMATERIALS USED IN SODIUM-ION BATTERIES

Nanomaterials possess distinct characteristics and engineering properties when observed at the nanoscale, setting them apart from the materials at micro-, meso-, and macroscales. Different chemical, mechanical, physical, and biological properties emerge at the nanoscale. To be classified as a nanomaterial, at least one dimension (length, width, or height) must fall within the range of 1–100 nanometers. Materials with all three dimensions in this nanoscale range are referred to as 0D materials. If two dimensions are in the nanoscale, the material is considered 1D, often taking the form of a tube or wore. Similarly, a material with one dimension in the nanoscale range is categorized as a 2D material, resembling a thin sheet or a piece of paper. The incorporation of 2D materials in the fabrication of NIBs proves to be an exceptional means of enhancing their performance and efficiency.

In the realm of NIBs, two-dimensional (2D) materials play diverse roles, serving as both anode and cathode materials. Examples of 2D anode materials include graphene, MXenes (e.g., titanium carbide MXene), and transition metal dichalcogenides (TMDs) like molybdenum disulfide (MoS_2) and for cathode materials, 2D materials encompass lithium nickel cobalt manganese oxide ($Li(NiCoMn)O_2$) and metal phosphates such as vanadium phosphates as shown in Figure 7.3.

Additionally, 2D materials find utility in solid electrolytes, like sodium titanate nanosheets and layered TMDs, and as coatings and protective layers, such as graphene oxide or hexagonal boron nitride (hBN), employed within NIBs. The incorporation of 2D materials in NIBs offers a promising avenue to amplify overall performance and efficiency, paving the way for advancements in energy storage technology.

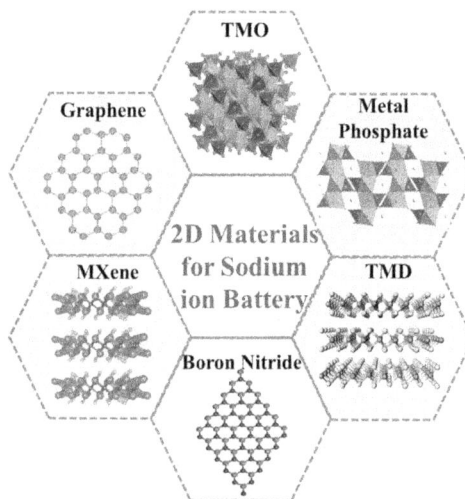

FIGURE 7.3 2D Materials for sodium-ion batteries (NIB).

7.4 APPLICATION OF 2D MATERIALS IN VARIOUS COMPONENTS OF SODIUM-ION BATTERIES

7.4.1 ANODE MATERIALS

The anode is a vital component of a battery, serving as the negative electrode typically made of oxidizing metals [15]. Its primary function is to transfer electrons to the cathode materials through the electrolyte. In rechargeable batteries, the anode materials significantly impact energy storage, and their selection depends on factors like electrical conductivity, physical and chemical stability, compatibility with other battery components, high sodium storage capacity, and structural stability [16]. In LIBs, graphite is commonly used as the main anode material. However, its application in non-lithium-ion batteries (NIBs) is limited due to restricted interlayer spacing. To overcome this, extensive research explores alternative materials like hard carbon, metal oxides, and metal sulfides as potential anodes for non-aqueous rechargeable lithium-ion batteries (NIBs). Moreover, ongoing research focuses on developing novel 2D anode materials with improved energy density, performance, and sustainability for next-generation electrochemical devices. Some widely employed 2D materials as anodes in NIBs are discussed below.

7.4.1.1 Graphene

Graphene, a 2D material composed of single-layer carbon atoms arranged in a hexagonal lattice structure as shown in Figure 7.4 [17], has gained immense recognition and widespread application in various fields, including electronics, energy storage devices such as batteries and supercapacitors, and sensors.

As an active anode material, graphene proves advantageous in accommodating sodium ions during charging, resulting in enhanced capacity and cycling stability owing to its high electronic conductivity [17,19]. In research conducted by Wang et al., reduced graphene oxide (rGO) demonstrated remarkable cycling stability,

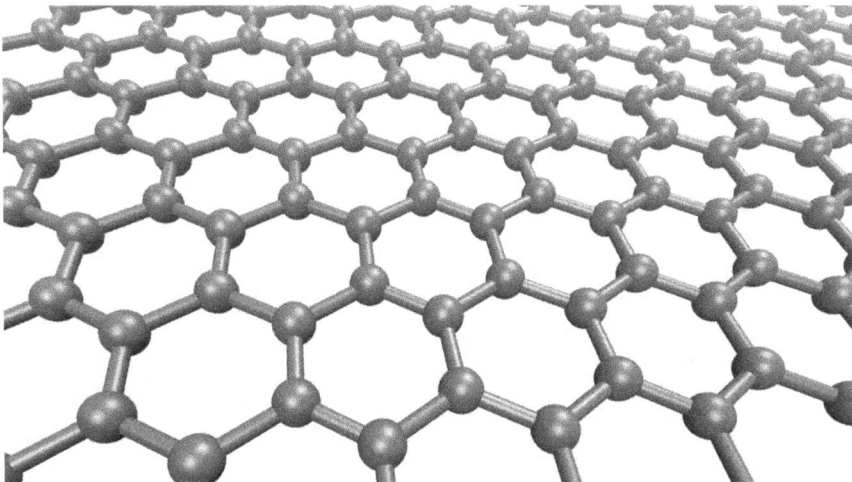

FIGURE 7.4 Schematic diagram of monolayer graphene lattice structure [18].

maintaining its capacity over multiple charge and discharge cycles as an anode material for NIBs [20].

Graphene exhibits great promise as a potential material for rechargeable NIBs. Dua et al. conducted an analysis of twin graphene's electrochemical performance as an anode material in NIBs, revealing exceptional characteristics in terms of sodium-ion storage capacity, rate capability, and cycling stability [21].

Chemical and physical treatments have been employed to improve the electrochemical performance of graphite-based batteries. Cabello et al.'s research study enhances our comprehension of the factors influencing the reliability and appropriateness of expanded graphite as an anode material for NIBs. Through the thermal expansion of natural graphite, they successfully increased the rate capacity and efficiency of NIBs to a significant extent [22]. Similarly, Wen et al. achieved excellent storage capacity, rate capability, and cycling stability for NIBs using treated graphite [23]. Overall, the utilization of graphene in NIBs offers the potential to advance energy storage technologies, enhancing their efficiency, durability, and environmental friendliness.

7.4.1.2 MXenes (Metal Carbide and Nitride)

MXenes have been recognized as a significant and valuable addition to the family of 2D materials. The schematic is shown in Figure 7.5 [24]. The discovery of MXenes,

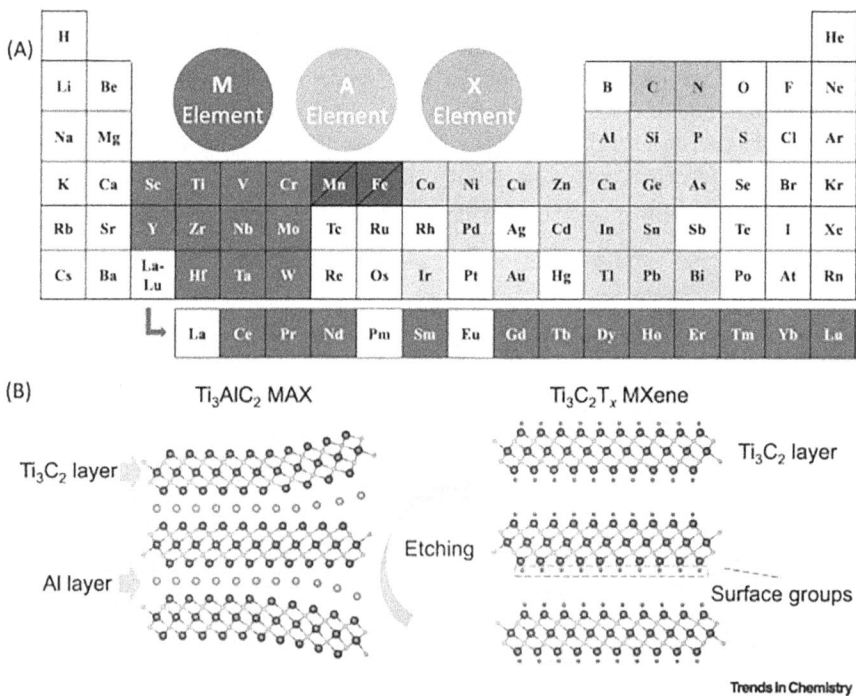

FIGURE 7.5 (A) The elements from the periodic table for MXene. (B) Schematic diagram of MXenes lattice structure [24].

which belong to the group of 2D carbides and nitrides, has sparked considerable interest among scientists and engineers primarily because other 2D materials often suffer from low mechanical properties and air instability.

MXenes offer promising alternatives and open new avenues for research and application in various fields. MXenes possess a metal-like electrical conductivity, rendering them highly suitable for utilization in energy storage and optoelectronic applications [25]. Notably, energy storage devices such as NIBs, LIBs, and other battery types exemplify user-friendly and environmentally conscious solutions. Given the abundant availability of sodium, NIBs hold the potential to replace traditional LIBs in applications like power grids, solar energy storage, and wind energy storage [26].

Continuous research endeavors are focused on investigating and enhancing the characteristics of MXenes, presenting exciting prospects for technological advancements across multiple domains. Xie et al.'s study showcases the potential of a composite paper combining MXene and carbon nanotubes as a highly promising material for high-capacity sodium-based energy storage devices. The composite paper exhibits a remarkable volumetric capacity, rendering it well-suited for sodium-based energy storage applications. Moreover, the porous structure of the composite paper facilitates efficient transport and storage of sodium ions, resulting in enhanced energy storage performance [27].

The electrochemical performance of MXenes in batteries can be influenced by several issues. One of these is the presence of compact multilayer structures and deficient delamination, which may reduce the number of accessible active sites for ion storage and impede fast ion diffusion within the MXene layers. Consequently, this can lead to decreased capacity, rate capability, and cycling stability in battery applications. To address these challenges and broaden the scope of MXene applications, Wu et al. introduced a method involving high-energy mechanical milling to achieve scale delamination of few-layer MXenes. This process has been found to significantly enhance the performance of MXenes in NIBs. Delaminated few-layer MXenes exhibit improved electrochemical properties, including enhanced capacity and cycling stability [28].

7.4.1.3 Transition Metal Dichalcogenides (TMDs)

TMDs, such as TiS_2, VS_2, CrS_2, CoTe2, $NiTe_2$, ZrS_2, NbS_2, and MoS_2, are presently captivating the attention of researchers as 2D materials. They are composed of transition metal atoms like molybdenum, tungsten, or titanium, arranged between chalcogen atoms like sulfur, selenium, or tellurium in a layered lattice structure [29]. The reason they are classified as 2D materials is their ultra-thin nature, consisting of only a few atomic layers as shown in Figure 7.6 [30].

In recent years, TMDs have garnered considerable interest owing to their distinctive properties and versatile applications across diverse fields. Their layered structure has been particularly noteworthy [31], as well as their semiconducting behavior, which enables electronic and optoelectronic applications. Additionally, TMDs exhibit good mechanical strength, making them viable candidates for structural applications. Their electrocatalytic properties have been demonstrated in the hydrogen evolution process [32,33], and they have also shown potential in energy storage applications such as batteries and supercapacitors [34].

(A)

IA																	0
1 H	IIA											IIIA	IVA	VA	VIA	VIIA	2 He
3 Li	4 Be											5 B	6 C	7 N	8 O	9 F	10 Ne
11 Na	12 Mg	IIIB	IVB	VB	VIB	VIIB		VIII			IB	IIB	13 Al	14 Si	15 P	16 S	17 Cl	18 Ar
19 K	20 Ca	21 Sc	22 Ti	23 V	24 Cr	25 Mn	26 Fe	27 Co	28 Ni	29 Cu	30 Zn	31 Ga	32 Ge	33 As	34 Se	35 Br	36 Kr
37 Rb	38 Sr	39 Y	40 Zr	41 Nb	42 Mo	43 Tc	44 Ru	45 Rh	46 Pd	47 Ag	48 Cd	49 In	50 Sn	51 Sb	52 Te	53 I	54 Xe
55 Cs	56 Ba	La-Lu	72 Hf	73 Ta	74 W	75 Re	76 Os	77 Ir	78 Pt	79 Au	80 Hg	81 Tl	82 Pb	83 Bi	84 Po	85 At	86 Rn
87 Fr	88 Ra	Ac-Lr	104 Rf	105 Db	106 Sg	107 Bh	108 Hs	109 Mt	110 Ds	111 Rg	112 Cn	113 Uut	114 Uuq	115 Uup	116 Uuh	117 Uus	118 Uuo

(B)

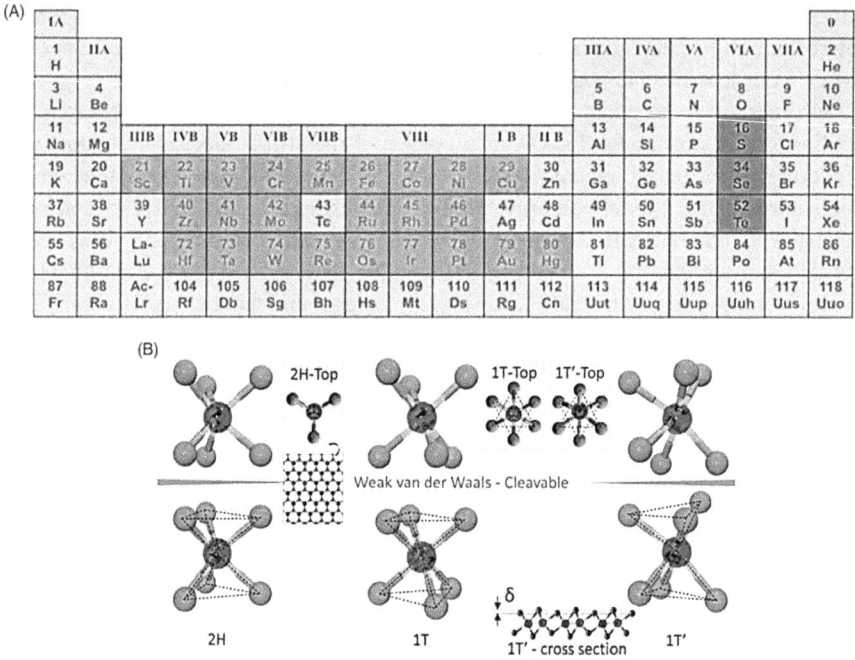

2H-Top   1T-Top   1T'-Top

Weak van der Waals - Cleavable

2H   1T   1T'

1T' - cross section

FIGURE 7.6 (A) The elements from the periodic table for transition metal dichalcogenides (TMDs). (B) Schematic diagram of TMDs lattice structure [30].

Extensive research is currently underway to investigate the synthesis, characterization, and manipulation of TMDs to fully unleash their potential and facilitate the development of innovative technologies. TMDs have garnered significant attention in energy storage and conversion devices due to their remarkable electrochemical stability, characterized by minimal degradation and corrosion [34]. In the realm of energy storage, NIBs have been gaining increasing interest from scientists and academics. This heightened attention is attributed to the various advantages they offer over traditional LIBs.

The production and disposal of LIBs can result in significant environmental impacts, mainly due to challenges associated with mining and recycling processes. To address these concerns, NIBs present a more environmentally friendly alternative. Unlike lithium, sodium is abundant and widely distributed, alleviating worries about resource scarcity. As a result, extensive research is underway to develop novel materials for producing the electrodes of NIBs, aiming to enhance their utilization and potentially surpass LIBs in various applications.

TMDs possess advantageous features such as high surface area and excellent electrical conductivity, making them a favorable choice for energy storage applications, including batteries [35]. In a study by Tang et al., the potential applications of heterostructures comprising TMDs and MXenes in NIBs and Na-O2 batteries were explored. The findings suggest that these heterostructures can significantly enhance the electrochemical performance of both battery systems, leading to improved rate capability, higher capacity, and enhanced stability. The combination of TMDs and MXenes in such heterostructures holds promise for advancing energy storage technologies [36].

Ongoing research is exploring TMDs as promising electrode materials for NIBs, owing to their favorable electrochemical performance. One notable study by David et al. focuses on the creation of a composite paper utilizing molybdenum disulfide (MoS2) and graphene as flexible NIB electrodes. The MoS2/graphene composite paper demonstrates significant enhancements in electrochemical performance, such as improved rate capability, doubled capacity, and exceptional cycling stability, showcasing its potential as a highly favorable material for NIB applications [37]. Mukherjee et al. investigate the potential application of exfoliated TMD nanosheets in supercapacitors and NIBs. These nanosheets demonstrate enhanced properties, such as high surface area and electrical conductivity, rendering them highly promising for energy storage purposes. The study suggests that incorporating these nanosheets could enhance the performance of supercapacitors and NIBs, leading to improved capacitance and storage capacity. Moreover, the research emphasizes the significance of nanosheet morphology in optimizing energy storage devices and underscores the promising prospects of TMD nanosheets in advancing energy storage technologies in the future [38].

7.4.2 Cathode Materials

The cathode, a vital component of battery cells, fulfills the essential function of gaining electrons, contrasting the anode's role of releasing them. As the positive electrode within the battery system, it is typically constructed using metals known for their excellent reducing properties [15]. The performance of cathode materials is contingent upon critical factors such as energy storage capacity, voltage range, and cyclic stability, which significantly influence the overall battery performance.

In the realm of NIBs, the ever-evolving field of 2D materials has seen promising advancements, with some materials finding extensive application as cathodes. These innovative 2D materials offer a host of advantages, including enhanced energy density, improved performance, and a sustainable outlook, contributing to the development of next-generation electrochemical devices. Several frequently utilized 2D materials serve as cathodes in NIBs, contributing to the advancement of this technology as follows.

7.4.2.1 Layered Transition Metal Oxides (TMO)

Layered transition metal oxides with the general formula [14] are recognized as highly promising cathode materials for LIBs with the general formula of $Na_xTMO_2 (x = 0 \sim 1)$ [39]. These oxides possess the capability to be tailored for utilization as NIB cathodes through the strategic substitution of some lithium ions with sodium ions, accommodating the larger size of sodium ions. As these oxides are employed as cathode materials in NIBs, they undergo a reversible process of sodium-ion insertion and extraction within their layered structure. This remarkable characteristic allows for efficient and reliable energy storage and release during the battery's charge and discharge cycles.

In the charging process of NIBs, a fascinating interplay takes place. Sodium ions are skillfully intercalated into the layers of the cathode material, causing oxidation for transition metal ions (Ni, Co, Mn, etc.) and reduction of oxygen. However, during discharging, the scene changes dramatically as the process reverses itself. Sodium ions elegantly de-intercalate from the cathode, while the transition metal ions and oxygen engage in their own synchronized performance of reduction and oxidation, respectively.

Alkali ions are tucked between layers of edge-sharing TMO_6 octahedra in layered transition metal oxides, as hypothesized by Delmas et al. [40,41] in the 1980s. When these TMO_6 octahedra are joined together by edge sharing, they create TMO_2 layers. Evolution of polymorphisms between TMO2 layers provides diverse intercalation sites for the alkali cation, leading to distinct kinds of layered oxides [5,42]. This is dependent on the stacking pattern of TMO_6 sheets along the c-axis. Delmas proposed dividing the layered oxide compounds of sodium (NaxMO2, M = transition metal) into two main classes: O2, O3, P2, and P3 as shown in Figure 7.7. The number denotes the total number of distinct transition metal interlayers in a repeating stacking unit perpendicular to the oxide layers, while the letter denotes whether the octahedral or prismatic sites occupied by alkali ions are represented by the letter 'O' or 'P'. Selecting the right layered metal oxides is crucial to the battery's electrochemical performance. The electrochemical behavior of a battery is determined by factors such as its Na concentration in its natural state, the stability of its individual layers, and the kinetics it experiences from its surroundings. Overall, Na ion sites in the prismatic configuration are more stable than those in the octahedral geometry.

The inclusion of Na vacancies in such a structure also results in increased interlayer distance between TMO_2 slabs due to electrostatic repulsion. Therefore, because of their closed-pack prismatic sites and wider inter-layer distance, P2 type

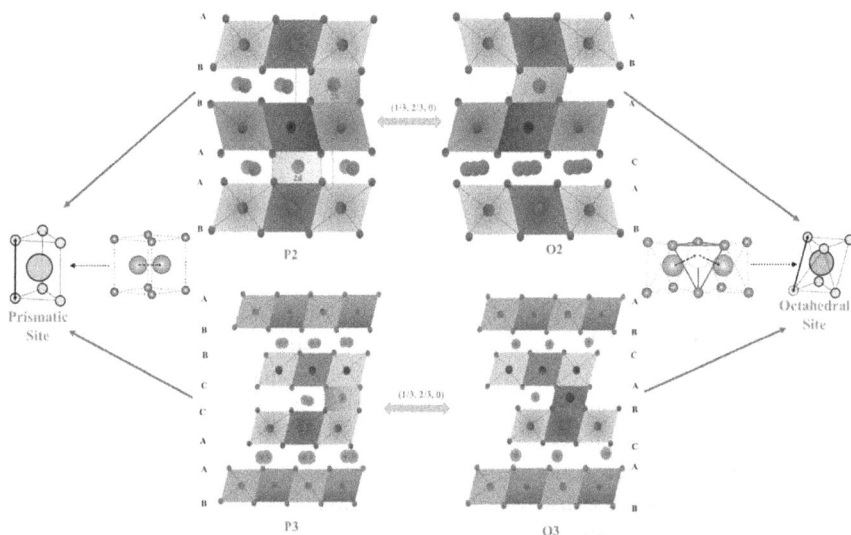

FIGURE 7.7 Crystal structure and sodium-ion migration path for O2, O3, P2, and P3-type cathode materials for SIBs.

frameworks allow for quicker charge/discharge of bigger Na^+ ions. However, in O3-type materials, Na^+ often follows a longer channel, reducing its diffusion rate. Because Na^+ ions need much more energy to migrate from the original octahedral site to the next octahedral site than P2-type ions do, they must first migrate to a face-shared neighboring tetrahedron site before moving on to the adjacent octahedral site across distinct planes. Since P2 type structures allow for more ion mobility than O3 type structures, they can easily acquire high-rate capabilities.

Layered transition metal oxides exhibit an array of impressive advantages in NIBs. The symphony of benefits includes achieving high energy density, maintaining stable cycling performance over multiple charge and discharge cycles, demonstrating exceptional rate capability, and retaining capacity with unwavering grace [41,43]. Among the notable performers on this stage of electrochemical innovation, sodium cobalt oxide ($NaCoO_2$) stands tall, accompanied by other remarkable examples of layered transition metal oxides, collectively contributing to the harmonious progress of NIB technology.

Zuo et al. conducted a comprehensive study on sodium-layered transition metal oxides (Na_xTMO_2, where TM represents transition metal/s), with a specific emphasis on Mn-based sodium-layered oxides. Their objective was to explore a cost-effective alternative to traditional LIB. The research findings revealed a significant cost difference between NIB and LIB, offering valuable insights into the potential feasibility of utilizing Mn-based sodium-layered oxides as an economically viable substitute for LiB [44]. This discovery holds promise in advancing the development of more affordable energy storage solutions.

In the pursuit of advancing NIBs and devising more efficient and reliable energy storage solutions, the incorporation of dopants becomes indispensable. Doping entails introducing foreign atoms or ions into the crystal lattice of materials to optimize their properties. In an illuminating study conducted by Shao et al., the electrochemical performance of layer-structured Ti-doped O3-$Na_{1-x}Cr_{1-x}Ti_xO_2$ cathode material for NIBs was thoroughly investigated. Through meticulous examination of three distinct doping levels ($x = 0, 0.03, 0.05$) of titanium within the cathode structure, the research revealed compelling outcomes. The Ti-doped O3 $Na_{1-x}Cr_{1-x}Ti_xO_2$ cathodes showcased remarkable electrochemical properties, including enhanced capacity, excellent rate capability, and superior cycling stability. These compelling findings underscore the immense potential of these materials as highly promising cathode candidates, capable of powering efficient and high-performance NIBs [45].

Furthermore, Zhang et al. demonstrated a novel approach by introducing rare earth elements, particularly iridium, into manganese-based Na-rich materials (Mn-based $Na_{1.2}Mn_{0.4}Ir_{0.4}O_2$) used as layered cathodes for NIBs. The strategic inclusion of iridium establishes a strong covalent bond with oxygen atoms, effectively suppressing the release of oxygen during battery operation [46]. This innovative modification holds significant promise for enhancing the stability and performance of NIBs, making them a potential candidate for future energy storage solutions.

Continuing efforts in research are dedicated to elevating the performance of these materials and innovating new cathode materials with superior electrochemical properties. The objective is to create highly efficient and sustainable NIBs, capable of effectively overcoming challenges such as weakened structural stability and capacity degradation during extensive cycling periods. By addressing these issues, these advancements pave the way for more reliable and long-lasting energy storage solutions.

7.4.2.2 Metal Phosphates

The fascinating potential of 2D metal phosphates, particularly vanadium phosphates, has opened up exciting opportunities for the advancement of NIB technology. With their layered crystal structure resembling that of graphene and possessing unique electrochemical properties, metal phosphates, including vanadium phosphates, have emerged as promising 2D materials for NIBs. Within this 2D structure, metal atoms are arranged in well-defined layers, each enveloped by oxygen and phosphorus atoms as shown in Figure 7.8 [47].

The stacking of these layers is facilitated by weak van der Waals forces. Notably, their abundant natural availability, cost-effectiveness, exceptional surface area, compositional flexibility, and high-rate capacity make them highly promising candidates as the next-generation materials for rechargeable batteries [48].

Embracing these 2D metal phosphates presents an avenue for significant advancements in NIB technology, propelling the future of sustainable energy storage solutions. The remarkable long-term structural stability of metal phosphates positions them as the ideal choice for electrode fabrication [48,49]. In a notable study, Fang et al. harnessed the impressive olivine NaFePO4 microsphere structure

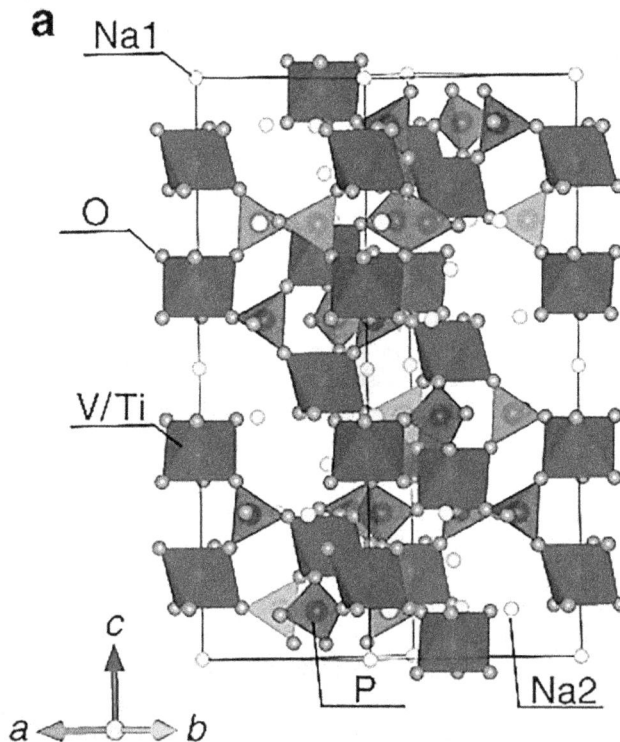

FIGURE 7.8 Schematic diagram of metal phosphates ($Na_2VTi(PO_4)_3$) lattice structure [47].

to significantly enhance the cathode's performance in NIBs. The utilization of this high-performance microsphere structure led to remarkable improvements in specific capacity, rate capability, and cycling stability, showcasing its potential as a valuable advancement in battery technology [50]. According to Fang et al., the fully sodiated NaVOPO4 cathode exhibits outstanding stability, high capacity, and prolonged cycling lifespan, showing great potential as a reliable choice for high-energy and long-lasting NIB applications [51].

Moreover, the continuous research and development in the field of metal phosphates hold significant potential for the discovery of novel cathode materials with even further improved performance characteristics. The pursuit of these advancements is expected to unlock broader applications in diverse energy storage devices, paving the way for more efficient and sustainable energy solutions for the future.

7.4.3 Solid Electrolytes

Solid electrolytes present a promising substitute for conventional liquid electrolytes in NIBs. They offer a host of advantages, notably improved safety, enhanced stability, and the potential to pave the way for the development of high-energy-density and long-lasting battery solutions.

Among the myriad of 2D materials, boron nitride (BN) stands out as an exceptionally promising candidate for serving as a solid electrolyte in advanced battery technologies. Its unique properties and characteristics make it a compelling choice to revolutionize the landscape of energy storage and unlock new possibilities for the future of battery technology.

In its fascinating two-dimensional (2D) form shown in Figure 7.9 [52], BN showcases a layered crystal structure similar to other renowned 2D materials like graphene, while also demonstrating exceptional thermal and mechanical stability.

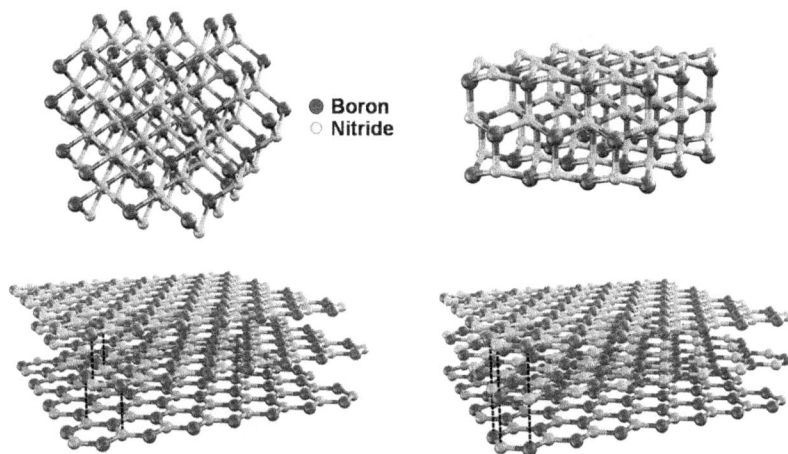

FIGURE 7.9 Schematic diagram of different crystal structures of boron nitride [53].

This remarkable combination of properties positions BN as a highly promising candidate for deployment as a 2D solid electrolyte, offering exciting prospects for the advancement of NIBs. Embracing BN in this role opens the door to the development of safer, more efficient, and longer-lasting energy storage systems, setting the stage for transformative innovations in the field of electrochemical energy storage.

The pursuit of solid electrolytes for NIBs remains an active and dynamic field of research, driven by the relentless efforts of scientists. They are continuously exploring cutting-edge materials and refining existing ones to push the boundaries of performance and unlock the full commercial potential of solid-state NIBs. The quest for superior energy storage solutions is paving the way for groundbreaking advancements in this exciting domain.

7.5   CONCLUSION

This chapter aims to enhance the understanding of 2D materials in NIBs, specifically focusing on graphene, TMD, and MXenes as anode materials, and transition metal oxides and metal phosphates as cathode materials. 2D materials have shown great potential in improving battery performance. Graphene's exceptional electrical conductivity and mechanical strength make it a promising anode material, leading to enhanced rate capability and capacity. Similarly, MXenes and TMDs exhibit excellent electrical conductivity and high surface areas, making them potential anode materials. However, layered transition metal oxides and metal phosphates offer advantages like high energy density, stable cycling performance, and excellent rate capability as cathode materials. Doping with elements like titanium and iridium further enhances their electrochemical properties. Additionally, boron nitride acts as an excellent solid electrolyte material, ensuring enhanced safety and stability. Utilizing 2D materials in NIBs holds great promise for advancing energy storage technologies, paving the way for more efficient, durable, and sustainable battery solutions. Continued research and optimization of these materials will undoubtedly drive the future of NIBs and contribute to the next generation of energy storage systems.

CONFLICTS OF INTEREST

The authors declare no conflicts of interest related to this book chapter.

REFERENCES

1. J. Y. Hwang, S. T. Myung, and Y. K. Sun, "Sodium-ion batteries: present and future," *Chem. Soc. Rev.*, 46,3529–3614, 2017.
2. A. Chandra, A. Chandra, and R. S. Dhundhel, "Electrolytes for sodium ion batteries: a short review," *Indian J. Pure Appl. Phys.*, 58, 113–119, 2020.
3. J. Wu, Z. X. Shen, and W. Yang, "Redox mechanism in Na-ion battery cathodes probed by advanced soft X-ray spectroscopy," *Front. Chem.*, 8, 1–16, Sep. 2020.
4. W. Li, "Investigation on the promising electrode materials for rechargeable sodium ion batteries," PhD Thesis, 2015.

5. T. Wang, D. Su, D. Shanmukaraj, T. Rojo, M. Armand, and G. Wang, "Electrode materials for sodium-ion batteries: considerations on crystal structures and sodium storage mechanisms," *Electrochem. Energy Rev.*, 1, 200–237, 2018.

6. P. K. Nayak, L. Yang, W. Brehm, and P. Adelhelm, "From lithium-ion to sodium-ion batteries: advantages, challenges, and surprises," *Angew. Chem. Int. Ed.*, 57, 102–120, 2018.

7. J. W. Somerville, "Understanding the structural and electrochemical behavior of high energy density layered oxide cathodes for sodium and lithium ion batteries," PhD Thesis, University of Oxford. 2019.

8. E. Peled, "The electrochemical behavior of alkali and alkaline earth metals in nonaqueous battery systems—the solid electrolyte interphase model," *J. Electrochem. Soc.*, 126, 2047–2051, Dec. 1979.

9. A. Manthiram, "A reflection on lithium-ion battery cathode chemistry," *Nat. Commun.*, 11, 1–9, 2020.

10. M. Armand, and P. Touzain, "Graphite intercalation compounds as cathode materials," *Mater. Sci. Eng.*, 31, 319–329, Dec. 1977.

11. M. B. Armand, "Intercalation electrodes," *NATO Conference Series, (Series) 6: Materials Science*, 2, 145–161, 1980.

12. M. V. Reddy, A. Mauger, C. M. Julien, A. Paolella, and K. Zaghib, "Brief history of early lithium-battery development," *Materials (Basel)*, 13, 1–9, 2020.

13. A. Mauger, C. M. Julien, J. B. Goodenough, and K. Zaghib, "Tribute to Michel Armand: from rocking chair – Li-ion to solid-state lithium batteries," *J. Electrochem. Soc.*, 167, 7, 070507, 2020.

14. C. Liu, Z. G. Neale, and G. Cao, "Understanding electrochemical potentials of cathode materials in rechargeable batteries," *Mater. Today*, 19, 109–123, 2016.

15. Dragonfly Energy. (n.d.). Anode vs. cathode: what's the difference? Dragonfly Energy. Retrieved from https://dragonflyenergy.com/anode-vs-cathode/

16. K. V. Galloway, and N. M. Sammes, "Fuel cells – solid oxide fuel cells I anodes," *Encyclopedia of Electrochemical Power Sources.* Elsevier, pp. 17–24, Jan. 2009.

17. R. Raccichini, A. Varzi, S. Passerini, and B. Scrosati, "The role of graphene for electrochemical energy storage," *Nat. Mater.*, 14, 271–279, 2015.

18. A. Armano, and S. Agnello, "Two-dimensional carbon: a review of synthesis methods, and electronic, optical, and vibrational properties of single-layer graphene," *C—J Carbon Res.*, 5, 67, 2019.

19. Y. M. Chang, H. W. Lin, L. J. Li, and H. Y. Chen, "Two-dimensional materials as anodes for sodium-ion batteries," *Mater. Today Adv.*, 6, 100054, 2020.

20. Y. X. Wang, S. L. Chou, H. K. Liu, and S. X. Dou, "Reduced graphene oxide with superior cycling stability and rate capability for sodium storage," *Carbon N. Y.*, 57, 202–208, 2013.

21. H. Dua, J. Deb, D. Paul, and U. Sarkar, "Twin-graphene as a promising anode material for Na-ion rechargeable batteries," *ACS Appl. Nano Mater.*, 4, 4912–4918, 2021.

22. M. Cabello et al., "On the reliability of sodium co-intercalation in expanded graphite prepared by different methods as anodes for sodium-ion batteries," *J. Electrochem. Soc.*, 164(14), A3804, 2017.

23. Y. Wen *et al.*, "Expanded graphite as superior anode for sodium-ion batteries," *Nat. Commun.*, 5(1), 4033, 2014.

24. Z. Lin, H. Shao, K. Xu, P. L. Taberna, and P. Simon, "MXenes as high-rate electrodes for energy storage," *Trends Chem.*, 2(7), 654–664, 2020.

25. M. Naguib, M. W. Barsoum, and Y. Gogotsi, "Ten years of progress in the synthesis and development of MXenes," *Adv. Mater.*, 33(39), 2103393, 2021.

26. M. K. Aslam, T. S. AlGarni, M. S. Javed, S. S. A. Shah, S. Hussain, and M. Xu, "2D MXene materials for sodium ion batteries: a review on energy storage," *J. Energy Storage*, 37, 102478, 2021.

27. X. Xie *et al.*, "Porous heterostructured MXene/carbon nanotube composite paper with high volumetric capacity for sodium-based energy storage devices," *Nano Energy*, 26, 513–523, 2016.

28. Y. Wu, P. Nie, J. Wang, H. Dou, and X. Zhang, "Few-layer MXenes delaminated via high-energy mechanical milling for enhanced sodium-ion batteries performance," *ACS Appl. Mater. Interfaces*, 9(45), 39610–39617, 2017.

29. A. Eftekhari, "Tungsten dichalcogenides (WS2, WSe2, and WTe2): Materials chemistry and applications," *J. Mater. Chem. A*, 5(35), 18299–18325, 2017.

30. M. Wu *et al.*, "Synthesis of two-dimensional transition metal dichalcogenides for electronics and optoelectronics," *InfoMat.*, 3(4), 362–396, 2021.

31. X. Chia, A. Y. S. Eng, A. Ambrosi, S. M. Tan, and M. Pumera, "Electrochemistry of nanostructured layered transition-metal dichalcogenides," *Chem. Rev.*, 115(21), 11941–11966, 2015.

32. A. Datar, M. Bar-Sadan, and A. Ramasubramaniam, "Interactions between transition-metal surfaces and MoS2 monolayers: implications for hydrogen evolution and CO2 reduction reactions," *J. Phys. Chem. C*, 124(37), 20116–20124, 2020.

33. L. Yang, P. Liu, J. Li, and B. Xiang, "Two-dimensional material molybdenum disulfides as electrocatalysts for hydrogen evolution," *Catalysts.*, 7(10), 285, 2017.

34. M. B. Askari *et al.*, "Transition-metal dichalcogenides in electrochemical batteries and solar cells," *Micromachines*, 14, 691, 2023.

35. Z. Shi, H. Huang, C. Wang, M. Huo, S. H. Ho, and H. S. Tsai, "Heterogeneous transition metal dichalcogenides/graphene composites applied to the metal-ion batteries," *Chem. Eng. J.*, 447, 137469, 2022.

36. C. Tang, Y. Min, C. Chen, W. Xu, and L. Xu, "Potential applications of heterostructures of TMDs with MXenes in sodium-ion and Na-O2 batteries," *Nano Lett.*, 19(8), 5577–5586, 2019.

37. L. David, R. Bhandavat, and G. Singh, "MoS2/graphene composite paper for sodium-ion battery electrodes," *ACS Nano*, 8(2), 1759–1770, 2014.

38. S. Mukherjee *et al.*, "Exfoliated transition metal dichalcogenide nanosheets for super-capacitor and sodium ion battery applications," *R. Soc. Open Sci.*, 6(8), 190437, 2019.

39. C. Tian, F. Lin, and M. M. Doeff, "Electrochemical characteristics of layered transition metal oxide cathode materials for lithium ion batteries: surface, bulk behavior, and thermal properties," *Acc. Chem. Res.*, 51(1), 89–96, 2018.

40. C. Delmas, C. Fouassier, and P. Hagenmuller, "Structural classification and properties of the layered oxides," *Physica B+C*, 99, 81–85, Jan. 1980.

41. C. Delmas, J. J. Braconnier, C. Fouassier, and P. Hagenmuller, "Electrochemical intercalation of sodium in NaxCoO2 bronzes," *Solid State Ionics*, 3–4, 165–169, Aug. 1981.

42. P. He, H. Yu, D. Li, and H. Zhou, "Layered lithium transition metal oxide cathodes towards high energy lithium-ion batteries," *J. Mater. Chem.*, 22, 3680–3695, Mar. 2012.

43. Y. Li Lu *et al.*, "Recent advances of electrode materials for low-cost sodium-ion batteries towards practical application for grid energy storage," *Energy Storage Mater.*, 7, 130–151, Jan. 2017.

44. W. Zuo, A. Innocenti, M. Zarrabeitia, D. Bresser, Y. Yang, and S. Passerini, "Layered oxide cathodes for sodium-ion batteries: storage mechanism, electrochemistry, and techno-economics," *Acc. Chem. Res.*, 56(3), 284–296, 2023.

45. Y. Shao, Z. Feng Tang, J. Ying Liao, and C. Hua Chen, "Layer-structured Ti doped O3-Na1−xCr1−xTixO2(x=0, 0.03, 0.05) with excellent electrochemical performance as cathode materials for sodium ion batteries," *Chin. J. Chem. Phys.*, 31(5), 673–676, 2018.

46. X. Zhang *et al.*, "Manganese-based Na-rich materials boost anionic redox in high-performance layered cathodes for sodium-ion batteries," *Adv. Mater.*, 2019.

47. D. Wang *et al.*, "Sodium vanadium titanium phosphate electrode for symmetric sodium-ion batteries with high power and long lifespan," *Nat. Commun.*, 2017.

48. Q. Cheng *et al.*, "Recent advances of metal phosphates-based electrodes for high-performance metal ion batteries," *Energy Storage Mater.*, 2021.

49. C. Masquelier, and L. Croguennec, "Polyanionic (phosphates, silicates, sulfates) frameworks as electrode materials for rechargeable Li (or Na) batteries," *Chem. Rev.*, 113(8), 6552–6591, 2013.

50. Y. Fang, Q. Liu, L. Xiao, X. Ai, H. Yang, and Y. Cao, "High-performance olivine NaFePO4 microsphere cathode synthesized by aqueous electrochemical displacement method for sodium ion batteries," *ACS Appl. Mater. Interfaces*, 7(32), 17977–17984, 2015.

51. Y. Fang *et al.*, "A fully sodiated NaVOPO4 with layered structure for high-voltage and long-lifespan sodium-ion batteries," *Chem*, 4, 1167–1180, 2018.

52. S. Mateti, I. Sultana, Y. Chen, M. Kota, and M. M. Rahman, "Boron nitride-based nanomaterials: synthesis and application in rechargeable batteries," *Batteries*, vol. 9, no. 7, p. 344, 2023.

53. S. Ogawa, S. Fukushima, and M. Shimatani, "Hexagonal boron nitride for photonic device applications: a review," *Materials*, 16(5), 2023.

54. Arnold, Stefanie, Wang, Lei, & Presser, Volker (2022). Dual-use of seawater batteries for energy storage and water desalination. *Small*, 18(43), 2107913, 10.1002/smll.202107913.

55. Slater, Michael D., Kim, Donghan, Lee, Eungje, & Johnson, Christopher S. (2012). Sodium-ion batteries. *Advanced Functional Materials*, 23, 947–958, 10.1002/adfm.201200691.

8 Two-Dimensional Materials for Flexible Batteries

Raj Kumar, Neetu Yadav, Mohit K Sharma, and Praveen Kumar

8.1 INTRODUCTION

The current climate emergency puts the need to develop energy materials (e.g., rechargeable batteries, solar cells, fuel cells, and catalysis) at the highest priority, as demonstrated by their inclusion in the United Nations (UN) Sustainable Development Goals [1]. The advancement of these materials toward powering a more sustainable and renewable future requires the understanding of physical and chemical processes triggered by external stimuli. One of the key challenges of the 21st century is the successful transition from fossil fuels to more sustainable renewable energies. In this context, electrochemical energy storage using rechargeable Li-ion batteries (LIBs) and supercapacitors is presently the technology of choice for the successful transition to renewable energy sources [2,3]. The future of LIBs looks promising, as they continue to be the dominant technology in the energy storage market and electric vehicles are the key technology to decarbonize road transport, a sector that accounts for 16% of global emissions (as per the International Energy Agency (IEA) report 2022) help achieving the Net Zero Emissions target by 2050 [4]. To meet the growing demand for energy storage, intensified research is required to develop next-generation LIBs with dramatically improved performances, for instance, high specific energy and power density, longer cycle lifetime, high mechanical stability, and safety.

The anode and cathode electrodes play an integral role in the efficient operation of LIBs. The existing carbon-based anodes are approaching their fundamental limits, intrinsically limited by the material properties [5,6], while standard cathode materials (lithium cobalt oxide($LiCoO_2$) and lithium iron phosphate $LiFePO_4$) are available which usually have a low capacity (than anode) but produce high voltages than anodes-based materials [7]. Figure 8.1 shows the theoretical specific capacity and electrochemical reduction potential for conventional cathode and anode electrodes [8]. To date, graphite is the only commercially available material used as standard anode in LIBs [5]. One of the drawbacks with graphite-based anodes is the low capacity, limited due to the intercalation process; it can accommodate a small fraction of Li-ions in its structure (6 carbon accommodates 1 Li). There is still

 DOI: 10.1201/9781003404729-8

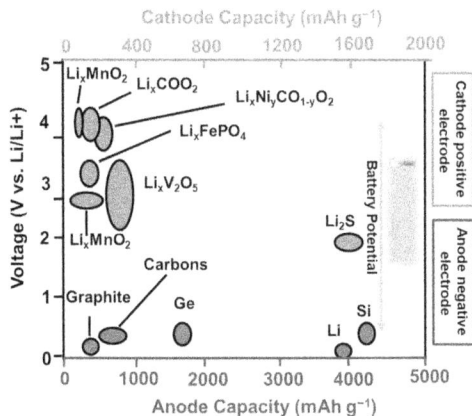

FIGURE 8.1 Theoretical gravimetric capacities and the electrochemical reduction potentials with respect to lithium metal for conventional anodes and cathode-based materials. (Adapted with permission from Ref. [8]. Copyright © 2009, Royal Society of Chemistry.)

plenty of space to search for new materials by replacing the standard graphite material with silicon (Si) and their composites toward high energy density batteries. Considering this, Si anodes have attracted much attention because of their exceptional theoretical specific capacity of 4200 mAh g^{-1}, which is about ten times higher than that of standard graphite anode 372 mAh g^{-1} but suffer from a rapid capacity fading and structural degradation over prolonged cycles [9–12]. Noteworthy is the lithiation mechanism which is distinct in the case of Si (Li$_x$Si$_x$ alloy formation) as compared with graphite (intercalation mechanism). By adding Si, we can significantly enhance the specific capacity (charge storage capacity) by storing more charge in a small volume. The most critical challenge for the practical implementation of Si-based anodes is the large volumetric change (up to 300%) that occurs during the charge/discharge cycles. The volume of Si material expands 300% during charge and contracts upon discharge process. This continuous volume expansion/contraction sequences result in structural degradation, as cracks, pulverization of Si, and potentially leading to the mechanical failure of the Si anode electrode. In addition, it leads to the unfavorable continuous solid electrolyte interface (SEI) formation limiting the migration of the Li-ions between the electrolyte solvent and the active material, which finally causes irreversible capacity loss [13–17]. The electrode swelling, rapid capacity fading, and mechanical fracture are the genuine issues that hamper the commercialization of Si as an anode in LIBs. Thus, intensive research efforts are undertaken to circumvent the inherent limitations of Si by designing hierarchical nanostructures and composites that would limit the degradation effects and maximize capacity retention over long-term cycling. In fact, the trend is to have increasingly complex multi-scale structures toward targeted functionalities [14–17].

Despite several discoveries and promising results on LIBs, still several challenges need to be addressed to further improve battery performance, including energy density, power density, cycle life, safety, and cost. One of the most effective pathways to realize high-performance batteries is to look beyond the scope of existing standard electrode material systems by searching and designing new electrode materials. The primary areas of focus are designing and implementing 2D materials such as graphene, transition metal dichalcogenides (TMDs), black phosphorus (BP), and other metal-oxide materials which can replace conventional bulk materials usually applied in LIBs and supercapacitors [18–20]. Another emerging field is all-solid-state batteries to overcome the safety issues without affecting the battery performance [21,22]. Indeed, the current trend is to have more hybrid or mixed composite 2D materials that can offer higher electronic conductivity, longer cycle life, high energy density, improved mechanical stability, and safety.

Recent advances using 2D materials for flexible batteries have shown promising results due to their unique mechanical, electrical, and chemical properties. Graphene is a 2D carbon material with unique properties such as high electrical conductivity, high specific surface area, and excellent mechanical strength. The flexible batteries based on graphene have shown high specific capacity, high rate capability, and excellent cycling stability [23,24]. However, the main difficulty lies in achieving high-quality large-area graphene layers that have limited its practical applications. Another exciting 2D material is BP which has a layered structure similar to graphene, but with a wider band gap and higher electron mobility. Various standard methods were utilized to synthesize BP such as liquid-phase exfoliation, chemical vapor deposition (CVD), and electrochemical deposition and have demonstrated high specific capacity, rate capability, and excellent cycling stability, making it a potential candidate for practical applications in flexible batteries [25,26]. TMDs are a family of 2D materials with unique electronic and optoelectronic properties, such as high carrier mobility, strong light-matter interactions, and tunable band gaps. TMDs-based flexible batteries are an excellent choice due to their scalability and thickness-dependent electrical/optical properties resulting in a high specific capacity, rate capability, and cycling stability [27–30].

Other 2D materials have also been investigated for flexible batteries that include MXenes [31], metal-oxide nanosheets [32], and layered double hydroxides [33]. The performance of 2D material-based flexible batteries depends on several factors, including thickness, growth quality, and composition, as well as the electrode architecture and the electrolyte used. To improve the performance and scalability of 2D material-based flexible batteries, researchers have been exploring new synthesis methods that are scalable and cost-effective for instance, roll-to-roll fabrication techniques [34], and developing novel architectures toward efficient batteries [35].

In the past years, 2D materials have been considered ideal for their implications in current collectors and electrodes in flexible batteries. Among them, graphene, TMDs, and MXenes-based materials have emerged as potential candidates for fabricating next-generation, highly efficient, and robust energy storage devices especially for Li-ion, Li-sulfur, magnesium-ion batteries, and supercapacitors [18–20]. In this chapter, we will discuss and highlight critical issues associated with conventional electrode materials and summarize the most promising results and applications of 2D materials to alleviate these issues of anode/cathode electrodes for LIBs. This chapter will also shed light on the current progress of various graphene/TMD-based electrodes in LIBs and provide a comprehensive picture of the state-of-the-art technologies, limitations, and challenges associated with these materials. It is difficult to encircle all aspects of 2D materials and their implications in various energy storage systems. Therefore, we restrict ourselves to anode/cathode systems using graphene and TMDs as one of the potential materials.

8.2 BASICS OF BATTERIES: COMPONENTS AND WORKING PRINCIPLE

The various components of a LIB are displayed in Figure 8.2. The basic structure of the LIBs consists of a positive electrode (cathode), a negative electrode (anode), an electrolyte, and a separator. The positive and negative electrodes are separated by a separator, a thin membrane that prevents the anode and cathode from coming into direct contact with each other and causing a short circuit, while allowing the flow of Li ions between the electrodes. The positive electrode is typically made of lithium cobalt oxide ($LiCoO_2$) or lithium iron phosphate-based materials. The standard negative electrode is made of graphite (a form of carbon). The electrolyte comprises a lithium salt (i.e., liquid $LiPF_6$) dissolved in organic solvents that facilitates the movement of lithium ions between the electrodes and blocks the electrons during the charge and discharge cycles. For the anode, Cu is generally used as a current collector while Al is applied as the current collector for cathodes. LIBs work by storing energy through a chemical reaction that takes place when the battery is charged. During the charging process, Li ions move from the positive electrode through the electrolyte to the negative electrode, where they are stored in the graphite material. When the battery is discharged, the reverse reaction occurs, and the Li ions move back to the positive electrode, generating an electric current [22,36,37].

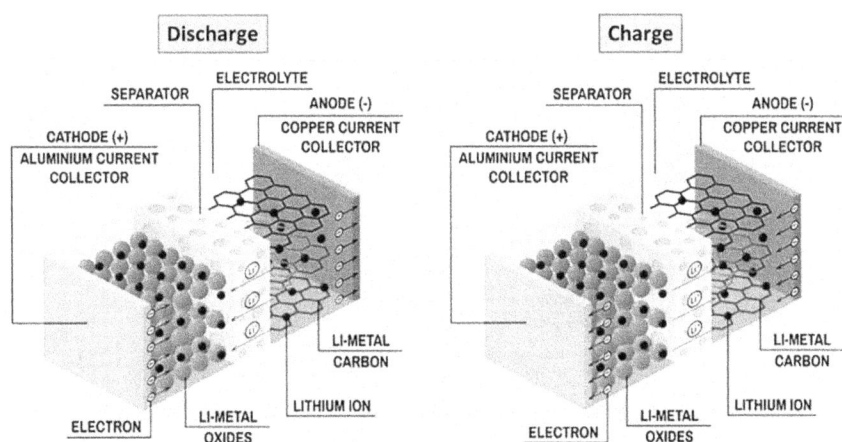

FIGURE 8.2 Components and working principle (charge/discharge process) of a Li-ion battery (Image Credit: https://blog.normagroup.com/en/how-batteries-live-long-in-electric-vehicles/).

The movement of Li ions between the anode and cathode through the electrolyte is facilitated by a difference in electric potential between the two electrodes. This potential difference is created by the chemical reactions (redox reactions) that occur at the electrodes during the charge/discharge process [36,37]. The overall performance of LIBs such as specific capacity and energy density strongly depends on the number of ions stored during the charging process, which strictly relies on the physical and chemical properties, microstructure, and composition of the material.

8.3 COMMON CHALLENGES WITH LIBS

There are several challenges with LIBs such as safety, mechanical stability, aging behavior, thermal stability, low energy density, and reduced capacity over prolonged charge/discharge cycles. Some of these issues have recently been highlighted and discussed in numerous review articles [18–20,22] and are summarized in Figure 8.3. The conventional electrode materials (graphite for anodes and mixed metal oxides and phosphates for cathodes) used in LIBs usually suffer from poor electrical properties that can lead to high polarization and low electrochemical performance [5]. These materials have limited capacity to store and transfer charge, resulting in lower energy and power densities and can undergo irreversible structural changes during cycling, leading to capacity loss and poor cycling stability. These challenges have driven researchers into developing new electrode materials with higher capacity, better electrical properties, and improved stability.

The anode is an essential component of LIBs, and while there have been significant efforts to develop high-capacity and stable anode materials, still many challenges lie ahead in their utilization for practical applications. The two most

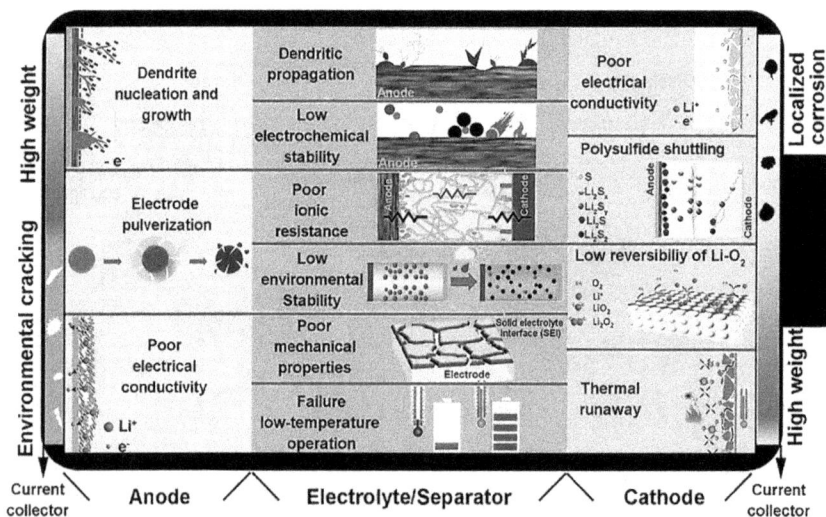

FIGURE 8.3 Common challenges in LIBs. Schematics show the typical challenges for current collectors, cathodes, anodes, separators, and electrolytes of LIBs. (Adapted with permission from Ref. [18]. Copyright © 2020, American Chemical Society.)

commonly used anode materials in LIBs are lithium metal and graphite [38,39]. Lithium metal has a high theoretical capacity of 3860 mA h/g and a low reduction potential of −3.04 V vs standard hydrogen electrode (SHE). However, it is highly reactive and prone to dendrite formation and short circuits upon cycling [40]. Graphite, on the other hand, has a low theoretical capacity of 372 mA h/g and a low Li^+ ion intercalation potential of 0–0.3 V vs Li/Li^+, which limits its usage toward high energy and high-power density. To overcome these limitations, Si, Ge, and their alloys have been extensively utilized to increase the capacity of the anode electrode [9–13,41–43]. In fact, Si has shown promising results (ten times higher capacity than graphite) but suffers from severe volume changes during cycling, leading to electrode pulverization, structural degradation in terms of unstable SEI formation and rapid capacity fading [13–17]. Overall, the development of high-capacity and stable anode materials remains a significant challenge and further research is needed to address the safety, stability, and performance issues associated with these materials.

On the other hand, cathode materials are a vital component of LIBs, and the limitations in their specific capacity, electrical conductivity, mechanical properties, and electrochemical stability can impede the overall performance of the battery [44,45]. The commonly used cathode materials in commercial applications include nickel-manganese-cobalt (NMC), lithium-cobalt-oxide ($LiCoO_2$), lithium-iron phosphate ($LiFePO_4$), and lithium-manganese spinel (LMO) [7,22,46,47]. However, these materials often have lower specific capacities than anodes, such as graphite, which can limit the overall capacity of the battery. Additionally, these materials may suffer from low thermal stability and fast capacity decay under certain conditions [48,49]. Consequently, major research is focused on developing cathode materials with higher specific capacity, improved safety, enhanced rate capability, and wide working voltage to improve the energy and power densities while reducing the size and cost of LIBs. Recent investigations reported that doping Mg in $LiFePO_4$ could increase the capacity by providing high voltage [50].

The safety of LIBs is of prime concern due to dendrite formation in LIBs and, therefore, the role of an electrolyte and the separator is critical that prevent the short-circuit between the cathode and anode electrodes. In LiBs, separators are typically made of organic or inorganic materials with controlled porosity that allows them to intake liquid electrolytes as ion-conducting media [22,51,52]. The existing LIBs are based on a liquid electrolyte and are a major safety concern resulting from side reactions during the charge and discharge cycles. In addition, the use of liquid electrolytes can lead to capacity fading and reduced battery performance due to the degradation of the active materials caused by the leakage or evaporation of the electrolyte [22]. This can result in a loss of ion storage sites and a reduction in the battery's capacity and energy density. Also, during the charge–discharge process, the dendrites can form on the anode and penetrate through the separator, causing a short circuit that can lead to thermal runaway and even explosion.

To address these safety and performance issues, researchers are developing solid-state electrolytes that can offer improved safety and stability. Solid-state electrolytes are non-flammable and can prevent dendrite formation and reduce the risk of short circuits. They can also improve the battery's energy density and reduce capacity fading by providing a more stable interface between the electrodes and the electrolyte [21,22]. While solid-state electrolytes have shown great promise, they still face several challenges. These include their relatively low ionic conductivity, high interface resistance with the electrodes, and difficulties in fabricating large-area, defect-free solid-state electrolyte membranes [21,22]. However, with continued research and development, solid-state electrolytes hold great potential for revolutionizing the safety and performance of batteries.

Other battery materials such as Li-O_2 [53–55] and Li-S [56–58] have also been studied extensively due to their high theoretical specific capacity and energy density. However, there are still several obstacles that need to be overcome to make them viable commercial technology. The major challenge with the Li-S battery is the insulating nature of sulfur and Li_2S, which can lead to poor electronic conductivity and ionic transport, resulting in low-rate capacity and high over-potential. Another significant issue with Li-S is the shuttle effect, where Li_2S_x species dissolve in the electrolyte and shuttle between the anode and cathode, resulting in the formation of insoluble Li_2S_2/Li_2S deposits on the Li anode. This leads to a loss of active material, a decrease in battery efficiency, and a short lifespan [56–58]. To overcome these issues, different approaches have been reported such as adding conductive carbon materials or introducing sulfur in a porous structure to enhance the surface area and facilitate the transport of lithium ions, adding polysulfide scavengers to the electrolyte, designing modified cathode structures, and developing advanced separator materials to suppress the shuttle effect [59]. Similarly, there are issues with Li-O_2 batteries, especially the formation of Li_2O_2 phase on the cathode electrode during the discharge process. The insulating nature of Li_2O_2 phase eventually clogs the cathode and reduces the battery's performance. Moreover, the low Coulombic efficiency and low reversibility are other factors that affect the performance of Li-O_2 batteries [53–55].

8.4   WHY 2D MATERIALS? ADVANTAGES OVER CURRENT MATERIALS

Two-dimensional (2D) materials are usually known to be van der Waals materials as they inherently contain substantial in-plane covalent bonds and frail out-of-plane van der Waals forces. Various 2D materials and their derivatives such as graphene, h-BN, BP, MXenes, metal oxides, and TMDs are displayed in Figure 8.4. Because of such a unique structure, these materials construct a stable structure. They can be reduced to an atomically thin layer, such as single-layer graphite, or graphene can be produced by mechanical exfoliation from highly ordered pyrolytic graphite. With unique electronic and chemical properties, these materials have enticed remarkable attention as one of the considerably encouraging electrode materials for rechargeable batteries [18–20]. These materials can satisfy the perennially growing demands of higher power and energy density, outstanding rate performance, and long cycling life. There are several advantages of 2D materials over conventional electrode materials which are summarized in Figure 8.5.

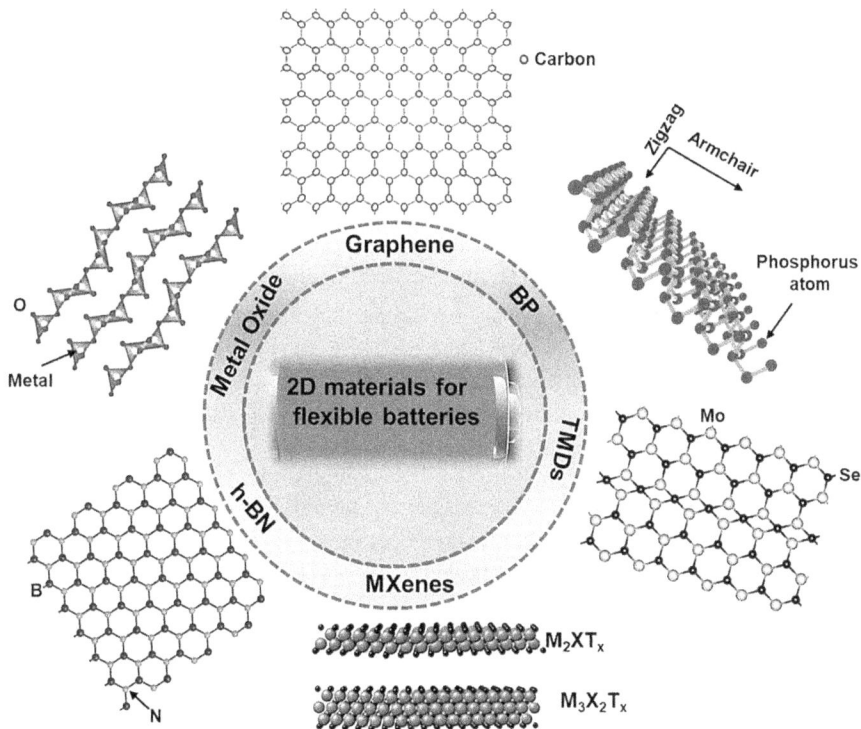

FIGURE 8.4 2D materials for flexible batteries.

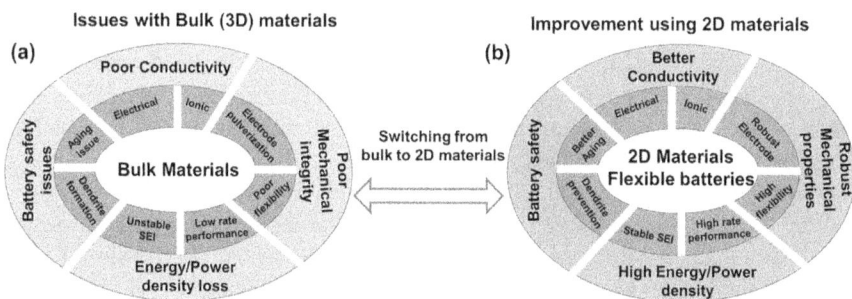

FIGURE 8.5 Issues with the conventional bulk battery materials and advantages of implementing 2D materials for highly efficient flexible batteries. (Adapted with permission from Ref. [18]. Copyright © 2020, American Chemical Society.)

1. High electrical conductivity and thermal stability: 2D materials possess excellent electrical conductivity, which is crucial for improving the rate capability and power density of batteries. Additionally, they exhibit high thermal stability, which is important for preventing thermal runaway and improving the safety of batteries [18–20,37].

2. Enhanced mechanical flexibility: 2D materials are inherently more flexible than 3D materials due to their thin, layered structure. This flexibility enables them to bend and twist without cracking or losing their electrochemical properties, making them ideal for use in flexible batteries for wearable and portable electronics [18–20,37].
3. Large surface area-to-volume ratio: Due to their thin, planar structure, 2D materials have a high surface area-to-volume ratio, which allows for a greater number of active sites to be exposed to the electrolyte. This results in improved Li-ion transfer and electrochemical performance such as higher capacity and faster charging/discharging rates [18–20,37].
4. Ultrathin layer: The ultrathin layer of 2D materials (typically one atom thick) makes them lightweight and comfortable to wear or carry, while still providing high energy density and power output [18–20,37].

The unique properties of 2D materials have the potential to revolutionize the field of flexible batteries owing to their excellent structural, chemical, and thickness-dependent properties. Research in this area continues to enable the development of more efficient, lightweight, and wearable energy storage devices and beyond.

8.5   2D MATERIALS: SYNTHESIS OF GRAPHENE/TMDS MATERIALS AND THEIR HETEROSTRUCTURES

Graphene, discovered by Andre Geim and Konstantin Novoselov in 2004 [60], opened up new opportunities in the materials research field with its excellent properties. The thinnest material, hexagonally arranged, sp^2 hybridized graphene, has a large specific area, high carrier mobility, outstanding optical transparency, a high thermal conductivity, and high Young's modulus. It is also the most transparent and thermally conductive material [61]. Sp^2 hybridized carbon bonds have two types of bonds: σ_{c-c} bonds (in-plane) and π bond s(out-of-plane). σ_{c-c} bonds are the strongest bonds, while π bonds provide weak interaction between the graphene layers. These π bonds provide a delocalized electron network which is responsible for electron conduction in graphene planes [62].

Graphene has unique properties, such as high carrier mobility, specific surface area, thermal conductivity, optical transmittance, and flexibility, with intrinsic tensile strength and Young's modulus [62]. Graphene is highly transparent to visible light, with a low absorbance of ~2.3% [63]. Single-layer graphene sheet has a high thermal conductivity of ~5000 $Wm^{-1}K^{-1}$ [64], and low stiffness of 20 N/m at room temperature [65]. The high Young's modulus (1 TPa) [66], and high tensile strength (42 N/m) [67] of graphene make it one of the strongest materials. Due to its high conductivity and large specific area, graphene has a low charge transfer resistance [68]. The water permeability of graphene was found to be much higher than conventional reverse osmosis membranes and, therefore, graphene also plays an important role in water purification [69].

Graphene can be synthesized by various methods including mechanical and liquid exfoliation, epitaxial growth, the CVD method, and graphene oxide reduction, but the yield in most of these production routes is typically very low

[70]. For large-scale production, Hummer's method [71], and modified Hummer's method [72] are preferable. The first three processes as mentioned above yield a relatively perfect structure and better electronic properties. However, Hummer's method is more popular owing to its low production cost and high yield but producing a less perfect structure of graphene.

Due to its hydrophilic nature, graphene oxide (GO) forms a stable colloidal suspension in water. Different methods of reducing graphene oxide partially restore the structure and properties of the material. The performance of the device in which reduced graphene oxide (rGO) is utilized is directly impacted by the properties of rGO, which are heavily dependent on the effectiveness of the reduction processes [73].

The thermal reduction of graphene oxide in a reducing and/or inert gas atmosphere is one method for reducing GO. There are two approaches to conducting thermal reduction: (1) rapid heating, and (2) slow heating. Rapid heating causes oxygen-containing functional groups to decompose into CO and CO_2 gases, which then suddenly expand to produce mechanical exfoliation of the structure. This causes high pressure between the layers. Thermal reduction is a useful way to create graphene in large quantities because it has both exfoliation and reduction as side effects. However, the primary causes of rapid heating are structural defects and small size. Due to the high energy generated by the decomposition of carbon dioxide, decomposed functional groups also remove carbon atoms from the structure during rapid heating. This distortion in the carbon planes causes graphene sheets to break into smaller sizes. Rapid heating results in a tiny, wrinkly, and structurally compromised graphene sheet. However, slow heating during reduction aids in the preservation of the structure of graphene by preventing the sheets from rapidly expanding. However, slow heating takes a long time to complete the process [74].

Studies have shown that the reduction of graphene oxide can be performed at high temperatures [75–77]. But if the rGO film is to be used on low melting substrates, e.g., glass and polymers, then in-situ reduction at high temperatures becomes more difficult. Graphene oxide can be reduced in a reducing atmosphere with or without inert gas. Hydrogen has a high reducing ability at lower temperatures, due to which graphene oxide can be reduced at low temperatures in hydrogen atmospheres.

In general, 2D TMDs synthesized through top-down methods are mainly used in the biomedical field, while 2D TMDs prepared through bottom-up methods are mostly applied in photoelectric devices and catalysis fields. Similar to the micro-mechanical exfoliation of graphene [78], MoS_2 flakes can be produced on a substrate using sticky tape. The starting material is bulk MoS_2, where some parts are peeled off with the tape and pressed into the substrate. Upon the release of the tape, some parts stay with the substrate rather than the tape due to the van der Waals force on the substrate. Repeating the process may produce flakes of MoS_2 with random shapes, sizes, and numbers of layers. This method offers 2D materials of the highest quality, enabling us to study their pristine properties and ultimate device performance.

Liquid phase exfoliation also starts with bulk MoS_2, producing flakes with random shapes, sizes, and number of layers. The quantities, however, are much larger, and the quality lower. Roughly, there are two routes to exfoliate MoS_2 in the solution. The first is exfoliation by mechanical means such as sonication, shearing, stirring, grinding, and bubbling [79–84]. The essence is purely physical, although some chemistry may still be involved [85–87]. For example, surfactants such as sodium deoxycholate (SDC) bile salt [85], and chitosan [86] may be added to the solution to prevent the exfoliated flakes from recombination. Electrolysis is employed to generate bubbles in some cases [87]. It is known that fine bubbles can squeeze into material interfaces and exfoliate them [88]. The yield of this method is dramatically improved over tape-assisted exfoliation, but the efficiency is still low for industrial applications. Therefore, a second route by atomic intercalation via solution chemistry [89], or electrochemistry [90], has been proposed. Lithium is typically used to intercalate between the MoS_2 layers and enlarge the interlayer spacing, easing the following exfoliation by mechanical treatment (e.g., sonication) [91].

8.6   RECENT DEVELOPMENTS IN GRAPHENE/TMDS COMPOSITES: APPLICATIONS IN BATTERIES AS ANODE/CATHODE EXAMPLES

Graphene is a promising choice for energy-storage devices and their manufacturing, particularly for its composite with metal oxide. The restacking problems are hampered by the composite nature of the electrode materials. The highly conductive nature of graphene, along with its porous structure, superior mechanical stability, and electrochemical stability, makes it a material with a lot of potential for use in energy-storage applications. Additionally, supercapacitors, fuel cells, batteries, and solar cells are among the devices that use graphene as an energy storage medium. For example, graphene can be used as the cathode and anode in the cell of lithium (Li) batteries, whereas the supports of proton-exchange membrane (PEM) fuel cells are constructed of graphene. In double-layer capacitors and pseudocapacitors, graphene-based electrodes are also utilized [92].

rGO substrates are promising for flexible high-energy-density LIBs due to their superior flexibility and lightweight benefits. Li et al. [93] constructed and demonstrated the performance of 3D rGO foam@S as a cathode and rGO@Li as an anode to evaluate their practical viability (for details, see Fig. 6 from ref. [93]). Notably, highly flexible lithium-sulfur batteries (LSBs) with such a particular topographic feature exhibit a high specific energy density of 754 Wh kg^{-1} and a power density of 377 W kg^{-1}.

In LIBs, graphite is currently employed as the active anode material. Graphite has a high capacity (372 mA g^{-1}) while producing a LiC_6 structure during lithium intercalation during charging due to its good electrical conductivity and high packing density [94]. For this aspect, laser-induced graphene (LIG) has been utilized as a new anode material for LIBs. It delivers higher capacity and power density than conventional regular graphite-based electrodes because of the following factors: i) LIG exhibits a high capacity as a binder-free anode because it has a large number of defects and graphitic edges, which can eventually boost the capacity. ii) Unlike graphite, which has a layered structure with closed pores, LIG has an open-pore structure that makes it simple for ions to move during the charge/discharge process resulting in high-power density. Shim et al. [95] reported on a simple method for creating LIGs on a Cu foil using dry transfer of LIGs created on a polyimide film. In order to create a LIG-based LIB anode, the LIG's natural 3D hierarchical porous structure was transferred to Cu foil while still being intact. Fast ion transport is made possible in the LIG anode by the graphitic edges, monolith structures, and macro/mesopores that make up LIG. As a result, a high mass loading of 3 mg cm^{-2} at a high rate of 20 A g^{-1} was able to provide a stable capacity of 114 mAh g^{-1}.

In-depth research has been done on LIG composites to enhance the anode performance. When employed alone as electrode materials, transition metal oxides (TMOs) are primarily constrained by low electrical/ionic conductivities and volume expansions. To overcome the limits of TMOs and the comparatively limited capacities of carbon materials, numerous attempts have been made to realize high-capacity electrodes employing composites of TMOs and carbon materials. Zhou et al. [96], for instance, presented a MnO_x-graphene composite that was made via a one-step laser scribing method. A copper current collector coated with a precursor layer was exposed to a CO_2 laser before being used as a MnO_x-LIG composite electrode. Additionally, nitrogen doping was carried out while laser scribing in a N_2 atmosphere. Doping increased the electrical conductivity and pseudocapacitance, which led to the MnO_x-N-doped LIG composite electrode with a high reversible capacity of 992 mAh g^{-1}.

High performance, exceptional flexibility, and mechanical stability are required for good batteries in a power supply. For greater energy density, it is also crucial to maintain the total mass of the battery's components as low as feasible. The binder and current collector make up the standard Li battery cathode material design. As a result, it produces a heavy and bulky battery architecture that affects the rate at which the specific capacity of LSB degrades. In light of this, Zhang et al. [97] introduce a free-standing hollow carbon nanofiber/rGO through straightforward vacuum filtering and heat treatment. The use of a freestanding cathode creates open pathways for ions and electrons, which leads to the LSB's high initial discharge capacity of 1318.4 mAh g^{-1}.

Low conductivities and sluggish kinetics, which result from a multielectron-transfer conversion reaction mechanism, are typical of high-capacity cathode materials made of metal fluorides and seriously hinder their commercialization by using a simple hydrothermal technique followed by successive freeze-drying, thermal reduction, and fluorination post-treatments. He et al. [98] fabricated a flexible free-standing FeF_3/chitosan pyrolytic carbon/reduced graphene oxide (FeF_3/C/rGO) sheet as an additive-free cathode. In addition to significantly reducing the Li+ diffusion pathway, the highly ordered rGO film serves as a matrix to restrain the complex interlamination reaction ($Fe^{3+} \leftrightarrow Fe^{2+} \leftrightarrow Fe$) between adjacent inter-layers with a spacing of approximately 30 nm. The FeF3/C/rGO film may attain an impressive capacity of up to 220 mAh g^{-1} over 200 cycles at 100 mA g^{-1} thanks to its free-standing structure, offering tremendous promise for wearable and flexible electronic devices.

Flexible power sources have a significant impact on the widespread promotion of wearable electronics. Their applications are still constrained despite withstanding urgent demands because of the low mass loading of active materials, poor durability, and relatively low energy density. Huang et al. [99], reported a synergistic interface bonding enhancement strategy to create flexible fiber-shaped composite cathodes to get around these issues. In this strategy, polypyrrole@sulfur (PPy@S) nanospheres are homogeneously implanted into the built-in cavity of self-assembled reduced graphene oxide fibers (rGOFs) using an easy microfluidic assembly method. The carbon and polymer interface in this architecture serves as a synergistic host for lithium polysulfides and sulfur nanospheres, which together enhance the chemical bonding at the interface and give the cathode superior mechanical flexibility, quick reaction kinetics, and good adsorption capacity. The PPy@S/rGOFs cathode consequently exhibits improved electrochemical performance and high-rate capabilities (5.8 mAh cm^{-2} at 0.2 A g^{-1}).

TMDs were considered a possible candidate for metal ion batteries (MIBs), particularly LIBs have a distinct physicochemical characteristic. Researchers have been working on the synthesis of various 2D TMDs and their composites over the past few decades by changing the physicochemical parameters to improve the electrochemical behavior of LIBs anodes [100,101]. Since then, there has been a significant growth in research on the electrochemical investigation of several TMDs, including MoS_2, WS_2, SnS_2, VS_2, and others [102,103]. Noteworthy is the TMDs' capacity to display numerous response processes for a single material that can open up new avenues for cutting-edge studies. Due to the preferred reaction mechanism, intermediate voltage for lithiation, and large interlayer spacing, MoS_2 is most investigated for battery applications [104]. Since MoS_2 has a wider interlayer spacing than graphite, metal ions can move through the layered structure more quickly. Additionally, it displays a high theoretical capacity for the maximum four-electron transfer reaction (670 mAh g^{-1}) [105]. In contrast to the graphite anode, which only experiences the intercalation reaction, MoS_2 undergoes intercalation followed by a conversion reaction during the metal ions (Li^+) insertion reaction. MoS_2 nanoplates made from the deformed graphene-like layer were produced by H. Hwang et al. [106], with a larger interlayer spacing (0.69 nm) than the bulk structure of the materials. The amorphous nature of the material and wide interlayer spacing enable faster Li-ion diffusion, which significantly improves its electrochemical performance as a LIB anode (reversible capacities at 1 C and 50 C are 917 mA h g^{-1} and 700 mA h g^{-1}, respectively).

In addition to the interlayer spacing, a number of additional elements are crucial in improving the electrochemical performance of MoS_2, including the impact of defects, the impact of pseudocapacity, and the presence of high surface reaction sites [107–109]. Barik et al. described the impact of defect engineering on electrochemical performance in one of their reports [108]. Theoretically, they determined the rate of adsorption and diffusion barrier of metal ions, and they discovered that the defective MoS_2 offers higher adsorption sites toward the metal ions with a low diffusion barrier compared to the pristine MoS_2, though they noted that defects in MoS_2 cannot affect the open circuit voltage of the electrode. Researchers frequently use graphene or rGO with MoS_2, which not only improves electrical conductivity but also limits volume expansion to improve electronic conductivity [103–105,110–113].

8.7 LIMITATIONS AND FUTURE PERSPECTIVES

In this book chapter, we have discussed current advances, challenges, and applications of numerous 2D materials in flexible energy storage devices. Despite significant advances, emerging novel 2D materials, particularly those beyond graphene, are still in their preliminary stages. Compared with the conventional LIBs, the field of flexible batteries using 2D materials is still in its infancy and has advanced dramatically in the last decade, from fundamental research to the development of next-generation battery technologies [18–20,37].

The importance of dimensionality in defining the intrinsic properties (high mechanical, electronic, and chemical properties) and possible uses of nanomaterials have obviously been emphasized. Graphene has been known for over a decade due to its exceptional electronic and mechanical properties but other 2D materials have also emerged as an alternative that will help us design desired characteristics for a particular application [23,24]. Still, there are several challenges that need to be resolved, for instance, restacking of 2D layers during the charge/discharge process that can affect the electrode performance and hence the capacity of the battery [18]. The scalability of the 2D materials to meet the industry demands is another factor that is still lacking. Additionally, the stability and degradation of 2D materials under various operating conditions, including mechanical stress, temperature, and exposure to air and moisture, is still an area of active research. Therefore, further research in the development of 2D by utilizing cost-effective methods with large scalability is required to fully realize their potential to meet industry demands.

Individual 2D materials have their own set of drawbacks limited due to their high production cost, poor electronic conductivity in the case of TMDs, single layer of MXenes is more susceptible toward stacking, and complex synthesis methods for materials like Xenes [18–20]. However, given the choice of 2D materials available, hybridization or mixing more than one material can provide a potential solution to circumvent issues associated with 2D materials [113]. For instance, graphene has been employed to enhance the electronic conductivity of metal oxides and TMDs with a goal to improve the overall conductivity of the system which eventually results in high performance toward electrochemical storage devices. Also, the aging behavior of the electrodes is another factor resulting due to inevitable side reactions in the process of charging and discharging of batteries. These side reactions can consume Li ions in the electrolyte and are eventually deposited on the anode electrode, which eventually passivates the active area of the electrode. The high surface area of 2D material makes them more susceptible to these side reactions. Therefore, high-end instruments such as advanced electron microscopy and other analytical tools (in situ XPS and Raman) are required to investigate and monitor these side reactions at meta level.

In terms of future perspective, there is a lot of excitement to unlock the potential of 2D materials for developing high-performance flexible batteries for a range of applications. Continued research and development efforts are likely to focus on improving the scalability, stability, and performance of 2D materials for use in flexible batteries, as well as exploring new applications and device architectures that can take advantage of their unique properties. Furthermore, the integration of 2D materials into existing battery architectures and the optimization of their electrochemical properties for specific applications is also an area of ongoing research. This includes improving the electrical conductivity and electrochemical stability of 2D materials and exploring novel device architectures that can take advantage of their unique properties.

ACKNOWLEDGMENTS

Dr. Praveen Kumar sincerely expresses gratitude for the financial support provided by the Shared Instrumentation Facility at the Colorado School of Mines, Golden, USA.

DECLARATION OF COMPETING INTEREST

The authors declare no conflicts of interest.

REFERENCES

1. The United States Sustainable Development Goals, https://www.un.org/sustainabledevelopment/.
2. Tarascon, J. M., and Armand, M. 2001. Issues and challenges facing rechargeable lithium batteries. *Nature* 414: 359–367.
3. Van Noorden, R. 2014. The rechargeable revolution: a better battery. *Nature* 507: 26–28.
4. The International Energy Agency (IEA) report. 2022. https://www.iea.org/reports/renewables-2022.
5. Lee, B. S., Oh, S. H., Choi, Y. J. et al. 2023. SiO-induced thermal instability and interplay between graphite and SiO in graphite/SiO composite anode. *Nat. Commun.* 14: 150.
6. Schweidler, S., Biasi, Ld., Schiele, A., Hartmann, P., Brezesinski, T., and Janek, J. 2018. Volume changes of graphite anodes revisited: a combined operando X-ray diffraction and in situ pressure analysis study. *J. Phys. Chem. C* 122: 8829–8835.
7. Manthiram, A. 2020. A reflection on lithium-ion battery cathode chemistry. *Nat. Commun.* 11: 1550.
8. Landi, J. B., Ganter, J. M., Cress, D. C., DiLeo, A. R., and Raffaelle, P. R. 2009. Carbon nanotubes for lithium ion batteries. *Energy Environ. Sci.* 2: 638. DOI: 10.1039/B904116H.
9. Chan, C., Peng, H., Liu, G. et al. 2008. High-performance lithium battery anodes using silicon nanowires. *Nat. Nanotechnol.* 3: 31–35.
10. Wu, H., and Cui, Y. 2012. Designing nanostructured Si anodes for high energy lithium-ion batteries. *Nano Today* 7: 414–429.
11. McDowell, M. T., Lee, S. W., Nix, W. D., and Cui, Y. 2013. 25th anniversary article: understanding the lithiation of silicon and other alloying anodes for lithium-ion batteries. *Adv. Mater.* 25: 4966–4985.
12. Obrovac, M. N., and Krause, L. J. 2007. Reversible cycling of crystalline silicon powder. *J. Electrochem. Soc.* 154: A103–A108.
13. Hanai, K., Liu, Y., Imanishi, N., Hirano, A., Matsumura, M., Ichikawa, T., and Takeda, Y. 2005. Electrochemical studies of the Si-based composites with large capacity and good cycling stability as anode materials for rechargeable lithium ion batteries. *J. Power Sources* 146: 156–160.

14. Vorauer, T., Kumar, P., Berhaut, C. L. et al. 2020. Multi-scale quantification and modeling of aged nanostructured silicon-based composite anodes. *Commun. Chem.* 3: 141.

15. Kumar, P., Berhaut, C. L., Zapata, D., De, E., Tardif, S., Pouget, S., Lyonnard, S., Jouneau, P.-H. 2020. Nano-architectured composite anode enabling long-term cycling stability for high-capacity lithium-ion batteries. *Small* 16: 1906812.

16. Karuppiah, S. et al. 2020. A scalable silicon nanowires-grown-on-graphite composite for high-energy lithium batteries. *ACS Nano* 14: 12006–12015.

17. Berhaut, C. L. et al. 2019. Multiscale multiphase lithiation and delithiation mechanisms in a composite electrode unraveled by simultaneous operando small-angle and wide-angle X-ray scattering. *ACS Nano* 13: 11538–11551.

18. Rojaee, R., and Shahbazian-Yassar, R. 2020. Two-dimensional materials to address the lithium battery challenges. *ACS Nano* 14: 2628–2658.

19. Xiao, Z. et al. 2021. Recent developments of two-dimensional anode materials and their composites in lithium-ion batteries. *ACS Appl. Energy Mater.* 4: 7440–7461.

20. Li, Y. et al. 2019. Emerging two-dimensional noncarbon nanomaterials for flexible lithium-ion batteries: opportunities and challenges. *J. Mater. Chem. A* 7: 25227.

21. Li, C. et al. 2021. An advance review of solid-state battery: challenges, progress, and prospects. *Sustain. Mater. Technol.* 29: e00297.

22. Sharma, S. K. et al. 2022. Progress in electrode and electrolyte materials: path to all-solid-state Li-ion batteries. *Energy Adv.* 1: 457.

23. Dai, C., Sun, G., Hu, L., Xiao, Y., Zhang, Z., and Qu, L. 2020. Recent progress in graphene-based electrodes for flexible batteries. *InfoMat* 2: 509–526.

24. Zhu, Y., Murali, S., Cai, W., Li, X., Suk, J. W., Potts, J. R., and Ruoff, R. S. 2010. Graphene and graphene oxide: synthesis, properties, and applications. *Adv. Mater.* 22: 3906–3924.

25. Zhang, Y., Ma, C., Xie, J., Ågren, H., and Zhang, H. 2021. Black phosphorus/polymers: status and challenges. *Adv. Mater.* 33: 2100113.

26. Xue, X.-X. et al. 2021. Black phosphorus-based materials for energy storage and electrocatalytic applications. *J. Phys. Energy* 3: 042002.

27. Choudhary, N. et al. 2018. Two-dimensional transition metal dichalcogenide hybrid materials for energy applications. *Nano Today* 19: 16–40.

28. Chhowalla, M., Liu, Z., and Zhang, H. 2015. Two-dimensional transition metal dichalcogenide (TMD) nanosheets. *Chem. Soc. Rev.* 44: 2584.

29. Lin, L. et al. 2019. Two-dimensional transition metal dichalcogenides in super-capacitors and secondary batteries. *Energy Storage Mater.* 19: 408–423.

30. Palchoudhury, S., Ramasamy, K., Han, J., Chena, P., and Gupta, A. 2023. Transition metal chalcogenides for next-generation energy storage. *Nanoscale Adv.* DOI: 10.103 9/D2NA00944G.

31. Shinde, P. A. et al. 2022. Two-dimensional MXenes for electrochemical energy storage applications. *J. Mater. Chem. A* 10: 1105.

32. Timmerman, M. A. et al. 2022. Metal oxide nanosheets as 2D building blocks for the design of novel materials. *Chem. Eur. J.* 26: 9084.

33. Sarfraz, M., and Shakir, I. 2017. Recent advances in layered double hydroxides as electrode materials for high-performance electrochemical energy storage devices. *J. Energy Storage* 13: 103–122.

34. Hempel, M. et al. 2018. Repeated roll-to-roll transfer of two-dimensional materials by electrochemical delamination. *Nanoscale* 10: 5522.

35. Xia, X., Yang, J., Liu, Y., Zhang, J., Shang, J., Liu, B., Li, S., and Li, W., 2023. Material choice and structure design of flexible battery electrode. *Adv. Sci.* 10: 2204875.

36. Cui, X. et al. 2020. Models based on mechanical stress, initial stress, voltage, current, and applied stress for Li-ion batteries during different rates of discharge. *Energy Storage* 2: e126.

37. Zhang, Y., Wang, L., Zhao, Y., and Peng, H. 2022. *Flexible Batteries*, Boca Raton: CRC Press/Taylor & Francis.

38. Roy, P., and Srivastava, S. K. 2015. Nanostructured anode materials for lithium ion batteries. *J. Mater. Chem. A* 3: 2454–2484.

39. Mahmood, N., Tang, T., and Hou, Y. 2016. Nanostructured anode materials for lithium ion batteries: progress, challenge and perspective. *Adv. Energy Mater.* 6: 1600374.

40. Deng, D. 2015. Li-ion batteries: basics, progress, and challenges. *Energy Sci. Eng.* 3: 385–418.

41. Abel, P. R. et al. 2013. Nanostructured Si(1-x)Gex for tunable thin film lithium-ion battery anodes. *ACS Nano* 7: 2249–2257.

42. Dominguez, D. Z. et al. 2022. (De)Lithiation and strain mechanism in crystalline Ge nanoparticles. *ACS Nano* 16: 9819–9829.

43. Bensalah, N., Matalkeh, M., Mustafa, N. K., and Merabet, H. 2020. Binary Si–Ge alloys as high-capacity anodes for Li-ion batteries. *Phys. Status Solidi A* 217: 1900414.

44. Khan, S. A., Ali, S., Saeed, K., Usman, M., and Khan, I. 2019. Advanced cathode materials and efficient electrolytes for rechargeable batteries: practical challenges and future perspectives. *J. Mater. Chem. A* 7: 10159–10173.

45. Ma, Y. 2018. Computer simulation of cathode materials for lithium ion and lithium batteries: a review. *Energy Environ. Mater.* 1: 148–173.

46. Rodrigues, M.-T. F. et al. 2017. Materials perspective on Li-ion batteries at extreme temperatures. *Nat. Energy* 2: 1–14.

47. Belharouak, I. et al. 2011. Electrochemistry and safety of Li4Ti5O12 and graphite anodes paired with LiMn2O4 for hybrid electric vehicle Li-ion battery applications. *J. Power Sources* 196: 10344.

48. Liu, C., Neale, Z. G., and Cao, G. 2016. Understanding electrochemical potentials of cathode materials in rechargeable batteries. *Mater. Today* 19: 109–123.

49. Nitta, N., Wu, F., Lee, J. T., and Yushin, G. 2015. Li-ion battery materials: present and future. *Mater. Today* 18: 252–264.

50. Xiaohui, Z. et al. 2012. Electrochemical properties of magnesium doped LiFePO$_4$ cathode material prepared by sol–gel method. *Mater. Res. Bull.* 47: 2819–2822.

51. Landesfeind, J. et al. 2016. Tortuosity determination of battery electrodes and separators by impedance spectroscopy. *J. Electrochem. Soc.* 163: A1373–A1387.

52. Wang, F., Li, L., Yang, X., You, J., Xu, Y., Wang, H., Ma, Y., and Gao, G. 2018. Influence of additives in a PVDF-based solid polymer electrolyte on conductivity and Li-ion battery performance. *Sustain. Energy Fuels* 2: 492–498.

53. Han, J. et al. 2017. Full performance nanoporous graphene based Li-O$_2$ batteries through solution phase oxygen reduction and redox additive mediated Li$_2$O$_2$ oxidation. *Adv. Energy Mater.* 7: 1601933.

54. Xia, C., Kwok, C. Y., and Nazar, L. F. 2018. A high-energy-density lithium-oxygen battery based on a reversible four-electron conversion to lithium oxide. *Science* 361: 777–781.

55. Kwak, W.-J., et al. 2020. Lithium–oxygen batteries and related systems: potential, status, and future. *Chem. Rev.* 120: 6626–6683.

56. Cao, R., Xu, W., Lv, D., Xiao, J., and Zhang, J. G. 2015. Anodes for rechargeable lithium-sulfur batteries. *Adv. Energy Mater.* 5: 1402273.

57. Chen, H. et al. 2019. Uniform high ionic conducting lithium sulfide protection layer for stable lithium metal anode. *Adv. Energy Mater.* 9: 1900858.

58. Liu, J. et al. 2018. Minimizing polysulfide shuttle effect in lithium-ion sulfur batteries by anode surface passivation. *ACS Appl. Mater. Interfaces* 10: 21965 –21972.
59. Ren, W., Ma, W., Zhang, S., and Tang, B. 2019. Recent advances in shuttle effect inhibition for lithium sulfur batteries. *Energy Storage Mater.* 23: 707–732.
60. Novoselov, K. S. et al. 2004. Electric field effect in atomically thin carbon films. *Science* 306: 666–669.
61. Deepak, T. G. et al. 2014. A review on counter electrode materials in dye-sensitized solar cells. *J. Mater. Chem. A* 2: 4474–4490.
62. Huang, X., Yin, Z., Wu, S., Qi, X. et al. 2011. Graphene-based materials: synthesis, characterization, properties, and applications. *Small* 7: 1876–1902.
63. Nair, R. R. et al. 2008. Fine structure constant defines visual transparency of graphene. *Science* 320: 1308.
64. Balandin, A. A. et al. 2008. Superior thermal conductivity of single-layer graphene. *Nano Lett.* 8: 902–907.
65. Nicholl, R., Conley, H., Lavrik, N. et al. 2015. The effect of intrinsic crumpling on the mechanics of free-standing graphene. *Nat. Commun.* 6: 8789.
66. Lee, C., Wei, X., Kysar, J. W., Hone. J. 2008. Measurement of the elastic properties and intrinsic strength of monolayer graphene. *Science* 321: 385–388.
67. Mokerov, V. G. et al. 2001. New quantum dot transistor. *Nanotechnology* 12: 552–555.
68. Bao, M., Zhang, C., Lahiri, D. et al. 2012. The tribological behavior of plasma-sprayed Al-Si composite coatings reinforced with nanodiamond. *JOM* 64: 702–708.
69. Cohen-Tanugi, D., and Grossman, J. C. 2012. Water desalination across nanoporous graphene. *Nano Lett.* 12: 3602–3608.
70. Roy-Mayhew, J. D., and Aksay, I. A. 2014. Graphene materials and their use in dye-sensitized solar cells. *Chem. Rev.* 114: 6323–6348.
71. Hummers, W. S. Jr., and Offeman, R. E. 1958. Preparation of graphitic oxide. *J. Am. Chem. Soc.* 80: 1339.
72. Marcano, D. C. et al. 2010. Improved synthesis of graphene oxide. *ACS Nano* 4: 4806–4814.
73. Pei, S., and Cheng, H.-M. 2012. The reduction of graphene oxide. *Carbon* 50: 3210–3228.
74. Pradhan, D., and Leung, K. T. 2008. Vertical growth of two-dimensional zinc oxide nanostructures on ITO-coated glass: effects of deposition temperature and deposition time. *J. Phys. Chem. C* 112: 1357–1364.
75. Becerril, H. A. et al. 2008. Evaluation of solution-processed reduced graphene oxide films as transparent conductors. *ACS Nano* 2: 463–470.
76. Wang, X., Zhi, L., and Mullen, K. 2008. Transparent, conductive graphene electrodes for dye-sensitized solar cells. *Nano Lett.* 8: 323–327.
77. Li, X., Wang, H., Robinson, J. T., Sanchez, H., Diankov, G., and Dai, H. 2009. Simultaneous nitrogen doping and reduction of graphene oxide. *J. Am. Chem. Soc.* 131: 15939–15944.
78. Vivid Lab Demonstration. Available online: https://www.youtube.com/watch?v=9I5d0YLZgec&feature=youtu.be (accessed on 30 June 2017).
79. Forsberg, V. et al. 2016. Exfoliated MoS_2 in water without additives. *PLoS ONE* 11: 0154522.
80. Gupta, A., Arunachalam, V., and Vasudevan, S. 2016. Liquid-phase exfoliation of MoS_2 nanosheets: the critical role of trace water. *J. Phys. Chem. Lett.* 7: 4884–4890.
81. Yao, Y., Lin, Z., Li, Z., Song, X., Moon, K.-S., and Wong, C.-P. 2012. Large-scale production of two-dimensional nanosheets. *J. Mater. Chem.* 22: 13494–13499.

82. Yu, Y., Jiang, S., Zhou, W., Miao, X., Zeng, Y., Zhang, G., and Liu, S. 2013. Room temperature rubbing for few-layer two-dimensional thin flakes directly on flexible polymer substrates. *Sci. Rep.* 3: 2697.

83. Varrla, E. et al. 2015. Large-scale production of size-controlled MoS$_2$ nanosheets by shear exfoliation. *Chem. Mater.* 27: 1129–1139.

84. Paton, K. et al. 2014. Scalable production of large quantities of defect-free few-layer graphene by shear exfoliation in liquids. *Nat. Mater.* 13: 624–630.

85. Zhang, M. et al. 2015. Solution processed MoS2-PVA composite for sub-bandgap mode-locking of a wideband tunable ultrafast Er:fiber laser. *Nano Res.* 8: 1522–1534.

86. Zhang, W. et al. 2015. one-step approach to the large-scale synthesis of functionalized MoS$_2$ nanosheets by ionic liquid assisted grinding. *Nanoscale* 7: 10210–10217.

87. Liu, N., Kim, P., Kim, J., Ye, J., Kim, S., and Lee, C. 2014. Large-area atomically thin MoS$_2$ nanosheets prepared using electrochemical exfoliation. *ACS Nano* 8: 6902–6910.

88. Liu, L. et al. 2016. A mechanism for highly efficient electrochemical bubbling delamination of CVD-grown graphene from metal substrates. *Adv. Mater. Interfaces* 3: 1500492.

89. Eda, G., Yamaguchi, H., Voiry, D., Fujita, T., Chen, M., and Chhowalla, M. 2011. Photoluminescence from chemically exfoliated MoS$_2$. *Nano Lett.* 11: 5111–5116.

90. Zeng, Z., Yin, Z., Huang, X., Li, H., He, Q., Lu, G., Boey, F., and Zhang, H. 2011. Single layer semiconducting nanosheets: high-yield preparation and device fabrication. *Angew. Chem. Int. Ed.* 50: 11093–11097.

91. Fan, X., Xu, P., Zhou, D., Sun, Y., Li, Y., Nguyen, M., Terrones, M., and Mallouk, T. 2015. Fast and efficient preparation of exfoliated 2H MoS$_2$ nanosheets by sonication-assisted lithium intercalation and infrared laser-induced 1T to 2H phase reversion. *Nano Lett.* 15: 5956–5960.

92. Olabi, A. G., Abdelkareem, M. A., Wilberforce, T. et al. 2021. Application of graphene in energy storage device – a review. *Renewable Sustainable Energy Rev.* 135: 110026.

93. Li, N., Zhang, K., Xie, K. Y., Wei, W. F., Gao, Y., Bai, M. H., Gao, Y. L., Hou, Q., Shen, C., Xia, Z. H., and Wei, B. Q. 2020. Reduced-graphene-oxide-guided directional growth of planar lithium layers. *Adv. Mater.* 32: 1907079.

94. Zhang, H. et al. 2021. Graphite as anode materials: fundamental mechanism, recent progress and advances. *Energy Storage Mater.* 36: 147–170.

95. Shim, H. C., Tran, C. V., Hyun, S., and In, J. B. 2021. Three-dimensional laser-induced holey graphene and its dry release transfer onto Cu foil for high-rate energy storage in lithium-ion batteries. *Appl. Surf. Sci.* 564: 150416.

96. Zhou, C. et al. 2020. Laser-induced MnO/Mn3O4/N-doped-graphene hybrid as binder-free anodes for lithium ion batteries. *J. Chem. Eng.* 385: 123720.

97. Zhang, Z., Wang, G., Lai, Y., and Li, J. 2016. A freestanding hollow carbon nanofiber/reduced graphene oxide interlayer for high performance lithium-sulfur batteries. *J Alloys Compd.* 663: 501–506.

98. He, D. et al. 2022. A flexible free-standing FeF3/reduced graphene oxide film as cathode for advanced lithium-ion battery. *J Alloys Compd.* 909: 164702.

99. Huang, L. et al. 2022.Synergistic interfacial bonding in reduced graphene oxide fiber cathodes containing polypyrrole@sulfur nanospheres for flexible energy storage. *Angew. Chem. Int. Ed.* 61: e202212151.

100. Stephenson, T. et al. 2014. Lithium ion battery applications of molybdenum disulfide (MoS2) nanocomposites. *Energy Environ. Sci.* 7: 209.

101. Pumera, M., Sofer, Z., and Ambrosi, A. 2014. Layered transition metal dichalcogenides for electrochemical energy generation and storage. *J. Mater. Chem. A* 2: 8981.

102. Santanu, M., Zhongkan, R., and Gurpreet, S. 2018. Beyond graphene anode materials for emerging metal ion batteries and supercapacitors. *Nano-Micro Lett.* 10: 70.

103. Chang, K., and Chen, W. 2011. ₗ-Cysteine-assisted synthesis of layered MoS2/graphene composites with excellent electrochemical performances for lithium ion batteries. *ACS Nano* 5: 4720.

104. Chang, K. et al. 2011. Graphene-like MoS2/amorphous carbon composites with high capacity and excellent stability as anode materials for lithium ion batteries. *J. Mater. Chem.* 21: 6251.

105. Cao, X., Shi, Y., Shi, W., Rui, X., Yan, Q., Kong, J., and Zhang, H. 2013. Preparation of MoS2-coated three-dimensional graphene networks for high-performance anode material in lithium-ion batteries. *Small* 9: 3433–3438.

106. Hwang, H., Kim, H., and Cho, J. 2011. MoS2 nanoplates consisting of disordered graphene-like layers for high rate lithium battery anode materials. *Nano Lett.* 11: 4826.

107. Cook, J. B., Kim, H. S., Yan, Y., Ko, J. S., Robbennolt, S., Dunn, B., and Tolbert, S. H. 2016. Mesoporous MoS2 as a transition metal dichalcogenide exhibiting pseudocapacitive Li and Na-ion charge storage. *Adv. Energy Mater.* 6: 1501937.

108. Barik, G., and Pal, S. 2019. Defect induced performance enhancement of monolayer MoS2 for Li- and Na-ion batteries. *J. Phys. Chem. C* 123: 21852.

109. Zhang, S. et al. 2015. Constructing highly oriented configuration by few-layer MoS2: toward high-performance lithium-ion batteries and hydrogen evolution reactions. *ACS Nano* 9: 12464.

110. Chang, K., and Chen, W. 2011. In situ synthesis of MoS2/graphene nanosheet composites with extraordinarily high electrochemical performance for lithium ion batteries. *Chem. Commun.* 47: 4252.

111. Kong, D., He, H., Song, Q., Wang, B., Lv, W., et al. 2014. Rational design of MoS2@graphene nanocables: towards high performance electrode materials for lithium ion batteries. *Energy Environ. Sci.* 7: 3320.

112. Han, S., Zhao, Y., Tang, Y., Tan, F., Huang, Y., Feng, X., and Wu, D. 2015. Ternary MoS_2/SiO_2/graphene hybrids for high-performance lithium storage. *Carbon* 81: 203–209.

113. Lu, X., Jin, X., and Sun, J. 2015. Advances of graphene application in electrode materials for lithium-ion batteries. *Sci. China Technol. Sci.* 58: 1829–1840.

9 Two-Dimensional Materials for Supercapacitor Applications

Rapaka S Chandra Bose, T S Varun,
Srikanth Ponnada, Lakshman Kumar Anisetty,
Kirankumar Venkatesan Savunthari,
Demudu Babu Gorle, and Rakesh K Sharma

9.1 INTRODUCTION

Supercapacitors are one of the most promising energy storage devices, owing to their advantages such as high-power density, high capacitive retention, fast charge discharge rate, and extended life cycle. Based on their energy storage techniques, supercapacitors are classified as electrical double-layer capacitors (EDLCs) or pseudocapacitors. The charge storage process in EDLCs is based on ion electrosorption and the formation of an electrochemical double layer (EDL); meanwhile, the charge storage mechanism in pseudocapacitors is based on faradic reactions. Supercapacitors' key constituents are electrodes and electrolyte materials, current collectors, separators, and sealants, all of which play a vital role in the electrochemical performance of supercapacitor devices. The selection of electrode materials and their design, in particular, are critical for improving the electrochemical properties of supercapacitor devices. To achieve specific capacitance, energy and power density, and cycle stability, the electrodes must have a high specific surface area, corrosion resistance, conductivity, and thermal and chemical stability [1]. They should also be low-cost and environmentally sustainable. In general, supercapacitor electrode materials are divided into three categories: carbon-based materials, transition metal oxides, and conducting polymers. While conducting polymers and transition metal oxides are known as pseudocapacitive materials due to their ability to provide fast-reversible redox reactions, the carbon-based materials are known as EDLCs [2,3]. Recently, two-dimensional (2D) materials have shown a significant potential to become a promising candidate for electrode material. These materials are highly suited for supercapacitor applications due to their unique physical and chemical behavior, good electrical and mechanical properties, and large surface area. The use of 2D materials as supercapacitor

DOI: 10.1201/9781003404729-9

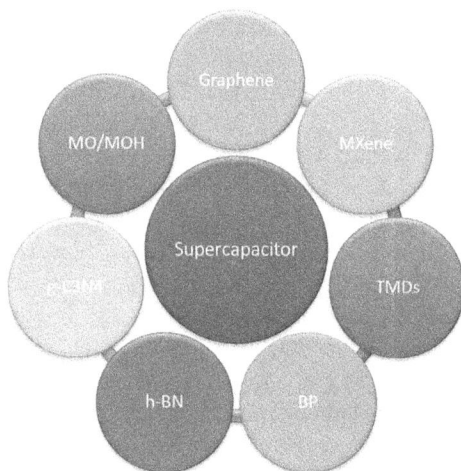

FIGURE 9.1 Various types of two-dimensional nanomaterials for supercapacitor applications presented in this chapter.

electrodes has resulted in better electrochemical features such as high specific capacitance, higher energy and power density, outstanding capacitive properties, and a high charging-discharging rate. This chapter presents the supercapacitor applications of 2D materials such as nanostructured carbon-based materials, MXene, transition metal dichalcogenides (TMDs), black phosphorus (BP), hexagonal boron nitride (h-BN), carbon nitride, and metal oxides/hydroxides [4–9] (Figure 9.1).

9.1.1 Nanostructured Carbon-Based Materials

Carbon-based materials are the most commonly employed as supercapacitor electrodes because of their unique chemical and physical features such as lower cost, larger surface areas, higher porosity, more sophisticated synthesis procedures, and availability [10]. Depending on the carbon materials utilized, an EDL forms between the electrode and the electrolyte interface in the charge storage process. Due to the higher specific surface area of carbon-based materials, the electrode material has a greater ability to accumulate the charge at the electrode/electrolyte interface and thereby increase the capacitance value. Aside from their specific surface area, the material's pore shape and structure, surface functionality, pore size distribution, and electrical conductivity are important aspects that influence electrochemical performance [11,12]. Carbon nanotubes, activated carbon, graphene, and carbon aerogels are examples of electrode materials. Because of its unique structure and amazing capabilities, graphene has been the topic of intensive research in supercapacitor applications among diverse carbon-based materials [13,14]. It is made of two-dimensional honeycomb crystals with sp^2-hybridized carbon sheets stacked in a hexagonal lattice. It has a theoretical specific surface area of roughly 2630 m^2g^{-1}, which is substantially higher than that of single-walled carbon nanotubes and other activated carbons. As a result, it is a good candidate for the EDLC [15–18]. In addition, graphene's high electrical conductivity of about 10^6 S cm^{-1}, corrosion resistance in aqueous electrolytes, excellent cyclic stability, good thermal conductivity of about 5000 W m^{-1}K^{-1}, high Young's modulus (1 TPa), excellent optical transmittance (97.7%), and intrinsic carrier mobility of about 2×10^5 cm^{2}V^{-1} s^{-1} are key features that make it an ideal material for energy storage devices [16,19]. Its high intrinsic carrier mobility increases charge transfer during the repetitive charge/discharge cycling process, which improves supercapacitor performance [15]. Individual graphene nanosheets, however, aggregate/restack during the production process due to strong interplanar contacts and mild van der Waals interactions between the graphene sheets [20,21]. It reduces the surface area of the graphene, limiting electrolyte ion transport and conduction and hence decreasing supercapacitive performance [21]. To address the shortcomings of graphene-based electrode materials, promising techniques such as heteroatom doping and chemical bonding were implemented. Preparation of hybrid materials or composites with highly conductive materials such as conductive polymers and metal oxides provides a synergistic effect of both parent and additive materials, overcoming the poor ion transport of carbon materials and preventing the aggregation of graphene nanosheets in order to increase the active surface area and diffusion of the electrolyte, thereby improving overall electrochemical perform-ance [19,21]. Chemical vapor deposition (CVD), chemical reduction, and physical and chemical exfoliation, liquid-phase exfoliation, epitaxial growth, and organic synthesis have all been used in the manufacture of graphene and its derivatives [18] (Figure 9.2).

Rounan et al. [22] investigated an electrostatic self-assembly approach for nitrogen and oxygen co-doping porous carbon nanosheets using graphene oxide as a template with cationic polyacrylamide and subsequent KOH activation. This doping produced a specific capacitance of 325.5 F g^{-1} at a current density of 0.5 A g^{-1} and a capacitance retention of 66% at a scan rate of 50 mV s^{-1}. Furthermore, the energy

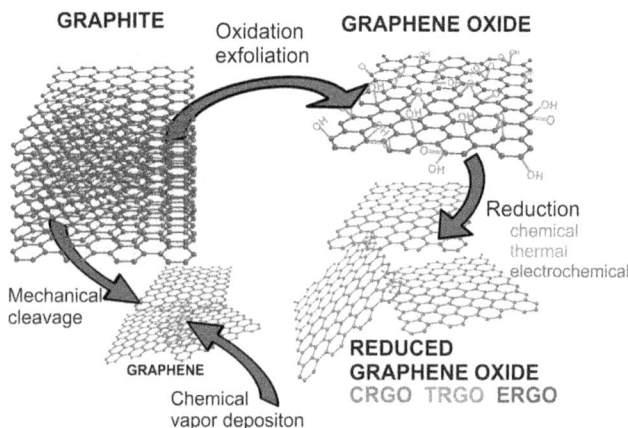

FIGURE 9.2 A schematic illustration of possible ways for preparation of graphene and rGO (ref. [18]). Image adapted with permission from [18]. Copyright © 2018 Elsevier. All rights reserved.

and power density of this composite material is 22.1 W h l^{-1}. Yan et al. [23] described the hierarchical structure of cobalt silicate coupled with reduced graphene oxide (CSO NN/rGO) as one of the promising electrode materials with high theoretical capacity and structural stability. They discovered that the electrode material significantly improves electrochemical behavior by optimizing electron/ion transportation and minimizing self-aggregation. It has a specific capacitance of 483 F g^{-1} at 0.5 A g^{-1} and a capacitance retention of 58% after 1000 cycles, which is substantially greater than the CSO NN. The presence of a hierarchical structure and the conductive character of rGO contribute to this improved electrochemical performance. Xu Zhang et al. [24] developed the 2D hybrid composite material NiCohydroxide nanowires/rGO. Because of the rapid movement of ions/electrons during electrochemical experiments, the presence of rGO in the composite material provides a conductive surface for the faradaic reaction, whereas nanowires have a rich redox reaction. Because of the synergistic impact of NiCo nanowires and rGO, excellent electrochemical performance of 1449 F g^{-1} at current density 1 Ag^{-1} with strong cycle stability in 6 M KOH electrolyte was achieved. Another study used the Ti$_3$C$_2$T$_x$/rGO composite as the electrode for a stretchable high-performance supercapacitor. Because of the strong mechanical and electrochemical performance of Ti$_3$C$_2$T$_x$, as well as the mechanical resilience of rGO, the resulting electrode obtained superior mechanical flexibility. Furthermore, the as-assembled symmetric supercapacitor demonstrated up to 300% stretchability and a specific capacitance of 18.6 mF cm^{-2} [25]. Zhang et al. used an electrode-assisted plasma electrolysis approach to prepare rGO@NiO composite supercapacitor electrode materials. At 1 Ag^{-1} current density, the as-prepared rGO@NiO composite supercapacitor has a specific capacitance of 1093 F g^{-1}. Furthermore, it has good cyclic stability, retaining 87% capacitance and 90.6% Coulombic efficiency after 5000 cycles [26]. Athira et al. investigated the influence of an anionic surfactant (SLS) on a polyaniline-wrapped reduced graphene oxide (SPGO) composite synthesized using an in situ polymerization process. In a 1 M H$_2$SO$_4$ electrolyte, a prototype symmetric supercapacitor based on an SPGO hybrid composite achieves a maximum specific capacitance of 531 Fg^{-1} at a current density of 0.2 Ag^{-1}. Furthermore, after 5000 cycles, the SLS addition on a PANI-rGO (PGO) hybrid achieved 98% Coulombic efficiency, which was significantly greater than PGO. The synergistic effect of SLS, PANI, and rGO with their associated Faradaic and double-layer mechanisms improves electrochemical performance and stability, making it an attractive option for practical energy storage applications [27]. Several constraints were impacted by the pure conducting polymers and metal oxides, including low electrical conductivity, poor cycle performance, and low energy and power density. Combining these elements with carbon compounds such as graphene allows the composite to benefit from all of the ingredients while overcoming the disadvantages of individual materials. Table 9.1 contains a tabular comparison of recently reported graphene/graphene-based nanocomposite materials.

9.1.2 MXene-Based Materials

Yury Gogotsi [28] synthesized the MXenes material for the first time in 2011 from the MAX phase. The MAX phase is defined as M$_{n+1}$AX$_n$, where M is a transition

TABLE 9.1

Electrochemical Performance of Supercapacitor Electrodes Fabricated by 2D Materials

Electrode Material	Specific Capacity	Current Density	Energy Density	Power Density	Cyclic stability	Ref.
Graphene-Based Materials						
N/O co-doped graphene	325.5 F g^{-1}	0.5 A g^{-1}	22.1 W h l^{-1}	22.1 W h l^{-1}	66%	22
CSO NN/rGO	483 F g^{-1}	0.5 A g^{-1}	-	-	58%	23
NiCo NW/rGO	1449 F g^{-1}	1 A g^{-1}	-	-	-	24
Ti$_3$C$_2$T$_x$/rGO	18.6 mF cm^{-2}	-	-	-	-	25
NiO@rGO	1093 F g^{-1}	1 A g^{-1}	-	-	87%	26
Polyaniline wrapped rGO	531 F g^{-1}	0.2 A g^{-1}	-	-	98%	27
Mxene-Based Materials						
Ti$_3$C$_2$/NiCo$_2$S$_4$	1927 F g^{-1}	-	-	-	-	37
MnO$_{1.88}$/MnO$_2$/Ti$_3$C$_2$T$_x$	312 F g^{-1}	1 A g^{-1}	-	-	91%	38
Ti$_3$C$_2$T$_x$/PANI	556.2 F g^{-1}	0.5 A g^{-1}	-	-	91.6%	39
V$_2$CT$_x$	181.1 F g^{-1}	0.2 A g^{-1}	-	-	89.1%	40
Ni$_{1.5}$Co$_{1.5}$S$_4$/Ti$_3$C$_2$	166.7 mA h g^{-1}	1 A g^{-1}	49.8 W h Kg^{-1}	800 W Kg^{-1}	90%	41
V$_4$C$_3$	330 F g^{-1}	-	-	-	90%	42
Ti$_3$C$_2$T$_x$	-	-	-	-	1133.5%	43
CoF/Mxene	1268.25 F g^{-1}	1 A g^{-1}	-	-	97%	44
TMD-Based Materials						
SnS$_2$/MoS$_2$	466.6 F g^{-1}	1 A g^{-1}	115 W h kg^{-1}	2230 W kg^{-1}	88.2%	71
Mn-integrated MoS$_2$ NFs	430 F g^{-1}	10 A g^{-1}	48.9 W h kg^{-1}	5 kW kg^{-1}	77%	72
MoS$_2$/rGO	460 F g^{-1}	-	-	-	90%	57
2H-MoS$_2$ nanoflowers	382 F g^{-1}	1 A g^{-1}	-	-	97.5%	55

Material	Capacitance	Current density		Retention	Ref.
spherical flowerlike MoS_2	255 F g^{-1}	0.25 A g^{-1}	-	70%	56
MoS_2@G/AC	334 F g^{-1}	0.5 A g^{-1}	-	83.8%	58
BP-Based Materials					
R-BP/SPC	364.5 F g^{-1}	0.5 A g^{-1}	-	89%	77
h-BN-Based Materials					
h-BN/rGO	304 F g^{-1}	1 A g^{-1}	-	98%	81
g-C_3N_4-Based Materials					
g-C_3N_4@$ZnCo_2O_4$	157 mA h g^{-1}	4 A g^{-1}	-	90%	89
MO/MOH-Based Materials					
Mesoporous β-$Co(OH)_2$	605 F g^{-1}	-	-	94.4%	98
Co-Ni$(OH)_2$	1366 F g^{-1}	1.5 A g^{-1}	-	96.26%	99
$ZnFe_2O_4$-rGO	1419 F g^{-1}	-	-	93%	100
$NiCo_2O_4$/CNT	1786 F g^{-1}	0.5 A g^{-1}	-	-	101

metal (e.g., V, Mo, Sc, Nb, Ti, Cr, Zr, Hf), X is nitrogen and/or carbon, and A is a group 13 or 14 elements (e.g., Sn, Al, and Si) [29]. MXene is a new class of transition metal nitrides, carbides, and carbonitrides that are prepared by selectively removing a specific element from their multi-layer precursors, such as by etching the element A from the MAX phase [29–31]. MXene is thus characterized by the formula $M_{n+1}X_nT_x$ (n = 1–3), where Tx is a surface terminal (e.g., OH, O, F) [30,32,33]. Because of their unusual hydrophilicity and metallic conductivity, MXenes are a viable prospect in the field of energy storage, particularly as electrodes for supercapacitors [34]. Because of their capacity to store charges via pseudocapacitive mechanisms, MXenes such as Ti_2CT_x, $Ti_3C_2T_x$, V_2CT_x, $Nb_4C_3T_x$, Nb_2CT_x, $Ta_4C_3T_x$, and Ti_2NT_x can be employed as electrode materials for supercapacitors [35,36]. MXenes material displayed exceptional mechanical and chemical resilience in addition to having ultrahigh conductivity and a hydrophilic surface [34]. Pure MXenes have the disadvantages of easy oxidation, low electrical conductivity, and mechanical flexibility. The above-mentioned restrictions can be mitigated by preparing composites with other materials, which effectively combines the excellent features of the various materials contained in the composites. When MXene is combined with other materials, the additives can accommodate between the MXene layers, suppress MXene layer restacking, and provide a broad surface area for improved electrochemical performance. The composite materials allow for a nearly fivefold increase in specific capacitance over pure MXene (Figure 9.3).

Wenling Wu et al. [37] synthesized a hierarchical new electrode material by combining two-dimensional Ti_3C_2-Mxene nanosheets with one-dimensional nickel-cobalt sulfide ($NiCo_2S_4$) via electrostatic interaction between negatively charged

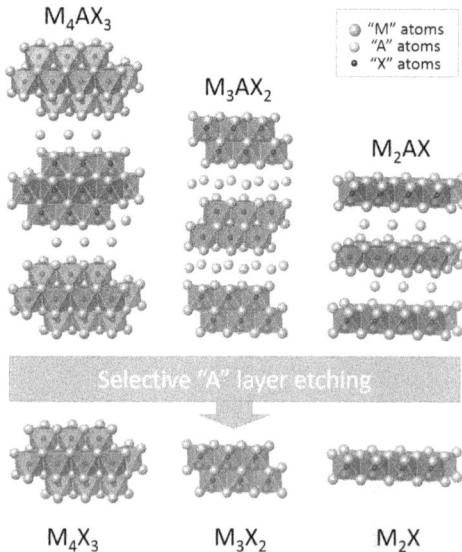

FIGURE 9.3 Structure of MAX phases and the corresponding MXenes. Image adapted with permission from GNU Free Documentation License Copyright © 2015 Wikipedia. All rights reserved.

MXene and positively charged $NiCo_2S_4$. Because of the conductive nature of MXene and the high pseudo-capacitance of $NiCo_2S_4$, the electrode materials exhibit exceptional electrochemical capabilities for the supercapacitor. At a scan rate of 2 mVs^{-1}, the Ti_3C_2-MXene/$NiCo_2S_4$ hybrid electrode material has an excellent specific capacitance of 1927 Fg^{-1} and good cyclic stability for 4000 cycles. The improved electrochemical property of the electrode is owing to the intrinsic features of both 2D and 1D materials, which also provide a shorter path for ion diffusion and electron transfer during an electrochemical phenomenon. Xinhua and colleagues [38] used an in situ growth and self-assembly to prepare a $MnO_{1.88}$/MnO_2/$Ti_3C_2T_x$ composite as a promising electrode material for supercapacitor applications. $Ti_3C_2T_x$ works as a molecular template as well as a reducing agent, reducing $KMnO_4$ to mixed valent MnO_2 and metastable $MnO_{1.88}$. The produced electrode was evaluated against electrochemical performances and demonstrated exceptional electrochemical behavior with specific capacitance of 312 F g^{-1} at current density of 1 Ag^{-1} and capacitance retention of 91% after 5000 charge-discharge cycles. All of the foregoing electrochemical features are attributable to the presence of mixed-valent manganese cations, an effective charge transport network in the composite, and a synergistic impact between $Ti_3C_2T_x$ and MnO_x. Huizhong Xu et al. [39] described a simple and efficient technique for producing $Ti_3C_2T_x$/Polyaniline (MXene/PANI) composite materials by chemical oxidative polymerization of aniline monomers on the surface of $Ti_3C_2T_x$ nanosheets under acidic conditions. The PANI formed a highly conductive porous structure on MXene, providing numerous pathways for electrolyte ion entry and lowering the composite material's charge transfer resistance. The presence of a highly conductive open structure, as well as efficient ion and electron transport, results in strong electrochemical characteristics. The composite materials have a maximum specific capacitance of 556.2 F g^{-1} at a current density 0.5 A g^{-1}, and a high cyclic stability of 91.6% after 5000 cycles. Hongtian et al. [40] investigated two-dimensional V_2CT_x MXene electrode material for supercapacitor fabrication by selectively etching Al layers from the V_2AlC MAX phase with NaF + HCl at 90 °C for 72 hours. The electrochemical property of the electrode system was tested using a standard three-electrode system using seawater as an electrolytic solution. Notably, the specific capacitance measured from a galvanometric charge-discharge curve for different current densities of 0.2, 0.5, 1.2, and 3 A g^{-1} was 181.1, 171.4, 91.3, and 11.8 F g^{-1}, respectively. After 5000 cycles, the electrode material demonstrates an excellent capacitance retention performance of 89.1%. As a result, the produced electrode material will be considered as a future electrode material for supercapacitors that use salt water as an electrolyte. Xingxing et al. [41] developed $Ni_{1.5}Co_{1.5}S_4$/Ti_3C_2 hybrid materials with varying MXene concentrations by growing $Ni_{1.5}Co_{1.5}S_4$ nanoparticles on few-layer Ti_3C_2 nanoflakes in situ using a simple one-step hydrothermal process. They tuned the mix of nickel and cobalt with MXene to enhance the supercapacitor behavior of composite materials. Because of the large surface area and improved electronic conductivity due to the synergistic effect of cobalt and nickel atoms, they have excellent electrochemical performances, with a specific capacitance of 166.7 mA hg^{-1} at current density 1 Ag^{-1} and a capacitance retention rate of 73.9% after a 20-fold increase in current density. Furthermore, the electrode material has a high energy and power density of 49.8 W h kg^{-1} and 800 W kg^{-1}, as well as cycling stability of 90% capacitance retention after 8000 cycles at 10 A g^{-1}. Based on the aforementioned features, it is possible to suggest a plausible method for building high-performance energy storage materials by multi-scale manipulation of compositions from atoms to components in the hybrid system. The vanadium carbide (V$_4$C$_3$) MXenes were prepared by employing HF to etch the intermediate Al layer. Etching the Al creates a channel for the ions to move from the electrolyte into the electrode, increasing the electrochemically active sites. The resulting electrode has a high specific capacitance (330 Fg^{-1} at 5 mVs^{-1}) and capacitive retention of 90% after 3000 cycles [42]. Tian et al. recently revealed for the first time that Ti$_3$C$_2$T$_x$-based supercapacitors in the H$_2$SO$_4$ electrolyte displayed an asymmetric voltage split between the positive and negative electrodes. They pointed out that the shift in the working potential range for the voltage window was caused by the positive electrode's partial oxidation behavior, which enables a greater capacitance to the negative electrode than the positive electrode. To improve the positive electrode's energy storage capacity, they proposed a high efficiency method in which a redox-active electrolyte was mixed with the standard H$_2$SO$_4$ electrolyte. They concluded that the presence of the redox electrolyte switched the storage mechanism from capacitive to battery, resulting in a better energy density. The mixed electrolyte supercapacitors demonstrated outstanding cyclic stability, retaining 113.5% of their initial capacitance after 10,000 cycles [43]. Ayman et al. reported the supercapacitor performance of an MXene/cobalt ferrite composite (CoF/MXene), which outperformed individual CoF and MXene in terms of electrochemical characteristics. In a 0.1 M KOH electrolyte, a specific capacitance of 1268.25 Fg^{-1} at a current density of 1 Ag^{-1} was obtained with 97% capacitance retention up to 5000 cycles [44]. Table 9.1 presents a comparative evaluation of MXene-based nanocomposite materials.

9.1.3 TRANSITION METAL DICHALCOGENIDE-BASED MATERIALS

TMDs are a type of two-dimensional material that contains transition metals from group IV to VI as well as chalcogens (S, Se, and Te). TMDCs have a generalized formula of MX$_2$ (M = transition metal; Ti, Zr, Hf, V, Nb, Ta, Mo, W, Tc, Re, Co, Rh, Ir, Ni, Pd, Pt), (X = chalcogens; S, Se, Te) [45]. Covalently bound sheets sandwich transition metal atoms between two layers of chalcogen atoms, forming stacks held together by weak van der Waals interactions [46]. Six chalcogen atoms coordinate one transition metal atom to form either octahedral or trigonal prismatic MX$_6$ polyhedrons. These MX$_6$ polyhedrons combine their edges to produce a single layer of TMDs [47]. So far, around 60 TMDCs have been identified. Among them, molybdenum disulfide (MoS$_2$), titanium disulfide (TiS$_2$), and tungsten disulfide (WS$_2$) have received a lot of attention. Among TMDs are semiconductors (MoS$_2$, WS$_2$, MoSe$_2$, WSe$_2$), metals (NbTe$_2$, TaTe$_2$), semimetals (MoTe$_2$), magnetic materials (CrSe$_2$, VS$_2$, VSe$_2$), topological insulators (WTe$_2$), and superconductors (NbSe$_2$) [48]. TMDs are interesting electrode materials for energy storage applications due to van der Waals interactions between layers and high surface areas due to sheet-like architectures. TMDs have the ability to store energy via both

Faradaic and non-Faradaic methods [49,50]. TMDs have a higher surface volume ratio, a graphene-like thin structure, an atomic-scale thickness, a short electron transport length, and a short ion diffusion path length, all of which help to accelerate the charge/discharge rate [49–51]. Due to their unique layered structure and good electrochemical characteristics, research has recently focused on creating various transition metal dichalcogenides-based supercapacitor electrodes [52–60]. The diverse synthesis methods of TMDs and their composites have resulted in a variety of morphologies that yield composites with increased surface area and porosity. Numerous reaction sites were formed as a result of the increased surface area, which aids in the development of supercapacitor performance. Because of its low cost, earth abundance, chemical stability, wide negative potential range, and high specific capacitance, TMDs have many applications [52,54,55]. Because of their semiconducting nature, TMDs have low electrical conductivity and are detrimental to electron mobilities and redox activities, severely restricting the charging/discharging rate and causing poor cycle stability. Further, the interactions between the 2D TMDs layers and their high surface energy increased the layers' restacking capability, reducing the number of active sites [61]. One effective method for addressing the aforementioned shortcomings and increasing TMD activity is to manufacture new functional hybrid materials composed of 2D TMDs and a highly conductive supporting material [62,63]. For example, multiple studies [62,64,65] have reported on the hybridization of 2D-TMDs with various carbon-based supporting materials to enhance active sites and facilitate charge transferability. Due to its electrical and increased physicochemical properties, the heterostructure hybrid of 2D-TMDs with graphene sheets has shown tremendous promise for a variety of applications in recent years. There are different approaches to synthesis the 2D-TMDs/Gr hybrids with specific characteristics, such as high specific surface area, unique morphology, thickness, and electroactive site numbers, according to their application purposes, including (a) chemical vapor deposition synthesis [66], (b) thermal and chemical reduction synthesis [67], (c) electrochemical synthesis [68], (d) microwave-assisted synthesis [69], (e) hydrothermal and solvothermal synthesis [70], etc. Conducting polymers are the best additions for improving specific capacitance in TMD composites. Figure 9.4 depicts the different crystal structures of monolayer TMDs.

Bo Wang et al. [71] used a simple hydrothermal procedure at 200 °C for 24 hours to prepare a 2D/2D SnS_2/MoS_2 heterojunction electrode material for supercapacitors. This electrode material is unique in that it prevents agglomeration and stacking during the electrochemical process. A typical three-electrode system with 0.5 M KOH electrolytic solution was used to investigate the electrochemical property. The electrode material has a specific capacitance of 466.6 F g^{-1} at 1 Ag^{-1} current density and a remarkable cycle stability of 88.2% capacitance retention at 4 A g^{-1}. In addition, the manufactured electrode has a very high energy density of 115 W h kg^{-1} and a power density of 2230 W kg^{-1}. Furthermore, for practical applications, they fabricated an all-solid symmetric supercapacitor with a specific capacitance of 98.8 F g^{-1} at a current density of 1 A g^{-1}. The good electrochemical performance with high specific capacitance could be attributed to the synergistic effect of nanostructure and electrical conductivity. Shib Shankar Singha et al. [72]

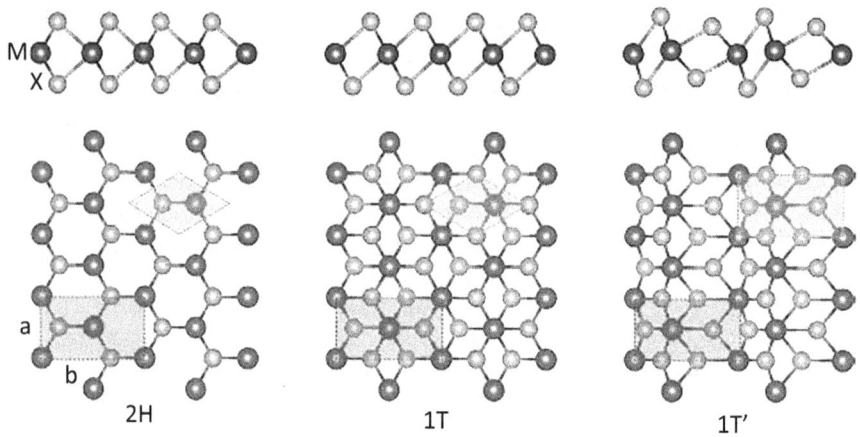

FIGURE 9.4 Three crystal structures of monolayer TMDs. The top schematics show cross-sectional views and the bottom schematics show basal plane views. The gray atoms are transition metal atoms and the red atoms are chalcogen atoms; in all three phases, a layer of transition metal atoms (M) is sandwiched between two chalcogenide layers (X). The semiconducting 2H phase has a trigonal prismatic structure, and the metallic 1T and semimetallic 1T' phases have octahedral and distorted octahedral structures, respectively. The gray shadow represents a rectangular computational cell with dimensions a×b, and the red shadow represents the primitive cell (ref. [48]). Image adapted with permission from [48]. Copyright © 2016 Springer Nature. All rights reserved.

used a simple one-step hydrothermal technique to prepare Mn-integrated MoS_2 NFs nanoflowers with varying manganese molar concentrations. Various physico-chemical approaches have been used to characterize the successful creation of electrode materials. The electrochemical measurement of the material was performed in an electrochemical cell using a typical three-electrode setup and 0.5 M Na_2SO_4 as an electrolyte. MoS_2 NFs have a specific capacitance of about 352 F g^{-1}, which was found to be enhanced to 430 F g^{-1} after the integration of a manganese atom on the surface of the MoS_2 NFs. The integration of manganese over MoS_2 increases the interlayer gap and specific surface area, which benefits the insertion or extraction of electrolyte ions during electrochemical measurements. Furthermore, the energy density is 48.9 W h kg^{-1} and the power density is 5 kW kg^{-1}, with exceptional capacity retention of up to 77% after 5000 cycles at a high current density of 10 A g^{-1}. Poulami Hota and colleagues [57] used a controlled temperature reflux approach to successfully generate amorphous MoS_2 ornamented on the surface of reduced graphene oxide. They discovered that the specific capacitance of the nanocomposite material varies with its size. As the size of the amorphous nanoparticles decreases, the specific capacitance of the composite materials gradually increases. When the size of the nanoporous material was around 50 nm, the specific capacitance was determined to be 270 F g^{-1}, but when the size of the particles was reduced to 5–7 nm, the specific capacitance increased to 460 F g^{-1}. After 5000 cycles, the retention in capacitance of materials was determined to be around 90%. All of the above electrochemical properties are affected by the size of the composite, which aids in the increase of active surface area with a large number of active sites, as well as better charge transport, which effectively enhances the storage capacity of the electrode materials. Mishra et al. have prepared 2H-MoS$_2$ nanoflowers using a one-step hydrothermal technique. At 1 Ag^{-1}, a specific capacitance of 382 Fg^{-1} was measured, with 97.5% retention after 4000 cycles. The authors concluded that the excellent electrochemical outputs were due to the 3D flowerlike form with exposed surface edges of distinct petals [55]. Gupta et al. discovered that spherical flowerlike MoS$_2$ particles produced by a simple hydrothermal method had a specific capacitance of 255 F g^{-1} at 0.25 A g^{-1} and a retention capacity of 70% after 1000 cycles [56]. Fu et al. discovered that MoS$_2$/graphene nanosheets embedded in activated carbon nanofibers (MoS$_2$@G/AC) manufactured through electrospinning, carbonization, and KOH activation had a high electrochemical performance. At current densities of 0.5 and 10 A g^{-1}, specific capacitances of 334 and 246.3 F g^{-1} were achieved, respectively. After 5000 charge/discharge cycles at 0.5 Ag^{-1}, the cyclic stability was found to be 83.8%. The high active surface area, adequate porosity, and synergy between the three separate components of the composite all contribute to the increased specific capacitance and stability of the MoS$_2$@G/AC nanofiber. The uniform distribution of the fiber network, with no beads or aggregation, provides for simple electron transfer [58]. Table 9.1 contains a detailed comparative examination of transition metal dichalcogenides and their composite materials.

9.1.4 BLACK PHOSPHORUS

Black phosphorus (BP) is one of the most promising 2D materials for future energy storage applications. White phosphorus, red phosphorus, and black phosphorus are the three major allotropes of phosphorus [73,74]. BP is the most stable of these allotropes, with phosphorene sheets held together by weak van der Waals forces and a structure comparable to graphene. In comparison to other 2D materials, BP has an outstanding theoretical capacity, high power density, large volumetric capacitance, strong carrier mobility, low redox potential, and direct controllable energy band gap [73–76]. Furthermore, BP has a high specific capacity of 2596 mA h g^{-1}, which is seven times greater than that of graphite. It also has a substantially larger surface area than graphene (2630 m^2 g^{-1}) because of the puckered shape and increased layer spacing [74]. Increased space between the BP layers causes ion diffusion within the layers to accelerate. These characteristics make the BP material a good electrode material for EDLCs (Figure 9.5).

Gopalakrishnan and Badhulika used a simple sonication procedure to create a hybrid red black phosphorus/sulfonated porous carbon (R-BP/SPC) composite. The resulting R-BP/SPC supercapacitor electrode has a maximum specific capacitance of 364.5 F g^{-1} at 0.5 A g^{-1} and a cyclic stability of 89%. After 10,000 cycles, only 11% of the capacitance is lost, demonstrating its usefulness in various super-capacitor systems [77]. Despite the fact that BP has several advantages, large-scale synthesis, high volume growth during charge/discharge cycles, and poor electronic conductivity limit its practical applicability. When used with other stable materials, it inhibits excessive volume expansion during repeated charge/discharge cycles and

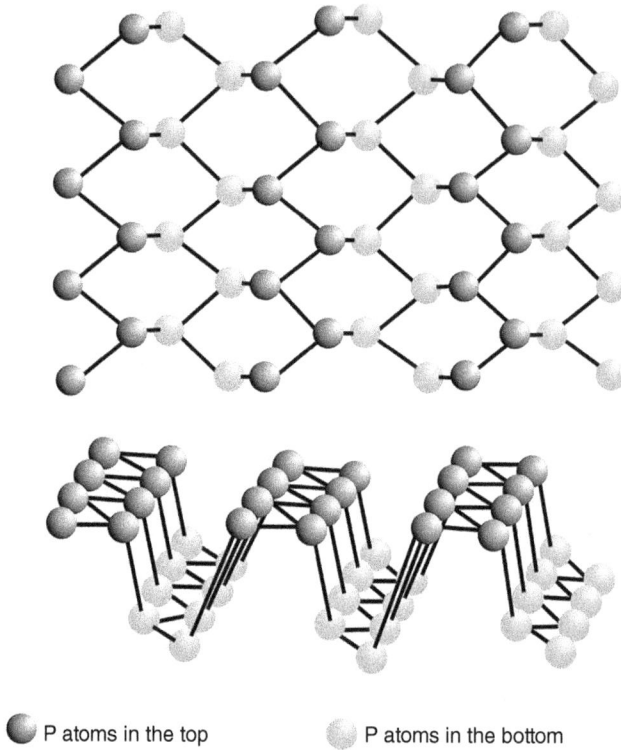

P atoms in the top   P atoms in the bottom

FIGURE 9.5   Schematic illustration of atomic structure of 2D-BP (ref. [73]). Image adapted with permission from [73]. Copyright © 2018 Springer Nature. All rights reserved.

boosts electrical conductivity. Moreover, in order to be used realistically, BP-based products require an efficient synthesis technique.

9.1.5 HEXAGONAL BORON NITRIDE

Hexagonal boron nitride (h-BN) is structurally identical to graphene, which has a honeycomb structure with sp^2 hybridization. It is made up of an equal number of boron and nitrogen atoms that are linked together by a strong covalent bond with a bond length of 1.45 Å and are coupled by weak van der Waals forces [78,79] (Figure 9.6). Boron nitride is available in crystalline and amorphous forms. In its crystalline form, it can have three major allotropes: hexagonal boron nitride (h-BN), sphalerite boron nitride (β-BN), and wurtzite boron nitride (γ-BN), the most stable of which is h-BN. It features a hexagonal structure and the P6$_3$/mmc space group [79]. Because of its unusual mechanical, thermal, and electronic qualities, two-dimensional hexagonal boron nitride has sparked increased interest in electronic applications, UV emitters, field emitters, and thermal radiators. Despite the fact that h-BN is structurally identical to graphene, its huge band gap (5.9 eV) has hampered its practical applicability due to poor electronic conductivity [79,80]. Many studies are still being conducted to increase the conductivity of h-BN, allowing it to be used

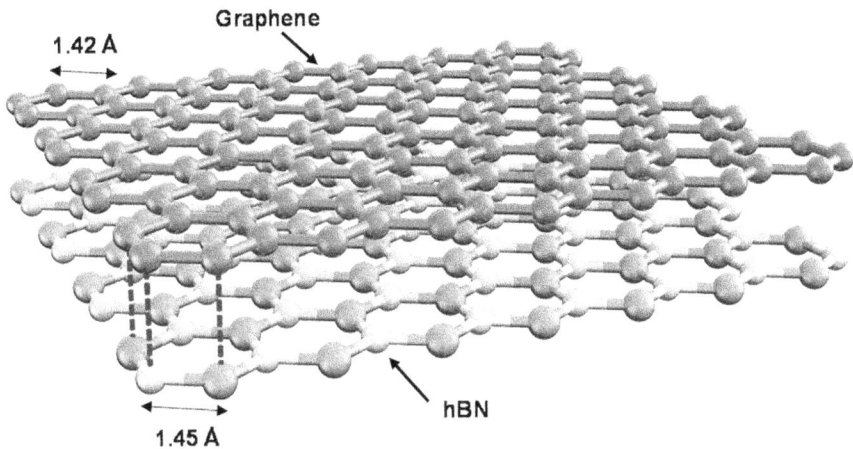

FIGURE 9.6 Lattice matching of graphene and hBN (ref. [78]). Image adapted with permission from [78]. Copyright © 2023 MDPI.

in energy storage devices such as supercapacitors and batteries. Researchers are currently concentrating on developing graphene-based h-BN nanocomposites with superior conductivity due to graphene [80].

Patil et al. described a simple method for synthesizing the 2D/3D heterostructure of h-BN and rGO, which demonstrated excellent capacitive performance in acidic and alkaline electrolytes. The van der Waals interaction between the h-BN and rGO creates an electroactive site for efficient ion and electron transport, resulting in fast reaction kinetics. The resulting h-BN/rGO heterostructure has a specific capacitance of 304 and 226 Fg^{-1} in alkaline and acidic conditions, respectively, at a current density of 1 Ag^{-1}, with 98% capacitance retention after 10,000 cycles [81]. Carbon atoms in the composite minimize the huge band gap of h-BN and introduce carbon conjugation, boosting electrical conductivity and electrochemical performance. Furthermore, the inclusion of different phases of h-BN and additives in the composite affects the supercapacitor electrodes' band structure and electronic work function. Furthermore, doping concentration can significantly improve the porosity of electrode materials, resulting in better specific capacitance and improved cycle stability. However, compared to other 2D materials, only few investigations on h-BN-based supercapacitors have been reported; thus, more research is needed.

9.1.6 CARBON NITRIDE

Carbon nitride (CN) is a two-dimensional layered material with a graphene-like structure that is emerging as a wonder material for energy storage and catalysis [82,83]. In general, seven C_3N_4 phases have been reported: α-C_3N_4, β-C_3N_4, cubic-C_3N_4, pseudocubic-C_3N_4, g-h-triazine, g-o-triazine, and g-h-heptazine. The graphitic carbon nitride (g-C_3N_4) phase is the most stable under ambient circumstances and is of particular relevance in the field of energy storage and conversion [84,85]. The g-C_3N_4 structure contains an N substituted graphitic network made up of sp^2-

hybridized carbon and nitrogen, resulting in a conjugated electronic structure [83,84]. In recent years, g-C$_3$N$_4$ has piqued the interest of many researchers due to its attractive properties, which include the presence of earth-abundant carbon and nitrogen, metal-free material, low cost, sustainability, simple synthesis conditions, exceptional chemical and thermal stability, excellent optical features, and appealing electronic properties, making it a fascinating candidate in the fields of photocatalysis and solar energy conversion, as well as the detection of toxic molecules [82–84]. The 2D graphitic carbon nitride (g-C$_3$N$_4$) exhibits outstanding physicochemical qualities, provides additional surface-active sites for electrochemical reactions, and shortens ion diffusion route lengths, allowing it to be used in supercapacitor applications in addition to graphene sheets. However, its low conductivity restricts its usage to some extent, which can be solved by combining it with other materials [84]. Carbon nitride has been extensively explored for its numerous applications; however, very few studies have been published on its use as supercapacitor electrodes [86–91]. Figure 9.7 represents the synthesis process of g-C$_3$N$_4$ by thermal polymerization of different precursors.

Sharma and Gaur investigated a composite of g-C$_3$N$_4$-supported ZnCo$_2$O$_4$ generated by a simple hydrothermal technique. The g-C$_3$N$_4$@ZnCo$_2$O$_4$ hybrid composite-based supercapacitor electrode demonstrated a specific discharge capacity of 157 mA h g^{-1} at 4 A g^{-1} and a retention rate of 90% after 2500 cycles. The symmetric supercapacitor designed using g-C$_3$N$_4$@ZnCo$_2$O$_4$//g-C$_3$N$_4$@ZnCo$_2$O$_4$ achieved a maximum specific capacity of 121 mA h g^{-1}, while maintaining 71% capacitance retention after 10,000 cycles [89]. The low electrical conductivity and

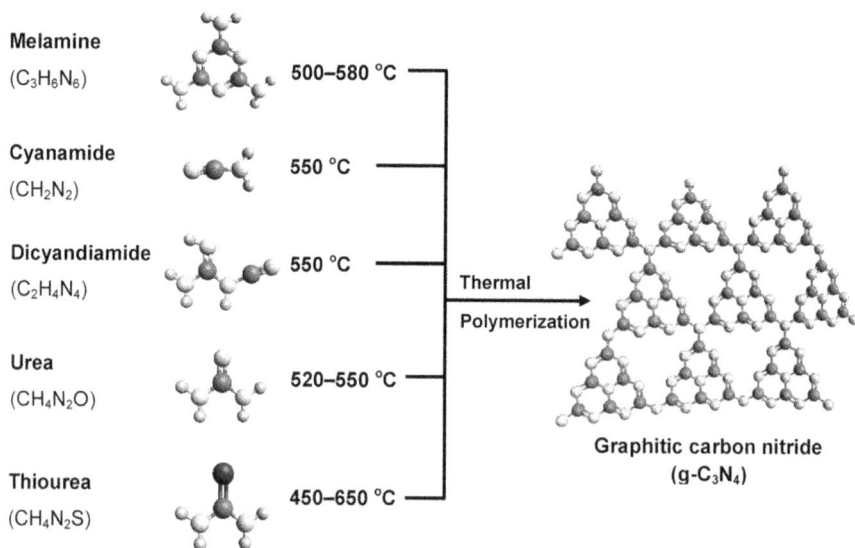

Melamine
(C$_3$H$_6$N$_6$)   500–580 °C

Cyanamide
(CH$_2$N$_2$)   550 °C

Dicyandiamide
(C$_2$H$_4$N$_4$)   550 °C   Thermal
                                    Polymerization

Urea
(CH$_4$N$_2$O)   520–550 °C

Thiourea
(CH$_4$N$_2$S)   450–650 °C

Graphitic carbon nitride
(g-C$_3$N$_4$)

FIGURE 9.7 Schematic illustration of the synthesis process of g-C$_3$N$_4$ by thermal polymerization of different precursors. See ref. [82] for further detail. Image adapted with permission from [82]. Copyright © 2016 American Chemical Society. All rights reserved.

tiny active surface area of g-C_3N_4 limit its practical use in energy storage devices. However, the g-C_3N_4 has a significant non-graphitic nitrogen content and a broad band gap of 2.7 eV, which limits its electrochemical performance and necessitates various procedures such as morphological tuning and doping with other conductive materials. Furthermore, van der Waals interactions of g-C_3N_4 produce single-layer restacking, reducing power and energy density. As a result, it has been discovered that hybridization with pseudocapacitive materials is more efficient.

9.1.7 METAL OXIDES/HYDROXIDES

The development of 2D materials for supercapacitor applications has progressed beyond graphene and currently includes 2D metal oxides/hydroxides. Atoms in 2D metal oxides/hydroxides are kept together by a strong covalent connection, and layers are held together by a weak van der Waals interaction. They have a high active surface area and great mechanical stability, flexibility, and well-defined surface termination, which are not possible in bulk, making them attractive candidates for next-generation battery and supercapacitor devices [92]. So far, a number of metal oxides and hydroxides exhibiting pseudocapacitive properties, such as RuO_2, V_2O_5, SnO_2, MnO_2, Co_3O_4, Fe_2O_3, $Ni(OH)_2$, and $Co(OH)_2$, have been effectively produced and investigated as electrode materials for energy storage devices [92–96]. RuO_2 is the ideal metal oxide for manufacturing supercapacitor electrodes; nevertheless, its cost and availability limit its practical applicability [92]. Due to the flexible tunability of metal ions, ability to keep multiple oxidation states, high reactive site, short ion diffusion path length, and improved reaction kinetics, 2D metal oxides/hydroxides have numerous superior properties for electrochemical processes [96,97]. Despite having high capacitive qualities, metal oxide/hydroxide has a restricted power density and cyclic stability due to weak electronic and ionic conductivities. The mixing of metal oxide/hydroxide with other materials such as graphene and CNT is a viable technique to improve supercapacitor performance.

Youssry et al. used different methodologies to report the electrochemical performance of $Co(OH)_2$ comprising two distinct phases. They used electrochemical deposition and chemical reduction to synthesize a homogeneous and evenly coated mesoporous β-$Co(OH)_2$ film supported on an FTO substrate, as well as mesoporous α-$Co(OH)_2$ powder. The scientists determined that the mesoporous β-$Co(OH)_2$ film had an outstanding specific capacitance of 605 F g^{-1} in 1 M KOH electrolyte and capacitance retention of 94.4% over 2000 cycles [98]. Vidhya et al. investigated a Ni-based mixed metal hydroxide composite ($Ni(OH)_2$ and Co-Ni $(OH)_2$) produced hydrothermally. At 1.5 A g^{-1}, the $Ni(OH)_2$ and Co-Ni$(OH)_2$ electrode materials exhibited pseudocapacitive behavior, with specific capacitances of 1038 and 1366 F g^{-1}, respectively. Because of the inclusion of Co, they discovered that the mixed metal hydroxide had higher capacitance than bare Ni $(OH)_2$ and had an excellent retention of 96.26% up to 2000 cycles [99]. Askari et al. examined an enhanced $ZnFe_2O_4$-rGO nanohybrid structure prepared by mixing a binary transition metal oxide based on zinc and iron with reduced graphene oxide. After 5000 charge/discharge cycles, the $ZnFe_2O_4$-rGO electrodes had a capacitance of 1419 F g^{-1} and 93% cyclic stability, making them an excellent contender for supercapacitor electrodes [100]. Cheng et al. reported a chemically synthesized core shell-structured metal hydroxide on CNT for high-performance supercapacitor electrodes. Because of the smaller crystal size of $NiCo_2O_4$ and the high conductivity of CNTs, the $NiCo_2O_4$/CNT composite has a higher specific capacitance of 1786 F g^{-1} at 0.5 A g^{-1} [101].

9.2   CHALLENGES

Supercapacitors will be promising energy storage devices in the future. However, there are a number of obstacles and roadblocks in practical application that should be considered in order to fully utilize the limitless applicability. The supercapacitor is found to have low specific capacitance and power density in the majority of instances. Researchers and R&D sectors are attempting to create novel materials that will pave the way for next-generation energy storage devices with adequate energy density, power density, and specific capacitance. Two-dimensional materials have significant advantages in the field of supercapacitor energy storage because they have a high specific surface area, excellent mechanical and electrical performances, and provide an ideal platform for the assembly of the thin film that is most suitable for supercapacitor construction. As a result, academics and the R&D sector are focusing on developing 2D nanomaterials with high redox potential that are better, more flexible, scalable, and have improved electrochemical properties. The unique physical and chemical features of 2D materials serve as a foundation for supercapacitive applications. However, there are several practical issues that limit the composite's commercial use in terms of supercapacitors. The chemical and thermodynamic stability of 2D materials in energy storage applications is a serious challenge. Mxenes are prone to oxidation in many aqueous electrolytes, which causes resistance and capacitance loss during cycling testing [102,103]. Sodium ascorbate has been used as an antioxidant to improve oxidation resistance. As a result, an effective mode of manufacturing electrode material and compatible electrolyte must be developed in order to improve oxidation resistance and its implications for long-term stability should be thoroughly investigated. Furthermore, the numerous opportunities for 2D nanostructures with fascinating features will make them an accessible material for future research that will benefit both industry and individuals.

9.3   CONCLUSION

This chapter covered recent advances in supercapacitor electrodes based on diverse 2D materials such as graphene, MXene, TMDs, carbon nitrides, h-BN, black phosphorus, and metal oxides/hydroxides. Because of the high surface area of 2D nanomaterials, an electrical double layer and the Faradaic charging process can be formed. Furthermore, the very porous structure of the 2D materials improves supercapacitor efficiency by offering a large number of surface-active sites, which shortens the ion diffusion path. Although 2D materials are the best candidate among other nanomaterials, some restrictions remain, such as high-quality large-scale production, integration with other electronic devices, and the construction of hybrid structures, among others. Individual 2D materials' poor electronic conductivity can be efficiently rectified by combining them with suitably conductive materials. This evaluation also includes recent literature studies on hybrid composite super-capacitive performance. Future research should, however, concentrate on effective hybridization mechanisms and structural performance connections of 2D materials. As a result of concentrated research in this field, scientists will be able to develop next-generation high-performance supercapacitors, a viable alternative to existing energy storage devices, electric vehicles, and solar panels.

ACKNOWLEDGMENTS

All authors would like to thank the Centre for Materials for Electronics Technology, Thrissur, India, Colorado School of Mines, USA, and the Indian Institute of Technology Jodhpur, India, for resource and technical support. The authors would also like to thank the Department of Science and Technology-India and University Grants Commission, India for funding support.

CONFLICTS OF INTEREST

The authors declare no competing interest.

REFERENCES

1. L. Xie, et al., Hierarchical porous carbon microtubes derived from willow catkins for supercapacitor applications, *J. Mater. Chem. A*, 4(5), 1637–1646 (Feb. 2016).
2. J. Li, X. Cheng, A. Shashurin, and M. Keidar, Review of electrochemical capacitors based on carbon nanotubes and graphene, *Sci. Res. J.*, 1, 1–13 (2012).
3. F. Shi, L. Li, C. D. Gu, and J. Tu, Metal oxide/hydroxide based materials for supercapacitors, *RSC Adv.*, 4, 41910–41921 (2014).
4. Y. Xue, Q. Zhang, W. Wang, H. Cao, Q. Yang, and L. Fu, Opening two-dimensional materials for energy conversion and storage: a concept, *Adv. Energy Mater.*, 7, 1602684 (2017).
5. S. Z. Butler, S. M. Hollen, L. Cao, Y. Cui, J. A. Gupta, H. R. Gutiérrez, T. F. Heinz, S. S. Hong, J. Huang, and A. F. Ismach, Progress, challenges, and opportunities in two-dimensional materials beyond graphene, *ACS Nano*, 7, 2898 (2013).
6. J. Cao, P. He, J. R. Brent, H. Yilmaz, D. J. Lewis, I. A. Kinloch, and B. Derby, Supercapacitor Electrodes from the in Situ Reaction between Two-Dimensional Sheets of Black Phosphorus and Graphene Oxide, *ACS Appl. Mater. Interfaces*, 10, 10330 (2018).
7. A. E. Del Rio Castillo, V. Pellegrini, H. Sun, J. Buha, D. A. Dinh, E. Lago, A. Ansaldo, A. Capasso, L. Manna, and F. Bonaccorso, Exfoliation of Few-Layer Black Phosphorus in Low-Boiling-Point Solvents and Its Application in Li-Ion Batteries, *Chem. Mater.*, 30, 506 (2018).
8. C. Rao, K. Gopalakrishnan, and U. Maitra, Comparative Study of Potential Applications of Graphene, MoS2, and Other Two-Dimensional Materials in Energy Devices, Sensors, and Related Areas, *ACS Applied Materials & Interfaces*, 7, 7809 (2015).
9. C. Tan, and H. Zhang, Two-dimensional transition metal dichalcogenide nanosheet-based composites, *Chem. Soc. Rev.*, 44, 2713 (2015).

10. Z. S. Iro, C. Subramani, and S. S. Dash, A brief review on electrode materials for supercapacitor, *Int. J. Electrochem. Sci.*, 11 (12), 10628–10643 (2016).

11. H. Yang, S. Kannappan, A. S. Pandian, J. H. Jang, Y. S. Lee, and W. Lu, Graphene supercapacitor with both high power and energy density, *Nanotechnology*, 28 (44), 2713–2731 (2017).

12. Z. Bo, S. Mao, Z. Jun Han, K. Cen, J. Chen, and K. Ostrikov, Emerging energy and environmental applications of vertically-oriented graphenes, *Chem. Soc. Rev.*, 44 (8), Royal Society of Chemistry, 2108–2121 (21 Apr. 2015).

13. M. Winter, and R. J. Brodd, What are batteries, fuel cells, and supercapacitors?, *Chem. Rev.*, 104 (10), 4245–4270 (2004).

14. A. Muzaffar, M. B. Ahamed, K. Deshmukh, and J. Thirumalai, A review on recent advances in hybrid supercapacitors: design, fabrication and applications, *Renewable Sustainable Energy Rev.*, 101, 123–145 (2019).

15. X. Zhang, H. Zhang, C. Li, K. Wang, X. Sun, and Y. Ma, Recent advances in porous graphene materials for supercapacitor applications, *RSC Adv.*, 4 (86), 45862–45884 (2014).

16. A. Yu, I. Roes, A. Davies, and Z. Chen, Ultrathin, transparent, and flexible graphene films for supercapacitor application, *Appl. Phys. Lett.*, 96 (25), 253105 (2010).

17. X. Cao, Y. Shi, W. Shi, G. Lu, X. Huang, Q. Yan, Q. Zhang, and H. Zhang, Preparation of novel 3D graphene networks for supercapacitor applications, *Small*, 7 (22), 3163–3168 (2011).

18. S. J. Rowley-Neale, E. P. Randviir, A. S. Abo Dena, and C. E. Banks, An overview of recent applications of reduced graphene oxide as a basis of electroanalytical sensing platforms, *Appl. Mater. Today*, 10, 218–226 (2018).

19. T. Palaniselvam, and J. B. Baek, Graphene based 2D-materials for supercapacitors, *2D Mater.*, 2 (3), 032002 (2015).

20. Y. Han, Y. Ge, Y. Chao, C. Wang, and G. G. Wallace, Recent progress in 2D materials for flexible supercapacitors, *J. Energy Chem.*, 27 (1), 57–72 (2018).

21. Y. Shao, M. F. El-Kady, L. J. Wang, Q. Zhang, Y. Li, H. Wang, M. F. Mousavi, and R. B. Kaner, Graphene-based materials for flexible supercapacitors, *Chem. Soc. Rev.*, 44 (11), 3639–3665 (2015).

22. R. Liu, Y. Wang, and X. Wu, Two-dimensional nitrogen and oxygen Co-doping porous carbon nanosheets for high volumetric performance supercapacitors, *Microporous Mesoporous Mater.*, 295, 109954 (2020).

23. Y. Cheng, Y. Zhang, H. Jiang, X. Dong, C. Meng, and Z. Kou, Coupled cobalt silicate nanobelt-on-nanobelt hierarchy structure with reduced graphene oxide for enhanced super capacitive performance, *J. Power Sources*, 448, 227407 (2020).

24. X. Zhang, Q. Fan, S. Liu, N. Qu, H. Yang, M. Wang, and J. Yang, A facile fabrication of 1D/2D nanohybrids composed of NiCo-hydroxide nanowires and reduced graphene oxide for high-performance asymmetric supercapacitors, *Inorg. Chem. Front.*, 7, 204 (2020).

25. Y. Zhou, K. Maleski, B. Anasori, J. O. Thostenson, Y. Pang, Y. Feng, K. Zeng, C. B. Parker, S. Zauscher, Y. Gogotsi, J. T. Glass, and C. Cao., Ti3C2Tx MXene-reduced graphene oxide composite electrodes for stretchable supercapacitors, *ACS Nano.*, 14, 3576–3586 (2020).

26. Y. Zhang, Y. Shen, X. Xie, W. Du, L. Kang, Y. Wang, X. Sun, Z. Li, and B. Wang, One-step synthesis of the reduced graphene oxide@ NiO composites for supercapacitor electrodes by electrode-assisted plasma electrolysis, *Mater. Des.*, 196, 109111 (2020).

27. A. R. Athira, S. Deepthi, and T. S. Xavier, Impact of an anionic surfactant on the enhancement of the capacitance characteristics of polyaniline-wrapped graphene oxide hybrid composite, *Bull. Mater. Sci.*, 44 (3), 1–10 (2021).

28. M. Naguib, M. Kurtoglu, V. Presser, J. Lu, J. Niu, M. Heon, L. Hultman, Y. Gogotsi, M. W. Barsoum., Two-dimensional nanocrystals produced by exfoliation of Ti3AlC2, *Adv. Mater.*, 23 (37), 4248–4253 (2011).

29. M. Naguib, J. Halim, J. Lu, K. M. Cook, L. Hultman, Y. Gogotsi, M. W. Barsoum., New two-dimensional niobium and vanadium carbides as promising materials for li-ion batteries, *J. Am. Chem. Soc.*, 135 (43), 15966–15969 (2013).

30. M. Ghidiu, M. Naguib, C. Shi, O. Mashtalir, L. M. Pan, B. Zhang, J. Yang, Y. Gogotsi, S. J. L. Billingebd and M. W. Barsoum et al., Synthesis and characterization of two-dimensional Nb4C 3 (MXene), *Chem. Commun.*, 50 (67), 9517–9520 (2014).

31. M. Naguib, O. Mashtalir, J. Carle, V. Presser, J. Lu, L. Hultman, Y. Gogotsi, M. W. Barsoum., Two-dimensional transition metal carbides, *ACS Nano*, 6 (2), 1322–1331 (2012).

32. B. Anasori, M. R. Lukatskaya, and Y. Gogotsi, 2D metal carbides and nitrides (MXenes) for energy storage, *Nat. Rev. Mater.*, 2(2), 1602, (2017).

33. M. Naguib, V. N. Mochalin, M. W. Barsoum, and Y. Gogotsi, 25th anniversary article: MXenes: a new family of two-dimensional materials, *Adv. Mater.*, 26 (7), 992–1005 (2014).

34. M. Ghidiu, M. R. Lukatskaya, M. Q. Zhao, Y. Gogotsi, and M. W. Barsoum, Conductive two-dimensional titanium carbide 'clay' with high volumetric capacitance, *Nature*, 516 (7529), 78–81 (2015).

35. A. Djire, A. Bos, J. Liu, H. Zhang, E. M. Miller, and N. R. Neale, Pseudocapacitive storage in nanolayered Ti2NTx MXene using Mg-ion electrolyte, *ACS Appl. Nano Mater.*, 2 (5), 2785–2795 (2019).

36. B. Anasori, Y. Xie, M. Beidaghi, J. Lu, B. C Hosler, Lars Hultman, P. R. C. Kent, Y. Gogotsi, M. W. Barsoum, Two-dimensional, ordered, double transition metals carbides (MXenes), *ACS Nano*, 9 (10), 9507–9516 (2015).

37. W. Wu, C. Zhao, J. Zhu, D. Niu, D. Wei, C. Wang, F. Wang, and L. Wang, Hierarchical materials constructed by 1D hollow nickel–cobalt sulfide nanotubes supported on 2D ultrathin MXenes nanosheets for high-performance supercapacitor, *Ceram. Int.*, 46, 12200 (2020).

38. X. Huang, X. Zhu, S. Luo, R. Li, N. Rajput, M. Chiesa, and V. Chan, MnO1.88/R-MnO2/Ti3C2(OH/F)x composite electrodes for high-performance pseudo-supercapacitors prepared from reduced MXenes, *New J. Chem.*, 46, 6583 (2020).

39. H. Xu, D. Zheng, F. Liu, W. Li, and J. Lin, Synthesis of an MXene/polyaniline composite with excellent electrochemical properties, *J. Mater. Chem. A*, 8, 5853 (2020).

40. H. He, Q. Xia, B. Wang, L. Wang, Q. Hu, and A. Zhou, Two-dimensional vanadium carbide (V2CTx) MXene as supercapacitor electrode in seawater electrolyte, *Chin. Chem. Lett.*, 31, 984 (2020).

41. X. He, T. Bi, X. Zheng, W. Zhu, and J. Jiang, Nickel cobalt sulfide nanoparticles grown on titanium carbide MXenes for high-performance supercapacitor, *Electrochim. Acta*, 332, 135514 (2020).

42. R. Syamsai, and A. N. Grace, Synthesis, properties and performance evaluation of vanadium carbide MXene as supercapacitor electrodes, *Ceram. Int.*, 46 (4), 5323–5330 (2020).

43. Y. Tian, C. Yang, Y. Luo, H. Zhao, Y. Du, L. B. Kong, and W. Que, Understanding MXene-based "symmetric" supercapacitors and redox electrolyte energy storage, *ACS Appl. Energy Mater.*, 3 (5), 5006–5014 (2020).

44. I. Ayman, A. Rasheed, S. Ajmal, A. Rehman, A. Ali, I. Shakir, and M. F. Warsi, CoFe2O4 nanoparticle-decorated 2D MXene: a novel hybrid material for super-capacitor applications, *Energy Fuels*, 34 (6), 7622–7630 (2020).

45. A. K. Geim, and I. V. Grigorieva, Van der Waals heterostructures, *Nature*, 499 (7459), 419–425 (2013).
46. K. S. Novoselov, A. Mishchenko, A. Carvalho, and A. H. Castro Neto, 2D materials and van der Waals heterostructures, *Science*, 353 (6298), aac9439 (2016).
47. T. Cao, J. Feng, J. Shi, Q. Niu, and E. Wang, MoS$_2$ as an ideal material for valleytronics: valley-selective circular dichroism and valley Hall effect, *Nat. Commun.*, 3 (887), 1–5 (2011).
48. Y. Li, K. A. Duerloo, K. Wauson, E. J. Reed., Structural semiconductor-to-semimetal phase transition in two-dimensional materials induced by electrostatic gating, *Nat. Commun.*, 7, 10671 (2016).
49. L. Lin, W. Lei, S. Zhang, Y. Liu, G. G. Wallace, and J. Chen, Two-dimensional transition metal dichalcogenides in supercapacitors and secondary batteries, *Energy Storage Mater.*, 19, 408–423 (2019).
50. N. S. Arul, and V. D. Nithya, Eds., *Two Dimensional Transition Metal Dichalcogenides: Synthesis, Properties, and Applications*, Springer (2019).
51. M. Mohan, K. N. Unni, and R. B. Rakhi, 2D organic-inorganic hybrid composite material as a high-performance supercapacitor electrode, *Vacuum*, 166, 335–340 (2019).
52. K. V. G. Raghavendra, R. Vinoth, K. Zeb, C. V. V. Muralee Gopi, S. Sambasivam, M. R. Kummara, I. M. Obaidat, and H. J. Kim, An intuitive review of supercapacitors with recent progress and novel device applications, *J. Energy Storage*, 31, 101652 (2020).
53. V. Shrivastav, S. Sundriyal, V. Shrivastav, U. K. Tiwari, and A. Deep, WS2/carbon composites and nanoporous carbon structures derived from zeolitic imidazole framework for asymmetrical supercapacitors, *Energy Fuels*, 35 (18), 15133–15142 (2021).
54. N. Joseph, P. M. Shafi, and A. C. Bose, Recent advances in 2DMoS2 and its composite nanostructures for supercapacitor electrode application, *Energy Fuels*, 34 (6), 6558–6597 (2020).
55. S. Mishra, P. K. Maurya, and A. K. Mishra, 2H–MoS2 nanoflowers based high energy density solid state supercapacitor, *Mater. Chem. Phys.*, 255, 123551 (2020).
56. H. Gupta, S. Chakrabarti, S. Mothkuri, B. Padya, T. N. Rao, and P. K. Jain, High performance supercapacitor based on 2D-MoS2 nanostructures, *Mater. Today: Proc.*, 26, 20–24 (2020).
57. P. Hota, M. Miah, S. Bose, D. Dinda, U. K. Ghorai, Y. K. Su, and S. K. Saha, Ultra-small amorphous MoS2 decorated reduced graphene oxide for supercapacitor application, *J. Mater. Sci. Technol.*, 40, 196–203 (2020).
58. H. Fu, X. Zhang, J. Fu, G. Shen, Y. Ding, Z. Chen, and H. Du, Single layers of MoS2/Graphene nanosheets embedded in activated carbon nanofibers for high-performance supercapacitor, *J. Alloys Compd.*, 829, 154557 (2020).
59. W. Yin, D. He, X. Bai, and W. W. Yu, Synthesis of tungsten disulfide quantum dots for high-performance supercapacitor electrodes, *J. Alloys Compd.*, 786, 764–769 (2019).
60. T. W. Lin, T. Sadhasivam, A. Y. Wang, T. Y. Chen, J. Y. Lin, and L. D. Shao, Ternary composite nanosheets with MoS2/WS2/graphene heterostructures as high-performance cathode materials for supercapacitors, *ChemElectroChem*, 5 (7), 1024–1031 (2018).
61. K. Leng, Z. Chen, X. Zhao, W. Tang, B. Tian, C. T. Nai, W. Zhou, and K. P. Loh, Phase restructuring in transition metal dichalcogenides for highly stable energy storage, *ACS Nano*, 10 (10), 9208–9215 (2016).
62. J. W. Jiang, and H. S. Park, Mechanical properties of MoS2/graphene heterostructures, *Appl. Phys. Lett.*, 105 (3), 033108 (2014).
63. M. Yousaf, A. Mahmood, Y. Wang, Y. Chen, Z. Ma, and R. P. S. Han, Advancement in layered transition metal dichalcogenide composites for lithium and sodium ion batteries, *J. Electr. Eng.*, 4(2), (2016).

64. P. S. Toth et al., Asymmetric MoS2/graphene/metal sandwiches: preparation, characterization, and application, *Adv. Mater.*, 28 (37), 8256–8264 (2016).

65. X. Zhang, et al., MoS2/carbon nanotube core-shell nanocomposites for enhanced nonlinear optical performance, *Chem. Eur. J.*, 23 (14), 3321–3327 (2017).

66. X. Liu, I. Balla, H. Bergeron, G. P. Campbell, M. J. Bedzyk, and M. C. Hersam, Rotationally commensurate growth of MoS2 on epitaxial graphene, *ACS Nano*, 10 (1), 1067–1075 (2016).

67. L. Jiang, B. Lin, X. Li, X. Song, H. Xia, L. Li, and H. Zeng, Monolayer MoS2-graphene hybrid aerogels with controllable porosity for lithium-ion batteries with high reversible capacity, *ACS Appl. Mater. Interfaces*, 8 (4), 2680–2687 (2016).

68. X. Li, L. Zhang, X. Zang, X. Li, and H. Zhu, Photo-promoted platinum nanoparticles decorated MoS2@graphene woven fabric catalyst for efficient hydrogen generation, *ACS Appl. Mater. Interfaces*, 8 (17), 10866–10873 (2016).

69. W. Qin, T. Chen, L. Pan, L. Niu, B. Hu, D. Li, J. Li, Z. Sun, MoS2-reduced graphene oxide composites via microwave assisted synthesis for sodium ion battery anode with improved capacity and cycling performance, *Electrochim. Acta*, 153, 55–61 (2015).

70. Y. Li, H. Wang, L. Xie, Y. Liang, G. Hong, and H. Dai, MoS2 nanoparticles grown on graphene: an advanced catalyst for the hydrogen evolution reaction, *J. Am. Chem. Soc.*, 133 (19), 7296–7299 (May 2011).

71. B. Wang, R. Hu, J. Zhang, Z. Huang, H. Qiao, L. Gong, and X. Qi, 2D/2D SnS2/MoS2 layered heterojunction for enhanced supercapacitor performance, *J. Am. Ceram. Soc.*, 103, 1088 (2020).

72. S. S. Singha, S. Rudra, S. Mondal, M. Pradhan, A. K. Nayak, B. Satpati, P. Pal, K. Das, and A. Singha, Mn incorporated MoS2 nanoflowers: A high performance electrode material for symmetric supercapacitor, *Electrochim. Acta*, 338, 135815 (2020).

73. H. You, Y. Jia, Z. Wu, F. Wang, H. Huang, and Y. Wang, Room-temperature pyro-catalytic hydrogen generation of 2D few-layer black phosphorene under cold-hot alternation, *Nat. Commun.*, 9, 2889 (2018).

74. J. Cheng, L. Gao, T. Li, S. Mei, C. Wang, B. Wen, W. Huang, C. Li, G. Zheng, H. Wang, and H. Zhang, Two-dimensional black phosphorus nanomaterials: emerging advances in electrochemical energy storage science, *Nano-Micro Lett.*, 12 (1), 1–34 (2020).

75. Y. Sui, J. Zhou, X. Wang, L. Wu, S. Zhong, and Y. Li, Recent advances in black-phosphorus-based materials for electrochemical energy storage, *Mater. Today*, 42, 117–136 (2021).

76. M. Wen, D. Liu, Y. Kang, J. Wang, H. Huang, J. Li, P. K. Chu, and X.-F. Yu, Synthesis of high-quality black phosphorus sponges for all-solid-state super-capacitors, *Mater. Horiz.*, 6 (1), 176–181 (2019).

77. A. Gopalakrishnan, and S. Badhulika, Facile sonochemical assisted synthesis of a hybrid red–black phosphorus/sulfonated porous carbon composite for high-performance supercapacitors, *Chem. Commun.*, 56 (52), 7096–7099 (2020).

78. S. Ogawa, S. Fukushima, and M. Shimatani, Hexagonal boron nitride for photonic device applications: a review, *Materials*, 16, 2005 (2023).

79. G. R. Bhimanapati, N. R. Glavin, and J. A. Robinson, 2D Boron Nitride: Synthesis and Applications, In *Semiconductors and Semimetals*, Elsevier, Vol. 95, 101–147 (2016).

80. T. Li, X. Jiao, T. You, F. Dai, P. Zhang, F. Yu, L. Hu, L. Ding, L. Zhang, Z. Wen, and Y. Wu, Hexagonal boron nitride nanosheet/carbon nanocomposite as a high-performance cathode material towards aqueous asymmetric supercapacitors, *Ceram. Int.*, 45 (4), 4283–4289 (2019).

81. I. M. Patil, S. Kapse, H. Parse, R. Thapa, G. Andersson, and B. Kakade, 2D/3D heterostructure of h-BN/reduced graphite oxide as a remarkable electrode Material for supercapacitor, *J. Power Sources*, 479, 229092 (2020).

82. W.-J. Ong, L.-L. Tan, Y. H. Ng, S.-T. Yong, and S.-P. Chai, Graphitic carbon nitride (g-C$_3$N$_4$)-based photocatalysts for artificial photosynthesis and environmental remediation: are we a step closer to achieving sustainability?, *Chem. Rev.*, 116 (12), 7159–7329 (2016).

83. Y. Wang, L. Liu, T. Ma, Y. Zhang, and H. Huang, 2D graphitic carbon nitride for energy conversion and storage, *Adv. Funct. Mater.*, 31 (34), 2102540 (2021).

84. M. Ghaemmaghami, and R. Mohammadi, Carbon nitride as a new way to facilitate the next generation of carbon-based supercapacitors, *Sustain. Energy Fuels*, 3(9), 2176–2204 (2019).

85. Y. Luo, Y. Yan, S. Zheng, H. Xue, and H. Pang, Graphitic carbon nitride based materials for electrochemical energy storage, *J. Mater. Chem. A*, 7 (3), 901–924 (2019).

86. R. R. Nallapureddy, M. R. Pallavolu, and S. W. Joo, Construction of functionalized carbon nanofiber–g-C3N4 and TiO2 spheres as a nanostructured hybrid electrode for high-performance supercapacitors, *Energy Fuels*, 35 (2), 1796–1809 (2021).

87. C. Shen, R. Li, L. Yan, Y. Shi, H. Guo, J. Zhang, Y. Lin, Z. Zhang, Y. Gong, and L. Niu, Rational design of activated carbon nitride materials for symmetric supercapacitor applications, *Appl. Surf. Sci.*, 455, 841–848 (2018).

88. R. Gonçalves, T. M. Lima, M. W. Paixão, and E. C. Pereira, Pristine carbon nitride as active material for high-performance metal-free supercapacitors: simple, easy and cheap, *RSC Adv.*, 8 (61), 35327–35336 (2018).

89. M. Sharma, and A. Gaur, Designing of carbon nitride supported ZnCo2O4 hybrid electrode for high-performance energy storage applications, *Sci. Rep.*, 10 (1), 1–9 (2020).

90. Y. Xu, L. Wang, Y. Zhou, J. Guo, S. Zhang, and Y. Lu, Synthesis of heterostructure SnO2/graphitic carbon nitride composite for high-performance electrochemical supercapacitor, *J. Electroanal. Chem.*, 852, 113507 (2019).

91. G. Dong, H. Fan, K. Fu, L. Ma, S. Zhang, M. Zhang, J. Ma, and W. Wang, The evaluation of super-capacitive performance of novel g-C3N4/PPy nanocomposite electrode material with sandwich-like structure, *Composites B: Eng.*, 162, 369–377 (2019).

92. N. Mahmood, I. A. De Castro, K. Pramoda, K. Khoshmanesh, S. K. Bhargava, and K. Kalantar-Zadeh, Atomically thin two-dimensional metal oxide nanosheets and their heterostructures for energy storage, *Energy Storage Mater.*, 16, 455–480 (2019).

93. C. Sun, W. Pan, D. Zheng, R. Yao, Y. Zheng, J. Zhu, and D. Jia, Low-cost MnO2 nanoflowers and La2O3 nanospheres as efficient electrodes for asymmetric supercapacitors, *Energy Fuels*, 34 (11), 14882–14892 (2020).

94. R. C. Rohit, A. D. Jagadale, J. Lee, K. Lee, S. K. Shinde, and D. Y. Kim, Tailoring the composition of ternary layered double hydroxides for supercapacitors and electrocatalysis, *Energy Fuels*, 35 (11), 9660–9668 (2021).

95. J. E. ten Elshof, H. Yuan, and P. Gonzalez Rodriguez, Two-dimensional metal oxide and metal hydroxide nanosheets: synthesis, controlled assembly and applications in energy conversion and storage, *Adv. Energy Mater.*, 6 (23), 1600355 (2016).

96. J. Mei, T. Liao, and Z. Sun, Two-dimensional metal oxide nanosheets for rechargeable batteries, *J. Energy Chem.*, 27 (1), 117–127 (2018).

97. H. T. Tan, W. Sun, L. Wang, and Q. Yan, 2D transition metal oxides/hydroxides for energy-storage applications, *ChemNanoMat*, 2 (7), 562–577 (2016).

98. S. M. Youssry, I. S. El-Hallag, R. Kumar, G. Kawamura, A. Matsuda, and M. N. El-Nahass, Synthesis of mesoporous Co(OH)2 nanostructure film via electrochemical

deposition using lyotropic liquid crystal template as improved electrode materials for supercapacitors application, *J. Electroanal. Chem.*, 857, 113728 (2020).

99. M. S. Vidhya, G. Ravi, R. Yuvakkumar, D. Velauthapillai, M. Thambidurai, C. Dang, and B. Saravanakumar, Nickel–cobalt hydroxide: a positive electrode for supercapacitor applications, *RSC Adv.*, 10 (33), 19410–19418 (2020).

100. M. B. Askari, P. Salarizadeh, M. Seifi, M. H. Ramezan zadeh, and A. Di Bartolomeo, ZnFe2O4 nanorods on reduced graphene oxide as advanced super-capacitor electrodes, *J. Alloys Compd.*, 860, 158497 (2021).

101. J. P. Cheng, B. Q. Wang, S. H. Gong, X. C. Wang, Q. S. Sun, and F. Liu, Conformal coatings of NiCo2O4 nanoparticles and nanosheets on carbon nanotubes for supercapacitor electrodes, *Ceram. Int.*, 47, 32727 (2021).

102. K. S. Kumar, N. Choudhary, Y. Jung, and J. Thomas, Recent Advances in Two-Dimensional Nanomaterials for Supercapacitor Electrode Applications, *ACS Energy Lett.*, 3, 482 (2018).

103. M. R. Lukatskaya, O. Mashtalir, C. E. Ren, Y. Dall'Agnese, P. Rozier, P. L. Taberna, M. Naguib, P. Simon, M. W. Barsoum, and Y. Gogotsi, Cation intercalation and high volumetric capacitance of two-dimensional titanium carbide, *Science*, 341, 1502 (2013).

Index

For Product Safety Concerns and Information please contact our EU
representative GPSR@taylorandfrancis.com
Taylor & Francis Verlag GmbH, Kaufingerstraße 24, 80331 München, Germany

www.ingramcontent.com/pod-product-compliance
Lightning Source LLC
Chambersburg PA
CBHW060554220326
41598CB00024B/3105